Mathematical Physics: An Overview

Volume I

Mathematical Physics: An Overview
Volume I

Edited by **Lucy Flynn**

NY RESEARCH PRESS
New York

Published by NY Research Press,
23 West, 55th Street, Suite 816,
New York, NY 10019, USA
www.nyresearchpress.com

Mathematical Physics: An Overview
Volume I
Edited by Lucy Flynn

International Standard Book Number: 978-1-63238-316-7 (Hardback)

Printed in China.

Contents

Preface

Mathematical physics is a branch which is related to the development of mathematical methodologies in the applications for problem solving in physics. The origins of this subject can be traced to a lot of eras from over the world. From Archimedes in Greece to Ptolemy in Egypt, Alhazen in Iraq and Al-Biruni in Persia, all these personalities inaugurated the field of Mathematical Physics in one way or the other. Prominent contributors in the 20th century include renowned scientists such as Albert Einstein, Isaac Newton, and Nicholas Copernicus among numerous others.

There are many branches of mathematical physics such as; geometrically advanced formulations of classical mechanics, partial differential equations, quantum theory, relativity and quantum relativistic theories and even statistical mechanics. It wouldn't be wrong to say that this field of study is sometimes idiosyncratic. Sometimes, components of mathematics which initially came into existence from the development of physics are not considered as a part of mathematical physics, while some times other closely related fields are considered a part of mathematical physics. For example, ordinary differential equation, simplistic geometry are most often referred to as purely mathematical disciplines, on the other hand dynamical systems and Hamiltonian mechanics are considered to be purely mathematical physics.

This book, through its balance between the conceptual and the practical, illustrates the development of mathematical methods suitable for applications and for the formulation of physical theories. I'd like to thank the authors and experts who have contributed to every chapter of this book. Not only do I appreciate their participation, but also their adherence as a group to the time parameters set for this publication. The outstanding contributions they have made resulted in a very easy book to edit. It was a very enjoyable experience and a privilege reading each contribution.

Editor

1

Conditional Probability, Three-Slit Experiments, and the Jordan Algebra Structure of Quantum Mechanics

Gerd Niestegge

Zillertalstrasse 39, 81373 Muenchen, Germany

Correspondence should be addressed to Gerd Niestegge, gerd.niestegge@web.de

Academic Editor: B. G. Konopelchenko

Most quantum logics do not allow for a reasonable calculus of conditional probability. However, those ones which do so provide a very general and rich mathematical structure, including classical probabilities, quantum mechanics, and Jordan algebras. This structure exhibits some similarities with Alfsen and Shultz's noncommutative spectral theory, but these two mathematical approaches are not identical. Barnum, Emerson, and Ududec adapted the concept of higher-order interference, introduced by Sorkin in 1994, into a general probabilistic framework. Their adaption is used here to reveal a close link between the existence of the Jordan product and the nonexistence of interference of third or higher order in those quantum logics which entail a reasonable calculus of conditional probability. The complete characterization of the Jordan algebraic structure requires the following three further postulates: a Hahn-Jordan decomposition property for the states, a polynomial functional calculus for the observables, and the positivity of the square of an observable. While classical probabilities are characterized by the absence of any kind of interference, the absence of interference of third (and higher) order thus characterizes a probability calculus which comes close to quantum mechanics but still includes the exceptional Jordan algebras.

1. Introduction

The interference manifested in the two-slit experiments with small particles is one of the best known and most typical quantum phenomena. It is somewhat surprising therefore that quantum mechanics rules out third-order interference. This was discovered by Sorkin [1] considering measures on the "sets of histories" with experimental set-ups like the well-known two-slit experiments but with three and more slits. He introduced the interference terms I_2 and I_3 and detected that, although the second-order interference is a typical quantum phenomenon ($I_2 \neq 0$), the third-order interference does not occur in quantum mechanics ($I_3 = 0$).

In the present paper, Sorkin's interference terms I_2 and I_3 are ported to the framework of quantum logics with unique conditional probabilities which was introduced by the author in [2, 3]. In [3] it was shown that each such quantum logic can be embedded in an order-unit space where a specific type of positive projections then represents the probability conditionalization similar to the Lüders-von Neumann quantum measurement process.

In this general framework, the identity $I_3 = 0$ is not automatically given, and its role in a reconstruction of quantum mechanics from a few basic principles or in an axiomatic access to quantum mechanics based on a few interpretable postulates is analysed in the paper. It is shown that the absence of third-order interference ($I_3 = 0$) has some important consequences. It entails the existence of a product in the order-unit space generated by the quantum logic, which can be used to characterize those quantum logics that can be embedded in the projection lattice in a Jordan algebra. Most of these Jordan algebras can be represented as operator algebras on a Hilbert space, and a reconstruction of quantum mechanics up to this point is thus achieved.

Besides the identity $I_3 = 0$, two further typical properties of quantum mechanics distinguishing it from more general theories are identified; these are a novel bound for quantum interference and a symmetry property of the conditional probabilities. This latter property was discovered by Alfsen and Shultz who used it as a postulate to derive the Jordan product for the quantum mechanical observables from it in their approach [4], and it was used in a similar way in [3], but a physical justification for it is hard to find. With the main result of the present paper, it can now be replaced by another postulate with a clearer physical meaning, namely, the absence of third-order interference ($I_3 = 0$).

The next two sections summarize those parts of [2, 3] which are relevant for the subsequent sections. In Section 4, the second- and third-order interference terms (I_2 and I_3) are considered and ported to the quantum logics with unique conditional probabilities. The bound for quantum interference and the symmetry property of the quantum mechanical conditional probabilities are studied in Sections 5 and 6. In Section 7, a useful type of linear maps is introduced, which is used in Section 8 to analyse the case $I_3 = 0$. A certain mathematical condition, the Jordan decomposition property, is outlined in Section 9 and then used in Section 10 to derive the product in the order-unit space from the identity $I_3 = 0$. Section 11 finally addresses the question under which further conditions the order-unit space becomes a Jordan algebra.

2. Quantum Logics with Unique Conditional Probabilities

A quantum logic is the mathematical model of a system of quantum events or propositions. Logical approaches use the name "proposition", while the name "event" is used in probability theory and will also be preferred in the present paper. The concrete quantum logic of standard quantum mechanics is the system of closed linear subspaces of a Hilbert space or, more generally, the projection lattice in a von Neumann algebra.

Usually, an abstract quantum logic is assumed to be an orthomodular partially ordered set and, very often, it is also assumed that it is lattice. For the purpose of the present paper, however, a more general and simpler mathematical structure without order relation is sufficient. Only an orthocomplementation, an orthogonality relation, and a sum operation defined for orthogonal events are needed. The orthocomplementation represents the logical negation, orthogonality means mutual exclusivity, and the sum represents the logical and operation in the case of mutual exclusivity. The precise axioms were presented in [2] and look as follows.

The quantum logic E is a set with distinguished elements 0 and $\mathbb{1}$, an orthogonality relation $\perp$ and a partial binary operation $+$ such that the following six axioms hold for e, f, $g \in E$:

(OS1) *if $e \perp f$, then $f \perp e$; that is, the relation $\perp$ is symmetric;*

(OS2) *$e + f$ is defined for $e \perp f$, and then $e + f = f + e$; that is, the sum operation is commutative.*

(OS3) *if $g \perp e$, $g \perp f$, and $e \perp f$, then $g \perp e + f$, $f \perp g + e$ and $g + (e + f) = (g + e) + f$; that is, the sum operation is associative;*

(OS4) *$0 \perp e$ and $e + 0 = e$ for all $e \in E$;*

(OS5) *for every $e \in E$, there exists a unique $e' \in E$ such that $e \perp e'$ and $e + e' = \mathbb{1}$;*

(OS6) *there exists $d \in E$ such that $e \perp d$ and $e + d = f$ if and only if $e \perp f'$.*

Then $0' = \mathbb{1}$ and $e'' = e$ for $e \in E$. Note that an orthomodular partially ordered set satisfies these axioms with the two definitions:

(i) *$e \perp f$ if and only if $f \leq e'$;*

(ii) *the sum $e + f$ is the supremum of e and f for $e \perp f$.*

The supremum exists in this case due to the orthomodularity.

A state is a map $\mu : E \to [0,1]$ such that $\mu(\mathbb{1}) = 1$ and $\mu(e + f) = \mu(e) + \mu(f)$ for orthogonal pairs e and f in E. Then $\mu(0) = 0$ and $\mu(e_1 + \cdots + e_k) = \mu(e_1) + \cdots + \mu(e_k)$ for mutually orthogonal elements $e_1, \ldots, e_k$ in E. Denote by S the set of all states on E. With a state μ and $\mu(e) > 0$ for an $e \in E$, another state ν is called a conditional probability of μ under e if $\nu(f) = \mu(f)/\mu(e)$ holds for all $f \in E$ with $f \perp e'$. Furthermore, the following axioms were introduced in [2].

(UC1) *If $e, f \in E$ and $\mu(e) = \mu(f)$ for all $\mu \in S$, then $e = f$.*

(UC2) *If $e \in E$ and $\mu \in S$ with $\mu(e) > 0$, there is one and only one conditional probability of μ under e.*

If these axioms are satisfied, E is called a UCP *space* named after the major feature of this mathematical structure which is the existence of the unique conditional probability, and the elements in E are called events. The unique conditional probability of μ under e is denoted by μ_e and, in analogy with classical mathematical probability theory, $\mu(f \mid e)$ is often written instead of $\mu_e(f)$ with $f \in E$. The above two axioms imply that there is a state $\mu \in S$ with $\mu(e) = 1$ for each event $e \neq 0$, that the difference d in (OS6) becomes unique, and that $e \perp e$ if and only if $e \perp \mathbb{1}$ if and only if $e = 0$ $(e \in E)$.

Note that the following identity which will be used later holds for convex combinations of states $\mu, \nu \in S$ $(0 < s < 1)$:

$$(s\mu + (1-s)\nu)_e = \frac{1}{s\mu(e) + (1-s)\nu(e)}\left(s\mu(e)\mu_e + (1-s)\nu(e)\nu_e\right). \tag{2.1}$$

A typical example of the above structure is the projection lattice E in a von Neumann algebra M without type I_2 part; $E = \{e \in M : e^* = e = e^2\}$. The conditional probabilities then have the shape:

$$\mu_e(f) = \mu(f \mid e) = \frac{1}{\mu(e)}\hat{\mu}(efe), \tag{2.2}$$

with $e, f \in E$, $\mu \in S$, and $\mu(e) > 0$. Note that $\widehat{\mu}$ on M is the unique positive linear extension of the state μ originally defined only on the projection lattice; this extension exists by Gleason's theorem [5] and its later enhancements to finitely additive states and arbitrary von Neumann algebras [6–9]. The linear extension $\widehat{\mu}$ does not exist if M contains a type I_2 part.

For the proof of (2.2), suppose that the state ν on E is a version of the conditional probability of the state μ under e and use the identity $f = efe + efe' + e'fe + e'fe'$. From $\nu(e') = 0$ and the Cauchy-Schwarz inequality applied with the positive linear functional $\widehat{\nu}$ it follows $0 = \widehat{\nu}(efe') = \widehat{\nu}(e'fe) = \widehat{\nu}(e'fe')$ and thus $\nu(f) = \widehat{\nu}(efe)$. By the spectral theorem, efe can be approximated (in the norm topology) by linear combinations of elements in $\{d \in E : d \perp e'\} = \{d \in E : d \leq e\}$ for which ν coincides with $\mu/\mu(e)$. The continuity of $\widehat{\nu}$ (due its positivity) then implies $\nu(f) = \widehat{\nu}(efe) = \widehat{\mu}(efe)/\mu(e)$. Therefore the conditional probability must have this shape and its uniqueness is proved. Its existence follows from $efe \geq 0$ and $efe = f$ for $f \leq e$, since then $\nu(f) := \widehat{\mu}(efe)/\mu(e)$ indeed owns all the properties of the conditional probability.

Equation (2.2) reveals the link to the Lüders-von Neumann quantum measurement process. The transition from a state μ to the conditional probability μ_e is identical with the transition from the state prior to the measurement to the state after the measurement where e represents the measurement result.

3. The Embedding of the Quantum Logic in an Order-Unit Space

A quantum logic with a sufficiently rich state space as postulated by (UC1) can be embedded in the unit interval of an order-unit space. In the present section, it will be shown that the existence and the uniqueness of the conditional probabilities postulated by (UC2) give rise to some important additional structure on this order-unit space, which was originally presented in [3].

A partially ordered real vector space A is an order-unit space if A contains an Archimedean order unit $\mathbb{1}$ [10–12]. The order unit $\mathbb{1}$ is positive and, for all $a \in A$, there is $t > 0$ such that $-t\mathbb{1} \leq a \leq t\mathbb{1}$. An order unit $\mathbb{1}$ is called Archimedean if $na \leq \mathbb{1}$ for all $n \in \mathbb{N}$ implies $a \leq 0$. An order-unit space A has a norm given by $\|a\| = \inf\{t > 0 : -\mathbb{1}t \leq a \leq \mathbb{1}t\}$. Each $x \in A$ can be written as $x = a - b$ with positive $a, b \in A$ (e.g., choose $a = \|x\|\mathbb{1}$ and $b = \|x\|\mathbb{1} - x$). A positive linear functional $\rho : A \to \mathbb{R}$ on an order-unit space A is bounded with $\|\rho\| = \rho(\mathbb{1})$ and, vice versa, a bounded linear functional ρ with $\|\rho\| = \rho(\mathbb{1})$ is positive.

The order-unit space A considered in the following is the dual space of a base-norm space V and, therefore, the unit ball of A is compact in the weak-*-topology $\sigma(A, V)$. For $\rho \in V$ and $x \in A$ define $\widehat{\rho}(x) := x(\rho)$; the map $\rho \to \widehat{\rho}$ is the canonical embedding of V in its second dual $V^{**} = A^*$. Then $\rho \in V$ is positive if and only if $\widehat{\rho}$ is positive on A.

For any set K in A, denote by $\overline{\text{lin}}\, K$ the $\sigma(A, V)$-closed linear hull of K and by $\overline{\text{conv}}\, K$ the $\sigma(A, V)$-closed convex hull of K. For a convex set K, denote by *ext* K the set of its extreme points which may be empty unless K is compact. A projection is a linear map $U : A \to A$ with $U^2 = U$ and, for $a \leq b$, define $[a, b] := \{x \in A : a \leq x \leq b\}$. Suppose that E is a subset of $[0, \mathbb{1}]$ in A such that

(a) $\mathbb{1} \in E$,

(b) $\mathbb{1} - e \in E$ if $e \in E$, and

(c) $d + e + f \in E$ if $d, e, f, d + e, d + f, e + f \in E$.

Define $e' := \mathbb{1} - e$ and call $e, f \in E$ orthogonal if $e + f \in E$. Then E satisfies the axioms (OS1),. . .,(OS6), and the states on E and the state space S can be considered as in Section 2.

Proposition 3.1. *Suppose that A is an order-unit space with order unit $\mathbb{1}$ and that A is the dual of the base-norm space V. Moreover, suppose that E is a subset of $[0, \mathbb{1}]$ satisfying the three above conditions (a), (b), (c) and that the following two conditions hold*

(i) *$A = \overline{\text{lin}}\, E$, and for each $\mu \in S$ there is a $\sigma(A, V)$-continuous positive linear functional $\widehat{\mu}$ on A with $\widehat{\mu}(e) = \mu(e)$ for $e \in E$.*

(ii) *For each $e \in E$ there is a $\sigma(A, V)$-continuous positive projection $U_e : A \to A$ such that $U_e\mathbb{1} = e, U_eA = \overline{\text{lin}}\{f \in E : f \leq e\}$ and $\widehat{\mu} = \widehat{\mu}U_e$ for $\mu \in S$ with $\mu(e) = 1$.*

Then E is a UCP space. The conditional probabilities have the shape $\mu(f \mid e) = \widehat{\mu}(U_e f)/\mu(e)$ for $e, f \in E$ and $\mu \in S$ with $\mu(e) > 0$.

Proof. For $e, f \in E$ with $e \neq f$ there is $\rho \in V_+$ with $\rho(e - f) \neq 0$. The restriction of $\rho/\rho(\mathbb{1})$ to E then yields a state $\mu \in S$ with $\mu(e) \neq \mu(f)$. Therefore (UC1) holds.

Suppose $e \in E$ and $\mu \in S$ with $\mu(e) > 0$. It is rather obvious that the map $g \to \widehat{\mu}(U_e g)/\mu(e)$ on E provides a conditional probability of μ under e. Now assume that ν is a further conditional probability of μ under e. Then $\nu(e) = 1$ and thus $\widehat{\nu} = \widehat{\nu}U_e$. From $U_e g \in \overline{\text{lin}}\{f \in E : f \leq e\}$ it follows that $\nu(g) = \widehat{\nu}(U_e g) = \widehat{\mu}(U_e g)/\mu(e)$ for $g \in E$. Therefore, (UC2) holds as well. □

Note that the linear extension $\widehat{\mu}$ in (i) is unique since $A = \overline{\text{lin}}\, E$. It will now be seen that the situation of Proposition 3.1 is universal for the quantum logics with unique conditional probabilities; that is, each UCP space has such a shape as described there.

Theorem 3.2. *Each UCP space E is a subset of the interval $[0, \mathbb{1}]$ in some order-unit space A with predual V as described in Proposition 3.1.*

Proof. Define $V := \{s\mu - t\nu : \mu, \nu \in S, 0 \leqslant s, t \in \mathbb{R}\}$, which is a linear subspace of the orthogonally additive real-valued functions on E, and consider for $\rho \in V$ the norm

$$\|\rho\| := \inf\{r \in \mathbb{R} : r \geq 0,\ \rho \in r \operatorname{conv}(S \cup -S)\}. \tag{3.1}$$

Then $|\rho(e)| \leq \|\rho\|$ for every $e \in E$. Let A be the dual space of the base-norm space V and let $\widehat{\mu}$ be the canonical embedding of $\mu \in V$ in $V^{**} = A^*$. If $\widehat{\mu}(x) \geq 0$ for all $\mu \in S$, the element $x \in A$ is called positive and in this case write $x \geq 0$. Equipped with this partial ordering, A becomes an order-unit space with the order unit $\mathbb{1} := \pi(\mathbb{1})$, and the order-unit norm of an element $x \in A$ is $\|x\| = \sup\{|\widehat{\mu}(x)| : \mu \in S\}$. With $e \in E$ define $\pi(e)$ in A via $\pi(e)(\rho) := \rho(e)$ for $\rho \in V$. Then $0 \leq \|\pi(e)\| \leq \mathbb{1}$, and $\pi(e + f) = \pi(e) + \pi(f)$ for two orthogonal events e and f in E. Moreover, A is the $\sigma(A, V)$-closed linear hull of $\pi(E)$.

Now the positive projection U_e will be defined for $e \in E$. Suppose $x \in A$ and $s\mu - t\nu \in V$ with $\mu, \nu \in S$ and $0 \leqslant s, t \in \mathbb{R}$. Then define $(U_e x)(s\mu - t\nu) := s\mu(e)\widehat{\mu}_e(x) - t\nu(e)\widehat{\nu}_e(x)$. Here, $\widehat{\mu}_e$ and $\widehat{\nu}_e$ are the canonical embeddings of the conditional probabilities μ_e and ν_e in A^*; they do not exist in the cases $\mu(e) = 0$ or $\nu(e) = 0$ and then define $\mu(e)\widehat{\mu}_e(x) := 0$ and $\nu(e)\widehat{\nu}_e(x) := 0$, respectively. It still has to be shown that U_e is well defined for $s\mu - t\nu = s'\mu' - t'\nu'$ with μ, μ', ν, $\nu' \in S$ and $0 \leq s, s', t, t'$. Then $s - t = (s\mu - t\nu)(\mathbb{1}) = (s'\mu' - t'\nu')(\mathbb{1}) = s' - t'$ and

$s + t' = s' + t$. If $s + t' = 0$, $s = s' = t = t' = 0$ and $U_e x$ is well defined. If $s + t' > 0$, then either $s\mu(e) + t'\nu'(e) = s'\mu'(e) + t\nu(e) = 0$ and $s\mu(e) = t'\nu'(e) = s'\mu'(e) = t\nu(e) = 0$, or $(s\mu + t'\nu')/(s + t') = (s'\mu' + t\nu)/(s' + t) \in S$ and calculating the conditional probability under e for both sides of this identity by using (2.1) yields $s\mu(e)\mu_e + t'\nu'(e)\nu'_e = s'\mu'(e)\mu'_e + t\nu(e)\nu_e$. In all cases, U_e is well defined.

If $\mu(e) = 1$ for $\mu \in S$, then $\mu = \mu_e$ and $\hat{\mu}(U_e x) = (U_e x)(\mu) = \hat{\mu}(x)$; that is, $\hat{\mu} = \hat{\mu}U_e$. Thus, $(U_e U_e x)(\mu) = \mu(e)\hat{\mu}_e(U_e x) = \mu(e)\hat{\mu}_e(x) = (U_e x)(\mu)$ for all $\mu \in S$ and hence for all $\rho \in V$. Therefore $U_e U_e = U_e$, that is, U_e is a projection. Its positivity, $\sigma(A, V)$-continuity as well as $U_e \mathbb{1} = \pi(e)$ and $U_e \pi(f) = \pi(f)$ for $f \in E$ with $f \le e$ follow immediately from the definition.

Therefore $\overline{\text{lin}}\{\pi(f) : f \in E, f \le e\} \subseteq U_e A$. Assume $U_e x \notin \overline{\text{lin}}\{\pi(f) : f \in E, f \le e\}$ for some $x \in A$. By the Hahn-Banach theorem, there is $\rho \in V$ with $\hat{\rho}(U_e x) \neq 0$ and $\rho(f) = 0$ for $f \in E$ with $f \le e$. Suppose $\rho = s\mu - t\nu$ with $\mu, \nu \in S$ and $0 \leqslant s, t \in \mathbb{R}$. Then $s\mu(f) = t\nu(f)$ for $f \in E$ with $f \le e$ and thus $s\mu(e)\mu_e(f) = t\nu(f)\nu_e(f)$. The uniqueness of the conditional probability implies $s\mu(e)\mu_e(f) = t\nu(f)\nu_e(f)$, that is, $\hat{\rho}(U_e f) = 0$ for all $f \in E$ and hence $\hat{\rho}U_e = 0$ which contradicts $\hat{\rho}(U_e x) \neq 0$. This completes the proof of Theorem 3.2 after identifying $\pi(E)$ with E. □

Lemma 3.3. *If e and f are events in a UCP space E with $e \le f$, then $U_e f = e = U_f e$ and $U_e U_f = U_f U_e = U_e$. If $e, f \in E$ are orthogonal, then $U_e f = 0 = U_f e$, $U_e U_f = U_f U_e = 0$ and $U_{e'} U_{f'} = U_{(e+f)'} = U_{f'} U_{e'}$.*

Proof. Suppose $e \le f$ and $\mu \in S$. Then $1 = \mu_e(e) \le \mu_e(f) \le 1$ for the conditional probability μ_e implies $\mu_e(f) = 1$ and hence $\hat{\mu}_e U_f = \hat{\mu}_e$. Therefore $\hat{\mu}U_e U_f = \mu(e)\hat{\mu}_e U_f = \mu(e)\hat{\mu}_e = \hat{\mu}U_e$ for all $\mu \in S$ and thus $U_e U_f = U_e$. The identity $U_f U_e = U_e$ immediately follows from (ii) in Proposition 3.1. Moreover $e = U_e \mathbb{1} = U_e U_f \mathbb{1} = U_e f$ and $e = U_e \mathbb{1} = U_f U_e \mathbb{1} = U_f e$.

Now assume that e and f are orthogonal. Then $e \leqslant f'$. Hence $e = U_e f' = U_e(\mathbb{1} - f) = e - U_e f$, and $U_e f = 0$. In the same way it follows that $U_f e = 0$. Therefore U_f vanishes on $U_e A = \overline{\text{lin}}\{d \in E : d \le e\}$ and $U_f U_e = 0$. The identity $U_e U_f = 0$ follows in the same way.

Moreover, $0 \le U_{e'} U_{f'} x \le U_{e'} U_{f'} \mathbb{1} = U_{e'} f' = U_{e'} \mathbb{1} - U_{e'} f = e' - f = (e + f)'$ for $x \in [0, \mathbb{1}]$. Therefore $\hat{\mu}U_{e'} U_{f'} = 0 = \hat{\mu}U_{(e+f)'}$ for $\mu \in S$ with $\mu((e + f)') = 0$. Now consider $\mu \in S$ with $\mu((e + f)') > 0$ and define $\nu := \hat{\mu}U_{e'} U_{f'} / \mu((e + f)') \in S$. From $(e + f)' \le e'$ and $(e + f)' \le f'$, it follows that $U_{e'} U_{f'} (e + f)' = (e + f)'$ and $\nu((e + f)') = 1$. Hence $\nu = \hat{\nu}U_{(e+f)'}$. From $U_{e'} U_{(e+f)'} = U_{(e+f)'} = U_{f'} U_{(e+f)'}$ it follows that $\nu = \hat{\mu}U_{(e+f)'} / \mu((e + f)')$ and thus $\hat{\mu}U_{e'} U_{f'} = \hat{\mu}U_{(e+f)'}$. This identity now holds for all states μ and therefore $U_{e'} U_{f'} = U_{(e+f)'}$. In the same way it follows that $U_{f'} U_{e'} = U_{(e+f)'}$. □

The projections U_e considered here are similar to but not identical with the so-called P-projections considered by Alfsen and Shultz in their noncommutative spectral theory [13]. A P-projection P has a quasicomplement Q such that $Px = x$ if and only if $Qx = 0$ (and $Qx = x$ if and only if $Px = 0$) for $x \ge 0$. If $U_e x = x$, then $U_{e'} x = U_{e'} U_e x = 0$ by Lemma 3.3, but $U_{e'} x = 0$ does not imply $U_e x = x$.

An element $P(\mathbb{1})$ with a P-projection P is called a projective unit by Alfsen and Shultz. In the case of spectral duality of a base-norm space V and an order-unit space A, the system of projective units in A is a UCP space if each state on the projective units has a linear extension to A (as with the Gleason theorem or in condition (i) of Proposition 3.1).

4. The Interference Terms I_2 and I_3

With two disjoint events e_1 and e_2, the classical conditional probabilities satisfy the rule $\mu(f \mid e_1 + e_2)\mu(e_1 + e_2) = \mu(f \mid e_1)\mu(e_1) + \mu(f \mid e_2)\mu(e_2)$. Only because violating this rule, quantum mechanics can correctly model the quantum interference phenomena observed in nature with small particles.

For instance, consider the two-slit experiment and let e_1 be the event that the particle passes through the first slit, e_2 the event that it passes through the second slit, and f the event that it is registered in a detector located at a fixed position somewhere behind the screen with the two slits. Then $\mu(f \mid e_1)\mu(e_1)$ is the probability that the particle is registered in the detector when the first slit is open and the second one is closed, $\mu(f \mid e_2)\mu(e_2)$ is the probability that the particle is registered in the detector when the second slit is open and the first one is closed, and $\mu(f \mid e_1+e_2)\mu(e_1+e_2)$ is the probability that the particle is registered in the detector when both slits are open. If the above rule were valid, it would rule out the interference patterns observed in the quantum physical experiments and correctly modelled by quantum theory. Therefore, a first interference term is defined by

$$I_2^{\mu,f}(e_1,e_2) := \mu(f \mid e_1 + e_2)\mu(e_1 + e_2) - \mu(f \mid e_1)\mu(e_1) - \mu(f \mid e_2)\mu(e_2), \tag{4.1}$$

where μ is a state, f any event, and e_1, e_2 an orthogonal pair of events. While $I_2^{\mu,f}(e_1,e_2) = 0$ in the classical case, it is typical of quantum mechanics that $I_2^{\mu,f}(e_1,e_2) \neq 0$.

With three orthogonal events e_1, e_2, e_3, a next interference term can be defined in the following way:

$$\begin{aligned} I_3^{\mu,f}(e_1,e_2,e_3) := {} & \mu(f \mid e_1 + e_2 + e_3)\mu(e_1 + e_2 + e_3) \\ & - \mu(f \mid e_1 + e_2)\mu(e_1 + e_2) - \mu(f \mid e_1 + e_3)\mu(e_1 + e_3) - \mu(f \mid e_2 + e_3)\mu(e_2 + e_3) \\ & + \mu(f \mid e_1)\mu(e_1) + \mu(f \mid e_2)\mu(e_2) + \mu(f \mid e_3)\mu(e_3). \end{aligned} \tag{4.2}$$

Similar to the two-slit experiment, now consider an experiment where the screen has three instead of two slits. Let e_k be the event that the particle passes through the kth slit ($k = 1,2,3$) and f again the event that it is registered in a detector located somewhere behind the screen with the slits. Then $\mu(f \mid e_k)\mu(e_k)$ is the probability that the particle is registered in the detector when the kth slit is open and the other two ones are closed, $\mu(f \mid e_i + e_j)\mu(e_i + e_j)$ is the probability that the particle is registered in the detector when the ith slit and the jth are open and the remaining third slit is closed, and $\mu(f \mid e_1 + e_2 + e_3)\mu(e_1 + e_2 + e_3)$ is the probability that the particle is registered in the detector when all three slits are open. The interference term $I_3^{\mu,f}(e_1,e_2,e_3)$ is the sum of these probabilities with negative signs in the cases with two slits open and positive signs in the cases with one or three slits open. The sum would become zero if these probabilities were additive in e_1, e_2, e_3 as they are in the classical case. However, this is not the only case; $I_3^{\mu,f}(e_1,e_2,e_3) = 0$ means that the detection probability with three open slits is a simple linear combination of the detection probabilities in the three cases with two open slits and the three cases with one single open slit. If the situation with three slits involves some new interference, the detection probability should not be such a linear combination and $I_3^{\mu,f}(e_1, e_2, e_3)$ should not become zero. The fascinating question now arises whether $I_3^{\mu,f}(e_1, e_2, e_3) = 0$ or not.

Considering experiments with three and more slits, this third-order interference term and a whole sequence of further higher order interference terms were introduced by Sorkin [1], but in another form. He used probability measures on "sets of histories". When porting his third-order interference term to conditional probabilities, it gets the above shape. The same shape is used by Barnum et al. [14] who adapted Sorkin's third-order interference term into an operational probabilistic framework. The further higher-order interference terms will not be considered in the present paper. It may be interesting to note that the absence of the third-order interference implies the absence of the interference of all orders higher than three.

Using the identity $\mu(f \mid e) = \widehat{\mu}(U_e f)/\mu(e)$, the validity of $I_3^{\mu,f}(e_1, e_2, e_3) = 0$ for all states μ is immediately equivalent to the identity

$$U_{e_1+e_2+e_3} f - U_{e_1+e_2} f - U_{e_2+e_3} f - U_{e_1+e_3} f + U_{e_1} f + U_{e_2} f + U_{e_3} f = 0. \tag{4.3}$$

If this will hold also for all events f, this means

$$I_3(e_1, e_2, e_3) := U_{e_1+e_2+e_3} - U_{e_1+e_2} - U_{e_2+e_3} - U_{e_1+e_3} + U_{e_1} + U_{e_2} + U_{e_3} = 0. \tag{4.4}$$

The term $I_3(e_1, e_2, e_3)$ does not any more depend on a state or the event f, but only on the orthogonal event triple e_1, e_2, e_3. Note that $I_3(e_1, e_2, e_3)$ is a linear map on the order-unit space A while $I_3^{\mu,f}(e_1, e_2, e_3) = \widehat{\mu}(I_3(e_1, e_2, e_3) f)$ is a real number.

This interference term shall now be studied in a von Neumann algebra where the conditional probability has the shape $\mu(f \mid e) = \widehat{\mu}(efe)/\mu(e)$ for projections e, $f$and a state μ with $\mu(e) > 0$, and hence $U_e f = efe$. Then

$$\begin{aligned} &U_{e_1+e_2+e_3} f - U_{e_1+e_2} f - U_{e_2+e_3} f - U_{e_1+e_3} f + U_{e_1} f + U_{e_2} f + U_{e_3} f \\ &\quad = (e_1 + e_2 + e_3) f (e_1 + e_2 + e_3) \\ &\quad\quad - (e_1 + e_2) f (e_1 + e_2) - (e_2 + e_3) f (e_2 + e_3) \\ &\quad\quad - (e_1 + e_3) f (e_1 + e_3) \\ &\quad\quad + e_1 f e_1 + e_2 f e_2 + e_3 f e_3 = 0. \end{aligned} \tag{4.5}$$

Therefore, in a von Neumann algebra and in standard quantum mechanics, the identity $I_3^{\mu,f}(e_1, e_2, e_3) = 0$ always holds, which was already seen by Sorkin. This becomes a first interesting property of quantum mechanics distinguishing it from the general quantum logics with unique conditional probabilities (UCP spaces). It is quite surprising that quantum mechanics has this property because there is no obvious reason why $I_3^{\mu,f}(e_1, e_2, e_3)$ should vanish while $I_2^{\mu,f}(e_1, e_2)$ does not. Likewise surprising are the bounds which quantum mechanics imposes on $I_2^{\mu,f}(e_1, e_2)$ and which will be presented in the next section.

5. A Bound for Quantum Interference

Suppose that E is a UCP space. By Theorem 3.2, it can be embedded in an order-unit space A such that Proposition 3.1 holds. For each event $e \in E$ define a linear map S_e on A by

$$S_e x := 2U_e x + 2U_{e'} x - x \quad (x \in A). \tag{5.1}$$

Then $S_e \mathbb{1} = \mathbb{1}$. Furthermore, $S_e S_{e'} x = S_{e'} S_e x = x$ for $x \in A$; that is, $S_{e'}$ is the inverse of the linear map S_e and S_e is a linear isomorphism. If it were positive, it would be an automorphism of the order-unit space A, but this is not true in general. It will now be studied what the positivity of the maps S_e would mean for the conditional probabilities.

Suppose $e, f \in E$. Then $0 \leqslant S_e f$ means $f \leqslant 2U_e f + 2U_{e'} f$, and this is equivalent to the following inequality for the conditional probabilities:

$$\mu(f) \leqslant 2\mu(f \mid e)\mu(e) + 2\mu(f \mid e')\mu(e') \tag{5.2}$$

or, equivalently,

$$I_2^{\mu,f}(e,e') \leq \mu(f \mid e)\mu(e) + \mu(f \mid e')\mu(e') \tag{5.3}$$

holding for all states μ. Exchanging f by f' yields from $0 \leqslant S_e f'$

$$\begin{aligned} 1 - \mu(f) = \mu(f') &\leqslant 2\mu(f' \mid e)\mu(e) + 2\mu(f' \mid e')\mu(e') \\ &= 2(1 - \mu(f \mid e))\mu(e) + 2(1 - \mu(f \mid e'))\mu(e') \\ &= 2 - 2\mu(f \mid e)\mu(e) - 2\mu(f \mid e')\mu(e') \end{aligned} \tag{5.4}$$

and thus a second inequality:

$$\mu(f) \geqslant 2\mu(f \mid e)\mu(e) + 2\mu(f \mid e')\mu(e') - 1 \tag{5.5}$$

or, equivalently,

$$I_2^{\mu,f}(e,e') \geq \mu(f \mid e)\mu(e) + \mu(f \mid e')\mu(e') - 1. \tag{5.6}$$

The above inequalities (5.3) and (5.6) introduce an upper bound and a lower bound for the interference term $I_2^{\mu,f}(e,e')$. How these bounds limit the interference is shown in Figure 1 using the inequalities (5.2) and (5.5) and not this specific interference term. The dashed diagonal line represents the classical case without interference, while the inequalities (5.2) and (5.5) allow the whole corridor between the two continuous lines and forbid the area outside this corridor.

In a von Neumann algebra, $U_e f = efe$ for projections e, $f$and hence $S_e x = 2exe + 2e'xe' - x = (e - e')x(e - e')$ with x in the von Neumann algebra. Therefore the maps S_e are positive and the above inequalities for the conditional probabilities ((5.2) and (5.5) illustrated

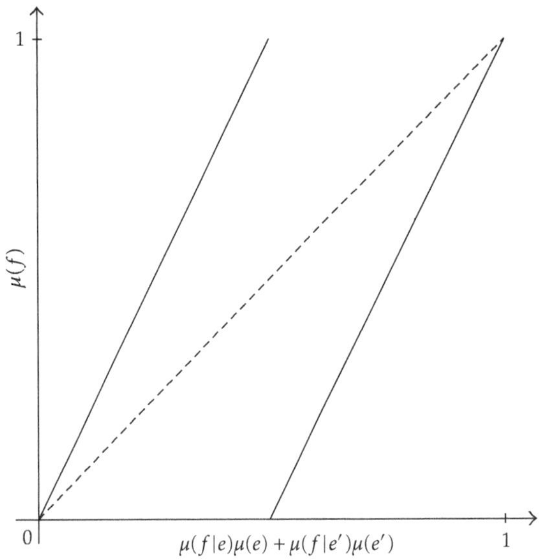

Figure 1

in Figure 1) and the resulting bounds for the interference term $I_2^{\mu,f}(e,e')$ (inequalities (5.3) and (5.6)) hold in this case. This is a second interesting property of quantum mechanics distinguishing it from other more general theories.

6. A Symmetry Property of the Quantum Mechanical Conditional Probabilities

Alfsen and Shultz introduced the following symmetry condition for the conditional probabilities in [4] and used it to derive a Jordan algebra structure from their noncommutative spectral theory:

$$\mu(f' \mid e)\mu(e) + \mu(f \mid e')\mu(e') = \mu(e' \mid f)\mu(f) + \mu(e \mid f')\mu(f'). \tag{A1}$$

This condition arose mathematically and a physical meaning is not immediately at hand. Alfsen and Shultz's interpretation was "*The probability of the exclusive disjunction of two system properties is independent of the order of the measurements of the two system properties.*" However, the exclusive disjunction is not an event or proposition.

The first summand on the left-hand side $\mu(f' \mid e)\mu(e)$ is the probability that a first measurement testing e versus e' provides the result e and that a second successive measurement testing f versus f' then provides the result f' (= "not f"). The second summand on the left-hand side and the two summands on the right-hand side can be interpreted in the same way with exchanged roles of e, e', f, f'.

Figure 2 displays a measuring arrangement consisting of two successive measurements M1 and M2; the first one tests e versus e' and the second one f versus f'. After the second one, the particle is let pass only in the two cases if the result of the first measurement is e and the result of the second one is f' or if the result of the first one is e' and the result of

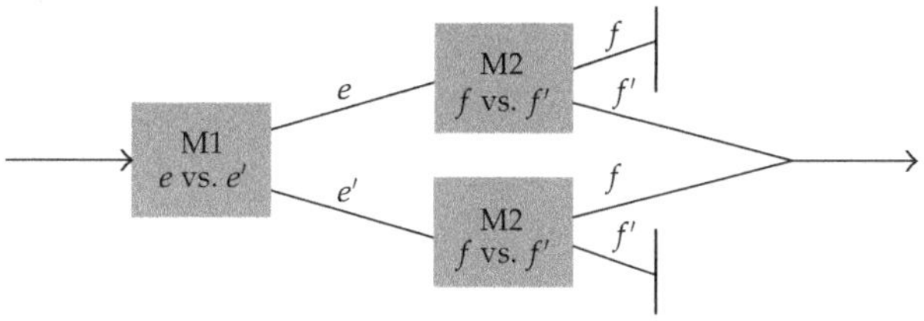

Figure 2

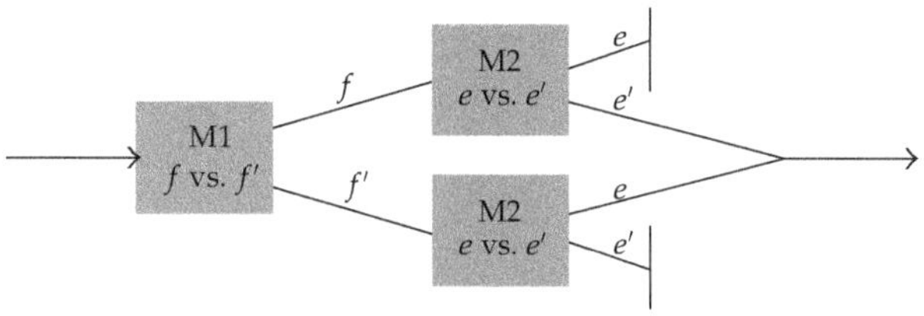

Figure 3

the second one is f. In the other two cases, if the result of the first measurement is e and the result of the second one is f or if the result of the first one is e' and the result of the second one is f', the particle is absorbed after the second measurement. Figure 3 displays the same measuring arrangement as Figure 2, but with exchanged roles of e and f.

Condition $(A1)$ now means that the probability that a particle passes through the measuring arrangement shown in Figure 2 is identical with the probability that a particle passes through the measuring arrangement shown in Figure 3.

The symmetry condition for the conditional probabilities $(A1)$ also plays a certain role in the study of different compatibility/comeasurability levels in [15].

Condition $(A1)$ will now be rewritten using the interference term $I_2^{\mu,f}(e_1, e_2)$ for two orthogonal events e_1 and e_2. To remove the dependence on f and the state μ, first define in analogy to (4.4)

$$I_2(e_1, e_2) := U_{e_1+e_2} - U_{e_1} - U_{e_2}, \tag{6.1}$$

which is a linear operator on the order-unit space A generated by an UCP space E. Note that $I_2^{\mu,f}(e_1, e_2) = \widehat{\mu}(I_2(e_1, e_2)f)$, recalling the identity $\mu(f \mid e)\mu(e) = \widehat{\mu}(U_e f)$ for states μ and events e and f. The validity of $(A1)$ for all states μ is immediately equivalent to the identity

$$U_e f' + U_{e'} f = U_f e' + U_{f'} e. \tag{6.2}$$

It implies

$$\begin{aligned} 0 &= U_e f' + U_{e'} f - U_f e' - U_{f'} e \\ &= e - U_e f + U_{e'} f - f + U_f e - U_{f'} e \\ &= I_2(f, f')e + 2U_f e - I_2(e, e')f - 2U_e f \end{aligned} \tag{6.3}$$

and hence

$$I_2(e,e')f - I_2(f,f')e = 2U_f e - 2U_e f. \tag{6.4}$$

Reconsidering the von Neumann algebras where $U_e f = efe$, (6.2) becomes $e(\mathbb{1} - f)e + (\mathbb{1} - e)f(\mathbb{1} - e) = f(\mathbb{1} - e)f + (\mathbb{1} - f)e(\mathbb{1} - f)$. Both sides of this equation are identical to $e + f - ef - fe$. Therefore $(A1)$ holds for all states and all events in a von Neumann algebra and in the standard model of quantum mechanics. This is a third interesting property of quantum mechanics distinguishing it from the general quantum logics with unique conditional probabilities (UCP spaces). However, it is a mathematical property without a clear physical reason behind it and the absence of third-order interference ($I_3 = 0$) is the more interesting property from the physical point of view.

7. The Linear Maps T_e

In addition to the U_e, a further useful type of linear maps T_e on the order-unit space A generated by a UCP space E will now be defined for $e \in E$:

$$T_e x := \frac{1}{2}(x + U_e x - U_{e'} x), \quad x \in A \tag{7.1}$$

In a von Neumann algebra, this becomes $T_e x = (ex + xe)/2$, which is the Jordan product of e and x. This is a first reason why some relevance is expected from the maps T_e on the order-unit space A.

A second reason is that $(A1)$ holds for all states μ if and only if $T_e f = T_f e$. This follows from (6.2). Thus, using the linear maps T_e, the symmetry condition $(A1)$ for the conditional probabilities is transformed to the very simple equation $T_e f = T_f e$. Actually, this is what brought Alfsen and Shultz to the discovery of $(A1)$.

Some further important characteristics of these linear maps will now be collected. Suppose $e \in E$ and $x \in A$. Using the above definition of T_e immediately yields $T_e x + T_{e'} x = x$. Moreover,

$$T_e x = \frac{1}{2}(x + U_e x - U_{e'} x) + U_e x + 0\, U_{e'} x, \tag{7.2}$$

is the spectral decomposition of T_e, and hence T_e has the three eigenvalues 0, 1/2, 1. Furthermore,

$$\begin{aligned} T_e U_e x &= \frac{1}{2}\left(U_e x + U_e^2 x - 0\right) = U_e x, \\ T_e U_{e'} x &= \frac{1}{2}\left(U_{e'} x + 0 - U_{e'}^2 x\right) = 0. \end{aligned} \tag{7.3}$$

Therefore

$$T_e^2 x = \frac{1}{2}(T_e x + T_e U_e x - T_e U_{e'} x) = \frac{1}{2}(T_e x + U_e x) \tag{7.4}$$

and thus

$$U_e = 2T_e^2 - T_e. \tag{7.5}$$

Originally, the T_e were derived from the U_e, and (7.5) means that the U_e can be reconstructed from the T_e.

The norm of the operator T_e will now be calculated. From $U_e[-\mathbb{1},\mathbb{1}] \subseteq [-e,e]$ and $U_{e'}[-\mathbb{1},\mathbb{1}] \subseteq [-e',e']$ it follows that

$$(U_e - U_{e'})[-\mathbb{1},\mathbb{1}] \subseteq [-\mathbb{1},\mathbb{1}] \tag{7.6}$$

and therefore $\|U_e - U_{e'}\| \leq 1$. Hence $\|T_e\| \leq 1$. Since $T_e e = e$ and $\|e\| = 1$ for $e \neq 0$, it follows that $\|T_e\| = 1$ unless $e = 0$ and $T_e = 0$.

Lemma 7.1. *If two events e and f in an UCP space E are orthogonal, then the linear maps T_e and T_f on the order-unit space A generated by E commute: $T_eT_f = T_fT_e$.*

Proof. By Lemma 3.3 the four projections $U_e, U_{e'}, U_f, U_{f'}$ commute pairwise. The linear maps T_e and T_f then commute by (7.1). □

An important link between the linear maps $T_e (e \in E)$ and Sorkin's interference term I_3 will be considered in the next sections.

8. Quantum Logics with $I_3 = 0$

The interference term $I_2^{\mu,f}(e_1,e_2)$ vanishes for all states μ and all events f if and only if $I_2(e_1,e_2) = U_{e_1+e_2} - U_{e_1} + U_{e_2} = 0$. The general validity of this identity for all orthogonal event pairs e_1, e_2 means that the map $e \to U_e$ is orthogonally additive in e. It will later be seen in Proposition 8.2 that the general validity of $I_3(e_1,e_2,e_3) = 0$ for all orthogonal event triples e_1, e_2, e_3 means that the map $e \to T_e$ is orthogonally additive in e. This will follow from the next lemma.

Lemma 8.1. *Suppose that E is a UCP space and that A is the order-unit space generated by E.*

(i) *If e_1, e_2, e_3 are three orthogonal events in E, then*

$$I_3(e_1,e_2,e_3) = U_{e_1+e_2+e_3} I_3\left((e_2+e_3)', e_2, e_3\right). \tag{8.1}$$

(ii) *If e and f are two orthogonal events in E, then*

$$T_e + T_f - T_{e+f} = \frac{1}{2} I_3\left(e, f, (e+f)'\right). \tag{8.2}$$

Proof. (i) $I_3((e_2+e_3)', e_2, e_3) = U_{\mathbb{1}} - U_{e_3'} - U_{e_2'} - U_{e_2+e_3} + U_{e_1} + U_{e_2} + U_{e_3}$ and then by Lemma 3.3,

$$\begin{aligned} U_{e_1+e_2+e_3} I_3((e_2+e_3)', e_2, e_3) &= U_{e_1+e_2+e_3} - U_{e_1+e_2} - U_{e_1+e_3} - U_{e_1+e_3} + U_{e_1} + U_{e_2} + U_{e_3} \\ &= I_3(e_1, e_2, e_3). \end{aligned} \tag{8.3}$$

(ii) With $e_1 := e, e_2 := f$ and $e_3 := (e+f)'$, it follows for $x \in A$:

$$\begin{aligned} T_{e+f}x &= \frac{1}{2}(x + U_{e_1+e_2}x - U_{e_3}x), \\ T_e x &= \frac{1}{2}(x + U_{e_1}x - U_{e_2+e_3}x), \\ T_f x &= \frac{1}{2}(x + U_{e_2}x - U_{e_1+e_3}x), \end{aligned} \tag{8.4}$$

and thus

$$\begin{aligned} T_e x + T_f x - T_{e+f}x &= \frac{1}{2}(x + U_{e_1}x - U_{e_2+e_3}x) + \frac{1}{2}(x + U_{e_2}x - U_{e_1+e_3}x) - \frac{1}{2}(x + U_{e_1+e_2}x - U_{e_3}x) \\ &= \frac{1}{2}(x - U_{e_1+e_2}x - U_{e_2+e_3}x - U_{e_1+e_3}x + U_{e_1}x + U_{e_2}x + U_{e_3}x) \\ &= \frac{1}{2} I_3(e_1, e_2, e_3)x. \end{aligned} \tag{8.5}$$

□

Proposition 8.2. *Suppose that E is a UCP space. Then the following three conditions are equivalent.*

(i) $I_3(e_1, e_2, e_3) = 0$ *for all orthogonal events* e_1, e_2, e_3 *in E.*

(ii) $I_3(e_1, e_2, e_3) = 0$ *for all orthogonal events* e_1, e_2, e_3 *in E with* $e_1 + e_2 + e_3 = \mathbb{1}$.

(iii) $T_{e+f} = T_e + T_f$ *for all orthogonal events e and f in E.*

Proof. The implication (i) ⇒ (ii) is obvious, the implication (ii) ⇒ (iii) follows from Lemma 8.1 (ii), and the implication (iii) ⇒ (i) from Lemma 8.1(i) and (ii) both. □

It now becomes clear that the symmetry condition $(A1)$ for the conditional probabilities implies the absence of third-order interference, that is, $I_3(e_1, e_2, e_3) = 0$ for all orthogonal events e_1, e_2, e_3. Recall that $(A1)$ implies the identity $T_e f = T_f e$ for all events e and f. Therefore, T_e is orthogonally additive in e and $I_3(e_1, e_2, e_3) = 0$ follows from Proposition 8.2.

9. Jordan Decomposition

In this section, the orthogonal additivity of T_e in e (by Lemma 7.1) will be used to define T_x for all x in the order-unit space A. The T_e are not positive and this involves some difficulties, which can be overcome when the real-valued bounded orthogonally additive functions on the UCP space E satisfy the so-called Jordan decomposition property or the stronger Hahn-Jordan decomposition property. These decomposition properties are named after the French mathematician Camille Jordan (1838–1922) who originally introduced the first one for functions of bounded variation and signed measures. The Jordan algebras are named after another person; this is the German physicist Pascual Jordan (1902–1980).

Consider functions $\rho : E \to \mathbb{R}$ which are orthogonally additive (i.e., $\rho(e+f) = \rho(e) + \rho(f)$ for orthogonal elements e and f in E) and bounded (i.e., $\sup\{|\rho(e)| : e \in E\} < \infty$). Let R denote the set of all these functions ρ on E. Then R comprises the state space S. The UCP space E is said to have the Jordan decomposition property if each $\rho \in R$ can be written in the form $\rho = s\mu - t\nu$ with two states μ and ν in S and nonnegative real number s and t. It has the ε-Hahn-Jordan decomposition property if the following stronger condition holds: for each $\rho \in R$ and every $\varepsilon > 0$ there are two states μ and ν in S, nonnegative real number s and t and an event e in E such that $\rho = s\mu - t\nu$ and $\mu(e) < \varepsilon$ as well as $\nu(e') < \varepsilon$. This ε-Hahn-Jordan decomposition property was studied in the framework of quantum logics by Cook [16] and Rüttimann [17]. The usual Hahn-Jordan decomposition property for signed measures is even stronger requiring that $\mu(e) = 0 = \nu(e')$.

Note that the projection lattices of von Neumann algebras have the Jordan decomposition property and the ε-Hahn-Jordan decomposition property, but this is not obvious. It is well known that these types of decomposition are possible for the bounded linear functionals on the algebra, but they are needed for the orthogonally additive real functions on the projection lattice. Bunce and Maitland Wright [18] showed that each such function on the projection lattice has a bounded linear extension to the whole algebra (this is the last step of the solution of the Mackey-Gleason problem which had been open for a long time), and then the decomposition of this extension provides the desired decomposition by considering the restrictions of the linear functionals to E.

Lemma 9.1. *Suppose that E is a UCP space with the Jordan decomposition property and $I_3(e_1, e_2, e_3) = 0$ for all orthogonal events e_1, e_2, e_3 in E. Then, for each x in the order-unit space A generated by E, the map $e \to T_e x$ from E to A has a unique $\sigma(A, V)$-continuous linear extension $y \to T_y x$ on A.*

Proof. Consider V as defined in the proof of Theorem 3.2. In the general case, only the inclusion $V \subseteq R$ holds, but the Jordan decomposition property ensures that $V = R$.

For each ρ in V and each x in A, define a function ρ^x on E by $\rho^x(e) := \widehat{\rho}(T_e x)$. By Proposition 8.2, ρ^x is orthogonally additive in e. Moreover, $|\rho^x(e)| \leqslant \|\rho\|\|T_e\|\|x\| \leqslant \|\rho\|\|x\|$ for $e \in E$. Thus ρ^x is bounded and lies in $R = V$. Let $\widehat{\rho}^x$ be its canonical embedding in $V^{**} = A^*$ and, with $y \in A$, consider the real-valued bounded linear map $\rho \to \widehat{\rho}^x(y)$ on V. It defines an element $T_y x$ in $V^* = A$ such that the map $y \to T_y x$ is linear as well as $\sigma(A, V)$-continuous on A and coincides with the original $T_e x$ for $y = e \in E$. □

10. The Product on the Order-Unit Space

Under the assumptions of the last lemma, a product can now be defined on the order-unit space A by $y \square x := T_y x$. It is linear in x as well as in y, but there is certain asymmetry concerning its $\sigma(A, V)$-continuity. The product $y \square x$ is $\sigma(A, V)$-continuous in $y \in A$ with $x \in A$ fixed and it is $\sigma(A, V)$-continuous in $x \in A$ with $y = e \in E$ fixed, but generally not with other $y \in A$. Moreover, $\mathbb{1} \square x = x$ and $y \square \mathbb{1} = y, e \square e = e$ and $\|e \square x\| \leq \|x\|$ for the elements x and

y in A and the events e in E. However, the inequality $\|y\Box x\| \leq \|y\|\|x\|$ is not yet available; this requires the ε-Hahn-Jordan decomposition property and will follow from the next lemma.

Lemma 10.1. *Suppose that E is a UCP space with the ε-Hahn-Jordan decomposition property. Then $[-\mathbb{1},\mathbb{1}]$ is identical with the $\sigma(A,V)$-closed convex hull of the set $\{e-e' : e \in E\}$ in A, and $[0,\mathbb{1}]$ coincides with the $\sigma(A,V)$-closed convex hull of E. Moreover, the extreme points of $[0,\mathbb{1}]$ lie in the $\sigma(A,V)$ closure of E.*

Proof. The inclusion $\overline{\text{conv}}\{e-e' : e \in E\} \subseteq [-\mathbb{1},\mathbb{1}]$is obvious, and it will now be shown that$[-\mathbb{1},\mathbb{1}] \subseteq \overline{\text{conv}}\{e-e' : e \in E\}$. Assume that an x exists in the interval $[-\mathbb{1},\mathbb{1}]$ which does not lie in the set $\overline{\text{conv}}\{e-e' : e \in E\}$. By the Hahn-Banach theorem, there is $\rho \in V$ with $\rho(x) > \sup\{\rho(e-e') : e \in E\}$. From $0 = (e-e')/2 + (e'-e)/2$ for any $e \in E$, it follows that $0 \in \text{conv}\{e-e' : e \in E\}$. Hence $\rho(x) > 0$ and $\rho \neq 0$.

Suppose $\varepsilon > 0$. Due to the ε-Hahn-Jordan decomposition property, there are two states μ and ν in S, nonnegative real number s and t and an event f in E such that $\rho = s\mu - t\nu$ and $\mu(f) < \varepsilon$ as well as $\nu(f') < \varepsilon$. Then

$$\begin{aligned} \rho(f'-f) &= s\mu(f') - s\mu(f) - t\nu(f') + t\nu(f) \\ &= s - 2s\mu(f) + t - 2t\nu(f') \\ &\geq s + t - 2\varepsilon(s+t) = (1-2\varepsilon)(s+t). \end{aligned} \tag{10.1}$$

Recall from the proof of Theorem 3.2 that $\|\rho\| = \inf\{r \in \mathbb{R} : r \geq 0 \text{ and } \rho \in r\,\text{conv}(S \cup -S)\}$. The case $s = t = 0$ cannot occur since this would imply $\rho = 0$. Therefore write

$$\rho = (s+t)\left(\frac{s}{s+t}\mu - \frac{t}{s+t}\nu\right), \tag{10.2}$$

to see that $\|\rho\| \leq s + t$.

Finally $\rho(x) \leq \|\rho\|\|x\| \leq \|\rho\| \leq s + t \leq \rho(f'-f)/(1-2\varepsilon)$. With ε small enough such that $(1-2\varepsilon)\rho(x) > \sup\{\rho(e-e') : e \in E\}$, an event f in E is found with $\rho(f'-f) \geq (1-2\varepsilon)\rho(x) > \sup\{\rho(e-e') : e \in E\}$, which is the desired contradiction with $e := f'$.

Thus $[-\mathbb{1},\mathbb{1}] = \overline{\text{conv}}\{e-e' : e \in E\}$. The map $x \to (x+\mathbb{1})/2$ is a $\sigma(A,V)$-continuous affine isomorphism mapping $[-\mathbb{1},\mathbb{1}]$ to $[0,\mathbb{1}]$ and $\{e-e' : e \in E\}$ to E; therefore $[0,\mathbb{1}] = \overline{\text{conv}}E$. The Krein-Milman theorem then yields that the extreme points of $[0,\mathbb{1}]$ lie in the $\sigma(A,V)$-closure of E. □

Lemma 10.2. *Suppose that E is a UCP space with the ε-Hahn-Jordan decomposition property and $I_3(e_1,e_2,e_3) = 0$ for all orthogonal events e_1, e_2, e_3 in E. Then $\|y\Box x\| \leq \|y\|\|x\|$ for all elements x, y in A.*

Proof. Suppose $e \in E$ and $x \in A$ with $\|x\| \leq 1$; that is, $x \in [-\mathbb{1},\mathbb{1}]$. Then $\|(e-e')\Box x\| = \|e\Box x - e'\Box x\| = \|T_e x - T_{e'} x\|$. Moreover, $T_e x - T_{e'} x = U_e x - U_{e'} x$ by (7.1) and $U_e x - U_{e'} x \in [-\mathbb{1},\mathbb{1}]$ by (7.6). Therefore, $\|(e-e')\Box x\| \leq 1$. This holds for all $e \in E$. Lemmas 9.1 and 10.1 then imply that $\|y\Box x\| \leq 1$ for all $y \in [-\mathbb{1},\mathbb{1}]$.

So far it has been seen that the Jordan decomposition property and the absence of third-order interference entail a product $y\Box x$ on the order-unit space A generated by the UCP space E which is linked to the conditional probabilities via the identity $U_e x = 2e\Box(e\Box x) - e\Box x$ holding for the events e and the elements x in A. This follows from (7.5).

The product $y\Box x$ is neither commutative nor associative and thus far away from the products usually considered in mathematical physics. The common product of linear operators is not commutative but associative and the Jordan produc $a \circ b := (ab + ba)/2$ is not associative but commutative. □

11. Jordan Algebras

The elements of a UCP space E represent the events or propositions; they can be considered also as observables with the simple discrete spectrum $\{0,1\}$ representing a yes/no test experiment. However, what is the meaning of the elements x in the order-unit space A generated by E? One might expect that they represent other observables with a larger and possibly nondiscrete spectrum and that $\widehat{\mu}(x)$ is the expectation value of the observable represented by x in the state μ. If it is assumed that they do so, one would also expect a certain behaviour.

First, one would like to identify the elements $x^2 = x\Box x$ and, more generally, x^n (inductively defined by $x^{n+1} := x\Box x^n$) in A with the application of the usual polynomial functions $t \to t^2$ or $t \to t^n$ to the observable.

Second, the expectation value of the square x^2 should be nonnegative; this means that $\widehat{\mu}(x^2) \geqslant 0$ for all x in A and for all states μ, and therefore $x^2 \geq 0$ in the order-unit space A.

Third, one would like to have the usual polynomial functional calculus allocating an element $p(x)$ in A to each polynomial function p such that $p_1(x)\Box p_2(x) = q(x)$ whenever p_1 and p_2 are two polynomial functions and the polynomial function q is their product. Since the product in A is not associative, $x^n\Box x^m$ do not need to be identical with x^{n+m}. If $x^n\Box x^m = x^{n+m}$ holds for all x in A and for all natural numbers n and m, A is called power-associative. The availability of the polynomial functional calculus for all elements in A means that A is power associative and vice versa.

The following theorem shows that these requirements make A a commutative Jordan algebra; that is, the product is Abelian and satisfies the Jordan condition $x\Box(x^2\Box y) = x^2\Box(x\Box y)$. The Jordan condition is stronger than power associativity and, in general, power associativity does not imply the Jordan condition.

Theorem 11.1. *Suppose that E is a UCP space with the ε-Hahn-Jordan decomposition property and with $I_3(e_1, e_2, e_3) = 0$ for all orthogonal events e_1, e_2, e_3 in E. Furthermore, assume that the order-unit space A generated by E, together with the multiplication □, is power associative and that $x^2 \geq 0$ for all x in A. Then the multiplication □ is commutative and A is a Jordan algebra. Moreover, E is a $\sigma(A,V)$-dense subset of $\{e \in A : e^2 = e\}$.*

Proof. Suppose $x, y \in A$. The positivity of the squares implies the Cauchy-Schwarz inequality $(\widehat{\mu}(x \Box y))^2 \leqslant \widehat{\mu}(x^2)\widehat{\mu}(y^2)$ with states μ. Then, with $y = \mathbb{1}$, $(\widehat{\mu}(x))^2 \leqslant \widehat{\mu}(x^2)$. Recall from the proof of Theorem 3.2 that $\|x\| = \sup\{|\widehat{\mu}(x)| : \mu \in S\}$. Therefore

$$\left\|x^2\right\| \leqslant \|x\|^2 = \sup\left\{ \left(\widehat{\mu}(x)\right)^2 : \mu \in S\right\} \leq \sup\left\{\widehat{\mu}\left(x^2\right) : \mu \in S\right\} = \left\|x^2\right\| \tag{11.1}$$

and hence

$$\left\|x^2\right\| = \|x\|^2. \tag{11.2}$$

Moreover,

$$\left\|x^2 + y^2\right\| = \sup\left\{\widehat{\mu}\left(x^2\right) + \widehat{\mu}\left(y^2\right) : \mu \in S\right\} \geq \sup\left\{\widehat{\mu}\left(x^2\right) : \mu \in S\right\} = \left\|x^2\right\|. \tag{11.3}$$

A second commutative product is now introduced on A by $x \circ y := (x\Box y + y\Box x)/2$. Note that, due to the power associativity, x^n is identical with the two products. Equipped with this new commutative product, A then becomes a Jordan algebra. This follows from a result by Iochum and Loupias [19] (see also [20]).

Since A is the dual of V, A with $\circ$ is a so-called JBW algebra [12]. The JBW algebras represent the Jordan analogue of the W^* algebras and these are the same as the von Neumann algebras but characterized by a different and more abstract set of axioms. The extreme points of $[0, \mathbb{1}]$ in a JBW algebra are the idempotent elements. Then, by Lemma 10.1, $E \subseteq \{e \in A : e^2 = e\} = \text{ext}[0, \mathbb{1}] \subseteq E$ and hence E is $\sigma(A, V)$ dense in $\{e \in A : e^2 = e\}$.

Suppose $e \in E$. If μ is any state on E with $\mu(e) > 0$, the map $f \rightarrow \widehat{\mu}(2e \circ (e \circ f) - e \circ f)/\mu(e)$ for $f \in E$ defines a version of the conditional probability (which was shown in [2]) and must coincide with $\widehat{\mu}(U_e f)/\mu(e)$. Therefore $U_e f = 2e \circ (e \circ f) - e \circ f$ and

$$e\Box f = T_e f = \frac{(f + U_e f - U_{e'} f)}{2} = e \circ f \tag{11.4}$$

for all e and f in E. The product on a JBW algebra is $\sigma(A, V)$ continuous in each component. It then follows in a first step that $e\Box y = e \circ y$ for all e in E and y in A, and in the second step that $x\Box y = x \circ y$ for all x and y in A. Note that the order of the two steps is important because of the asymmetry of the $\sigma(A, V)$ continuity of the product $x\Box y$. Thus finally $x\Box y = y\Box x$. $\Box$

When the starting point is the projection lattice E in a JBW algebra M (e.g., the self adjoint part of a von Neumann algebra with the Jordan product) without type I_2 part, the order-unit space A generated by E is the second dual $A = M^{**}$ of M. It contains M by the canonical embedding in its second dual but is much larger (unless M has a finite dimension); M is the norm-closed linear hull of E in A, while A is the $\sigma(A, V)$-closed linear hull of E.

A rich theory of Jordan algebras is available and most of them can be represented as a Jordan subalgebra of the self-adjoint linear operators on a Hilbert space. The major exception is the Jordan algebra consisting of the 3×3 matrices with octonionic entries, and the other exceptions are the so-called exceptional Jordan algebras which all relate to this one [12].

A reconstruction of quantum mechanics up to this point has thus been achieved from a few basic principles. The first one is the absence of third-order interference and the second one is the postulate that the elements of the constructed algebra exhibit a behaviour which one would expected from observables. The third one, the ε-Hahn-Jordan decomposition property, is less conceptional and more a technical mathematical requirement.

12. Conclusions

The combination of a simple quantum logical structure with the postulate that unique conditional probabilities exist provides a powerful general theory which includes quantum mechanics as a special case. It is useful for the reconstruction of quantum mechanics from a few basic principles as well as for the identification of typical properties of quantum mechanics that distinguish it from other more general theories. Three such properties have been studied: a novel bound for quantum interference, a symmetry condition for the conditional probabilities, and the absence of third-order interference ($I_3 = 0$); the third property has been the major focus of this paper.

In the framework of the quantum logics with unique conditional probabilities, the absence of third-order interference ($I_3 = 0$) has some important consequences. It entails the existence of a product in the order-unit space generated by the quantum logic, which can be used to characterize those quantum logics that can be embedded in the projection lattice in a Jordan algebra. Most of these Jordan algebras can be represented as operator algebras on a Hilbert space, and a reconstruction of quantum mechanics up to this point is thus achieved.

As the identity $I_2 = 0$ distinguishes the classical probabilities, the identity $I_3 = 0$ thus characterizes the quantum probabilities. It may be expected that there are other more general theories with $I_3 \neq 0$, and the quantum logics with unique conditional probabilities may provide an opportunity to establish them. For the time being, however, the projection lattices in the exceptional Jordan algebras are the only known concrete examples which do not fit into the quantum mechanical standard model but still have all the properties discussed in the present paper and do not exhibit third-order interference. Further examples can be expected from Alfsen and Shultz's spectral duality, but unfortunately all the known concrete examples of this theory are either covered by the Jordan algebras or do not satisfy the Gleason-like extension theorem (part (i) of Proposition 3.1). Besides the examples with $I_3 \neq 0$, it would also be interesting to find examples where the identity $I_3 = 0$ holds, but where the product on the order-unit space is not power associative or where the squares are not positive or where the symmetry condition ($A1$) for the conditional probabilities does not hold.

The possibility that no such examples exist is not anticipated but cannot be ruled out as long as no one has been found. It would mean that every quantum logic with unique conditional probabilities can be embedded in the projection lattice in a Jordan algebra and that third-order interference never occurs. Moreover, the further postulates concerning the behaviour of the observables (power associativity, positive squares) would then as well become redundant in the reconstruction of quantum mechanics. If this could be proved, the reconstruction process could be cleared up considerably.

In any case, it seems that third-order interference can play a central role in the reconstruction of quantum mechanics from a few basic principles, in an axiomatic access to quantum mechanics with a small number of interpretable axioms, as well as in the characterisation of the projection lattices in von Neumann algebras or their Jordan analogue, the JBW algebras, among the quantum logics. Mathematically, the symmetry property ($A1$) of the conditional properties (see Section 6) can play the same role, which was shown by Alfsen and Shultz [4] and by the author [3]. However, it has a less clear physical meaning than the third-order interference and, therefore, the approach of the present paper seem to be superior form a physical point of view.

Acknowledgments

The author would like to thank the Perimeter Institute and Philip Goyal for hosting the workshop "Reconstructing Quantum Theory" in August 2009 and Howard Barnum for the inspiring discussions during this workshop and for directing my attention to Sorkin's higher order interference.

References

[1] R. D. Sorkin, "Quantum mechanics as quantum measure theory," *Modern Physics Letters A*, vol. 9, no. 33, pp. 3119–3127, 1994.

[2] G. Niestegge, "Non-Boolean probabilities and quantum measurement," *Journal of Physics A*, vol. 34, no. 30, pp. 6031–6042, 2001.

[3] G. Niestegge, "A representation of quantum measurement in order-unit spaces," *Foundations of Physics*, vol. 38, no. 9, pp. 783–795, 2008.

[4] E. M. Alfsen and F. W. Shultz, "On non-commutative spectral theory and Jordan algebras," *Proceedings of the London Mathematical Society*, vol. 38, no. 3, pp. 497–516, 1979.

[5] A. M. Gleason, "Measures on the closed subspaces of a Hilbert space," *Journal of Mathematics and Mechanics*, vol. 6, pp. 885–893, 1957.

[6] E. Christensen, "Measures on projections and physical states," *Communications in Mathematical Physics*, vol. 86, no. 4, pp. 529–538, 1982.

[7] S. Maeda, "Probability measures on projections in von Neumann algebras," *Reviews in Mathematical Physics*, vol. 1, no. 2-3, pp. 235–290, 1989.

[8] F. J. Yeadon, "Measures on projections in W^*-algebras of type II_1," *The Bulletin of the London Mathematical Society*, vol. 15, no. 2, pp. 139–145, 1983.

[9] F. J. Yeadon, "Finitely additive measures on projections in finite W^*-algebras," *The Bulletin of the London Mathematical Society*, vol. 16, no. 2, pp. 145–150, 1984.

[10] E. M. Alfsen, *Compact Convex Sets and Boundary Integrals*, Springer, Berlin, Germany, 1971.

[11] L. Asimow and A. J. Ellis, *Convexity Theory and Its Applications in Functional Analysis*, Academic Press, New York, NY, USA, 1980.

[12] H. Hanche-Olsen and E. Størmer, *Jordan Operator Algebras*, Pitmann, Boston, Mass, USA, 1984.

[13] E. M. Alfsen and F. W. Shultz, *Non-Commutative Spectral Theory for Affine Function Spaces on Convex Sets*, vol. 172 of *Memoirs of the American Mathematical Society*, 1976.

[14] H. Barnum, J. Emerson, and C. Ududec, "Three slit experiments and the structure of quantum theory," *Foundations of Physics*, vol. 41, no. 3, pp. 396–405, 2011.

[15] G. Niestegge, "A hierarchy of compatibility and comeasurability levels in quantum logics with unique conditional probabilities," *Communications in Theoretical Physics*, vol. 54, no. 6, pp. 974–980, 2010.

[16] T. A. Cook, "The geometry of generalized quantum logics," *International Journal of Theoretical Physics*, vol. 17, no. 12, pp. 941–955, 1978.

[17] G. T. Rüttimann, "Jordan-Hahn decomposition of signed weights of finite orthogonality spaces," *Commentarii Mathematici Helvetici*, vol. 52, no. 1, pp. 129–144, 1977.

[18] L. J. Bunce and J. D. Maitland Wright, "The Mackey-Gleason problem," *Bulletin of the American Mathematical Society*, vol. 26, no. 2, pp. 288–293, 1992.

[19] B. Iochum and G. Loupias, "Banach-power-associative algebras and J-B algebras," *Annales de l'Institut Henri Poincaré*, vol. 43, no. 2, pp. 211–225, 1985.

[20] G. Niestegge, "A representation of quantum measurement in nonassociative algebras," *Foundations of Physics*, vol. 39, no. 2, pp. 120–136, 2009.

2

Combination Mode of Internal Waves Generated by Surface Wave Propagating over Two Muddy Sea Beds

Ray-Yeng Yang[1] and Hwung Hweng Hwung[2]

[1] *Tainan Hydraulics Laboratory, Research Center of Ocean Environment and Technology, Tainan Hydraulics Laboratory, National Cheng Kung University, No.1, Ta-Hsueh Road, Tainan 701, Taiwan*
[2] *Department of Hydraulic and Ocean Engineering, Research Center of Ocean Environment and Technology, National Cheng Kung University, No.1, Ta-Hsueh Road, Tainan 701, Taiwan*

Correspondence should be addressed to Ray-Yeng Yang, ryyang@mail.ncku.edu.tw

Academic Editor: Sanith Wijesinghe

When surface wave propagating over the two layer system usually induces internal wave in three different modes: they are external, internal and combination. In the present study, the nonlinear response of an initially flat sea bed, with two muddy sections, to a monochromatic surface progressive wave was investigated. From this theoretical result, it shows that a surface water wave progressing over two different muddy sections, the surface wave will excite two opposite-traveling short interfacial waves, forming a nearly standing wave at the interface of the fresh water and the muddy layer. Meanwhile, two opposite-outgoing "mud" waves each with very long wavelength will be simultaneously induced at the interface of two muddy sections. As a result, the amplitudes of the two short internal waves are found to grow exponentially in time. Furthermore, it will be much difficult to excite the internal waves when surface water wave progressing over two muddy sections with the large density gap.

1. Introduction

Generally, wave propagating over the two layer system usually induces internal wave in three different modes: they are external, internal, and combination, as shown in Figure 1. These internal waves dissipate rapidly with imaginary and real wave numbers of a similar magnitude and remove energy from the surface waves. Also, nonlinear damping mechanisms have been proposed based on wave-wave instabilities and interactions.

The present work is motivated by recent studies on the interaction between a progressive surface wave and the nearly standing subharmonic internal waves in a two-layer system. It is well known that the loading of progressive surface waves, a silty sediment

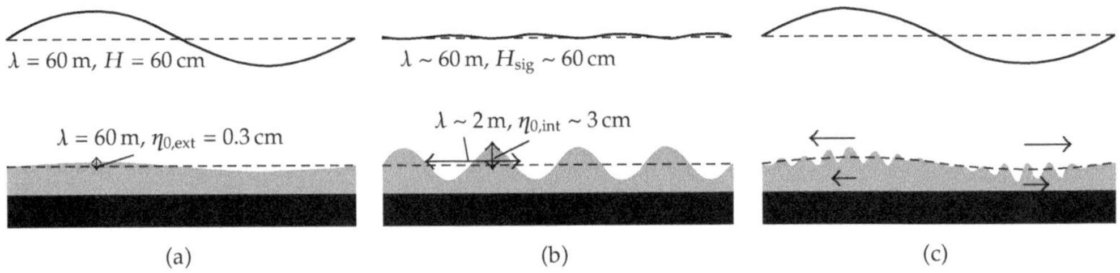

Figure 1: Schematic diagram of the two layer system showing (a) the external wave mode with the same frequency and wavelength for both the air-water and water-mud interface waves, (b) the internal wave mode with the same frequency, but much shorter wavelength for the water-mud interface waves, and (c) the combination of the two. The arrows in panel (c) indicate shear across the mud water interface, which could lead to a shear instability mechanism to generate the internal mode waves in each half cycle of the surface wave.

bed was repeatedly and extensively fluidized. The broad-based interest in understanding this phenomenon induces from the application to studies in sediment transport, wave attenuation, and the design of marine structures. The nonlinear response of an initially flat sea bed to a monochromatic surface progressive wave was studied by Wen [1] for using the multiple scale perturbation method. She found that two opposite-traveling subliminal internal "mud" waves are triggered and form a resonant triad with the surface wave. The resonant generation of internal waves on sediment bed was presented as a new mechanism of sediment suspension. Nonlinear wave interactions are considered to be as an important aspect of the dynamics of the oceans and the atmosphere. Of particular interest are resonant interactions, which are important in the redistribution of energy among wave modes with different spatial and temporal scales. Ball [2] used a second-order nonlinear resonance theory to analyze that linear growth of an internal wave could result from the interactions between two finite-amplitude surface waves. Watson et al. [3] extended the investigation by analyzing how a spectrum of surface waves could generate a corresponding spectrum of internal waves. Wen's [1] work was followed by Hill and Foda [4], who treated the problem in two dimensions for both an inviscid and a viscous lower layer. Hill and Foda [5] and Jamali [6] have presented theoretical and experimental studies of the resonant interaction between a surface wave and two oblique interfacial waves. Despite many similarities between the findings, there is one seemingly major difference. Hill and Foda's [5] analysis predicts only narrow bands of frequency, density ratio, and direction angle within which growth is possible. However, Jamali [6] predicted and observed wave growth over wide ranges of frequency and direction angle, and for all the density ratios that he investigated. Therefore, Jamali et al. [7] presented the study to investigate the contradictory results between the findings of Hill and Foda [5] and Jamali [6]. From their result, it is showed that the crucial difference between the two studies is in the dynamic interfacial boundary condition. The boundary condition used by Hill and Foda is missing a term proportional to the time derivative of the square of the velocity shear across the interface. When this missing term is included in the analysis, the theoretical predictions are consistent with the results of Jamali's [6] laboratory experiments. Both of the Hill and Foda's [5] and Jamali's [6] study found that the interfacial waves are short, have a frequency of nearly half that of the surface wave, and propagate in nearly opposite directions. The nonlinear response of an initially flat sea bed, with two muddy sections, to a monochromatic surface progressive wave was investigated in the present study. Based on an analysis similar to that of Hill and Foda's paper [5], the multiple-scale perturbation method was adopted, and the boundary value problem was

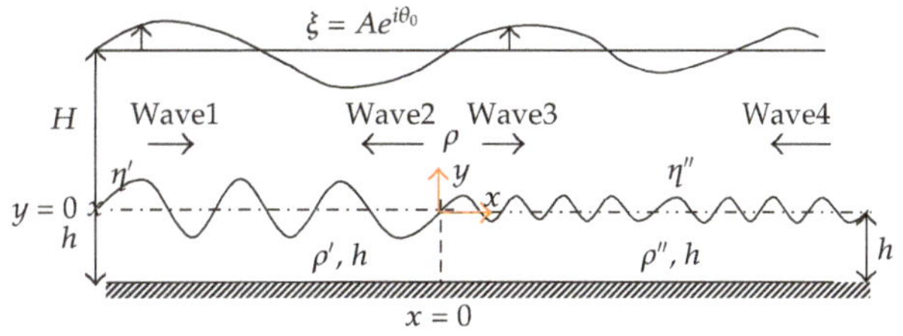

Figure 2: Configuration of the problem in the two-layer system with two muddy sections.

expanded in a power series of the surface-wave steepness. The linear harmonics and the conditions for resonance were obtained by the leading order, while the temporal evolution equations for the internal-wave amplitudes were investigated by a second-order analysis. It was found that result for equal density of two muddy sections is similar to that of Hill and Foda's paper [5]. Two opposite-traveling internal "mud" waves are selectively excited and form a resonant triad with the progressive surface wave. However, for a surface water wave progressing over two different muddy sections, the surface wave will also excite only two opposite-traveling short interfacial waves, forming a nearly standing wave at the interface of the fresh water and the muddy layer. Meanwhile, two opposite-outgoing "mud" waves each with very long wavelength will be simultaneously induced at the interface of two muddy sections. As a result, the amplitudes of the two short internal waves are found to grow exponentially in time.

2. Formulation

As shown in Figure 2, the origin of a two-dimensional Cartesian coordinate system is placed on the undisturbed interface between a surface layer of depth H and density ρ and a lower depth h at the interface between two different density sections, ρ and ρ''. The y-coordinate is defined as pointing vertically upward, and the density rations, $\gamma = \rho/\rho$ and $\gamma = \rho/\rho^{''}$, are assumed to be less than unity. To the leading order, the wave field is assumed to be made up of a linear progressive surface wave of amplitude A, wave number k, and frequency ω, propagating in the positive x-direction. Firstly, we assumed that the perturbation internal waves at left muddy section have amplitudes a_1 and a_2, wave numbers λ_1 and λ_2, frequencies σ_1 and σ_2, and propagate in the positive and negative x-directions, respectively. Then for the right muddy section, there are also the perturbation internal waves with amplitudes a_3 and a_4, wave numbers λ_3 and λ_4, frequencies σ_3 and σ_4, and move in the positive and negative x-directions, respectively. It is noted that a_1, a_2, a_3, a_4, and A are taken to be complex and $\lambda_1 \sim \lambda_4$ and $\sigma_1 \sim \sigma_4$ are all defined to have positive real values. For resonant interactions to occur in two muddy sections, the following resonance conditions are imposed on the four-handed wave numbers and frequencies:

$$\begin{aligned}
&\lambda_1 - \lambda_2 - \lambda_4 = k, \\
&\lambda_1 + \lambda_3 - \lambda_4 = k, \\
&\sigma_1 - \sigma_2 - \sigma_4 = \omega, \\
&\sigma_1 + \sigma_3 - \sigma_4 = \omega.
\end{aligned} \tag{2.1}$$

Expressing the flow-field in the two-layer inviscid system in terms of a velocity potential Φ, we assume that Φ satisfies Laplace's equation throughout the depth of the fluid

$$\nabla^2\Phi = 0, \qquad -h \le y \le H + \xi. \tag{2.2}$$

The displacement ξ of the free surface from its static elevation $y = H$ is given by

$$\xi = Ae^{i\theta_0} \tag{2.3}$$

in which the phase function $\theta_0 \equiv (kx - \omega t)$. The free surface will oscillate synchronously with ξ, with an amplitude A in response to the passage of the surface wave.

At the free surface, the usual kinematic and dynamic conditions are given by

$$\begin{aligned} \frac{D\xi}{Dt} &= \Phi y, \qquad y = H + \xi, \\ \Phi_t + g\xi + \frac{1}{2}\nabla\Phi \cdot \nabla\Phi &= 0, \qquad y = H + \xi. \end{aligned} \tag{2.4}$$

At the interface between the water and the slurry, there are similar conditions of continuity of pressure and vertical velocity. Therefore, on the left muddy section we have

$$\begin{aligned} \rho\left(\Phi_t^l + g\eta' + \frac{1}{2}\nabla\Phi^l \cdot \nabla\Phi^l\right)^+ &= \rho'\left(\Phi_t^l + g\eta' + \frac{1}{2}\nabla\Phi^l \cdot \nabla\Phi^l\right)^-, \quad y = \eta', \\ \frac{D\eta'}{Dt} &= \Phi_y^{l+} = \Phi_y^{l-}, \quad y = \eta'. \end{aligned} \tag{2.5}$$

Similar conditions of continuity of normal velocities and traction stresses are imposed at the disturbed interface on the right muddy section

$$\begin{aligned} \rho\left(\Phi_t^r + g\eta'' + \frac{1}{2}\nabla\Phi^r \cdot \nabla\Phi^r\right)^+ &= \rho''\left(\Phi_t^l + g\eta'' + \frac{1}{2}\nabla\Phi^r \cdot \nabla\Phi^r\right)^-, \quad y = \eta'', \\ \frac{D\eta''}{Dt} &= \Phi_y^{r+} = \Phi_y^{r-}, \quad y = \eta''. \end{aligned} \tag{2.6}$$

Meanwhile at the interface between two muddy mass of sections, conditions of continuity of pressure and continuity of water and mud are shown as followed

$$\begin{aligned} \left[\rho(\Phi_l)_x\right]_{y=\eta'}^{y=H+\xi} &= \left[\rho(\Phi_r)_x\right]_{y=\eta''}^{y=H+\xi}, \quad \text{at } x = 0, \text{ (forwater)}, \\ \left[\rho'(\Phi_l)_x\right]_{y=0}^{y=\eta'} &= \left[\rho''(\Phi_r)_x\right]_{y=0}^{y=\eta''}, \quad \text{at } x = 0, \text{ (formud)}. \end{aligned} \tag{2.7}$$

Note that the l and r superscripts denote the tiny interface between left muddy section and right muddy section, and + and – superscripts show evaluation just above and just below the interface between the water and the fluid-mud, respectively. And finally, at the bottom of mud layer, we have the no flux condition

$$\Phi_y = 0, \qquad y = -h. \tag{2.8}$$

With the system for the interface and free surface being assumed to be only weakly nonlinear interaction, Taylor's expansion at the interface and free surface is used to eliminate ξ, η', and η''. And by using the method of successive approximations, we have three free wave harmonics in the velocity potential Φ^l for left side ($x < 0$), that is we assume the following expansion for Φ^l:

$$\Phi^l = \varepsilon\varphi(y)e^{i(kx+\omega t)} + \varepsilon^2\left\{\psi(y)e^{i(\lambda_1 x+\sigma_1 t)} + \chi(y)e^{i(\lambda_2 x-\sigma_2 t)}\right\} + \Phi^l_{n.l.} + \text{c.c.} \tag{2.9}$$

The first three terms of the above expansion represent the three free harmonics; first term is the surface wave, and the following two terms are for the internal waves, $\Phi^l_{n.l.}$ represents the harmonics, and c.c. denotes complex conjugate. The expansion parameter is taken to be the steepness of the surface wave, $\varepsilon = kA$. Our analysis is restricted to the case of small internal waves, only so the internal wave harmonics appear at $O(\varepsilon^2)$ and not at $O(\varepsilon)$.

Similarly, the velocity potential Φ^r for fight side ($x > 0$) is expanded as follows:

$$\Phi^r = \varepsilon\varphi(y)e^{i(kx-\omega t)} + \varepsilon^2\left\{\psi(y)e^{i(\lambda_3 x-\sigma_3 t)} + \chi(y)e^{i(\lambda_4 x+\sigma_4 t)}\right\} + \Phi^r_{n.l.} + \text{c.c.} \tag{2.10}$$

3. Perturbation Solution

For the solution procedure, a standard perturbation analysis for a weakly nonlinear wave-field system is used. The solution procedure involves solving the above boundary value problem in an ordered sequence, by separating terms in the governing equations and the boundary conditions according to their order in ε and their phase. Substituting (2.9) and (2.10) in the governing equations and collecting the leading $O(\varepsilon)$ order, the approximate linear solutions for the interacting harmonics are obtained. For the dispersion relationship of surface wave,

$$\begin{gathered}\omega^4\left\{(\gamma'\gamma'')^{1/2}\coth(kh)\coth(kH) + 1\right\} - \omega^2(\gamma'\gamma'')^{1/2} \\ \times\{\coth(kH)+\coth(kh)\}gk + \left\{(\gamma'\gamma'')^{1/2} - 1\right\}g^2k^2 = 0.\end{gathered} \tag{3.1}$$

For the dispersion relationship for internal waves of two muddy sections, the following is obtained

$$\sigma_1^2 = \frac{g\left\{1-(\gamma'\gamma'')^{1/2}\right\}}{(\gamma'\gamma'')^{1/2}+\coth(\lambda_1 h)}, \quad x<0,$$

$$\sigma_2^2 = \frac{g\left\{1-(\gamma'\gamma'')^{1/2}\right\}}{(\gamma'\gamma'')^{1/2}+\coth|\lambda_2 h+\lambda_4 h|}, \quad x<0,$$

$$\sigma_3^2 = \frac{g\left\{1-(\gamma'\gamma'')^{1/2}\right\}}{(\gamma'\gamma'')^{1/2}+\coth(\lambda_1 h+\lambda_3 h)}, \quad x>0, \tag{3.2}$$

$$\sigma_4^2 = \frac{g\left\{1-(\gamma'\gamma'')^{1/2}\right\}}{(\gamma'\gamma'')^{1/2}+\coth|(\lambda_4 h)|}, \quad x>0.$$

Quadratic interactions between the above linear harmonics are analyzed at the second order, $O(\varepsilon^2)$. Since the homogeneous version of the boundary value problem had a nontrivial solution, the inhomogeneous problem has a solution only if the forcing terms are orthogonal to the homogeneous solution. Invoking solvability, through the use of Green's theorem, the desired evolution equations via the internal wave amplitudes are obtained. For simplicity, the amplitude equations—via solvability—may be straightforwardly shown as follows:

$$\frac{da_l}{dt} = i\alpha_l a_r A,$$
$$\frac{da_r}{dt} = i\alpha_r a_l A^*, \tag{3.3}$$

in which $a_l = a_1 + a_2 + a_4$ and $a_r = a_1 + a_3 + a_4$. α_l and α_r are the interaction coefficients. Taking cross differentiation of (3.3), the growth for amplitudes of the internal waves is governed by exponentials:

$$a_l, a_r \propto \exp\{\pm\sqrt{-\alpha_l\alpha_r}At\},$$

$$\alpha = \sqrt{-\alpha_l\alpha_r} = \frac{1}{2}|A|\left\{\frac{-g}{H}\left(1-2(\gamma'\gamma'')^{1/2}\frac{\sigma_2+\sigma_4}{\sigma_1+\sigma_3}+\frac{k}{\lambda_1+\lambda_3}\right)\right.$$
$$\left.\times\left(1-2(\gamma'\gamma'')^{1/2}\frac{\sigma_1+\sigma_3}{\sigma_2+\sigma_4}-\frac{k}{\lambda_2+\lambda_4}\right)(\lambda_1+\lambda_3)(\lambda_2+\lambda_4)\right\}^{1/2}. \tag{3.4}$$

Under the condition of α being purely imaginary, then the amplitudes a_l and a_r are not able to grow with time. Equivalently, the surface water wave will be stable to these internal wave perturbations. Only when α is real, the growth of the internal wave amplitudes will occur.

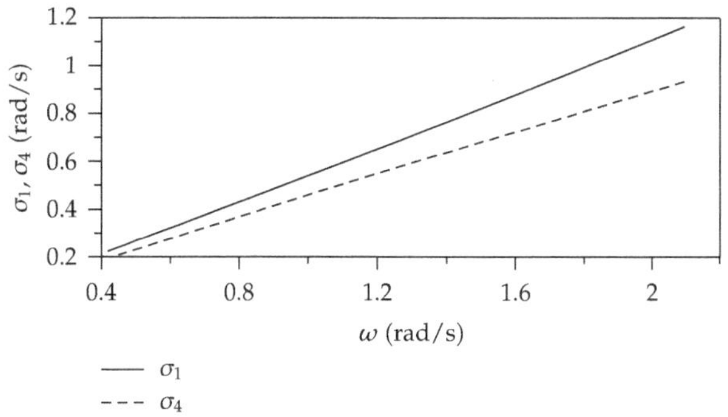

Figure 3: Two internal wave frequencies as functions of surface wave frequency ω, $H = 4\,\text{m}$, $\gamma' = \gamma'' = 0.83$, and $h = 0.1\,\text{m}$.

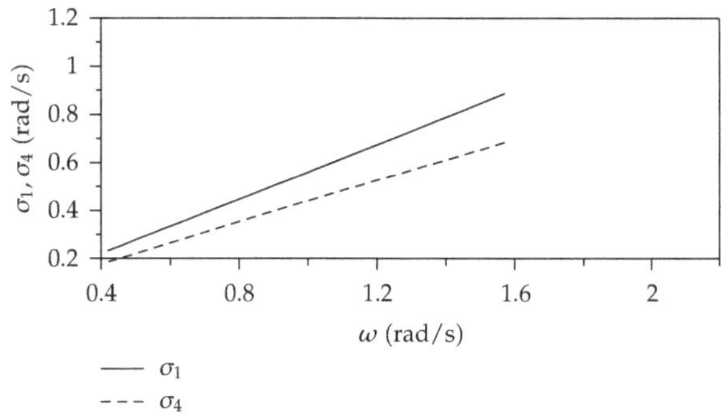

Figure 4: Two internal wave frequencies as functions of surface wave frequency ω, $H = 4\,\text{m}$, $\gamma' = \gamma'' = 0.83$, and $h = 0.3\,\text{m}$.

4. Results and Discussion

In an effort to develop a mechanism of the generation for internal waves of two muddy sections by surface wave, an inviscid second-order nonlinear resonant interaction has been analyzed. An example of the same density for two muddy sections, the internal-wave parameters as functions of ω for the cases of $\gamma' = \gamma'' = 0.83$, $H = 4\,\text{m}$, $h = 0.1\,\text{m}$ and $\gamma' = \gamma'' = 0.83$, $H = 4\,\text{m}$, $h = 0.3\,\text{m}$ is shown in Figures 3 and 4. The result of the present study is similar to that of Hill and Foda's paper [5]. It is obvious that there is a surface-wave frequency, a cut-off frequency, below which resonant triads do not exist. The cut-off frequency is a function of H, h, γ', and γ''. Furthermore, the critical frequency corresponds to these cases in which the internal waves in muddy layer are nearly subharmonic to the surface wave, that is, $\sigma_1 \approx \sigma_4 \approx \omega/2$, and are propagating in the same and opposite directions to the surface wave. Figure 5 shows the growth rate of internal waves as a function of the surface wave frequency for $\gamma'' = 0.83$, $H = 4\,\text{m}$, and $h = 0.1\,\text{m}$ and $0.3\,\text{m}$. The growth rate of internal waves will be suppressed when the depth of mud layer increases. That is because if the mud layer depth being too thick, the interaction among surface and interfacial internal waves weakens. Internal waves are also subdued. However, for a surface water wave progressing over two different muddy sections ($\gamma' \neq \gamma''$), the surface wave will also excite only two opposite-traveling interfacial short waves, forming a nearly standing wave at the interface of the fresh water and the muddy layer. Meanwhile, two opposite-outgoing

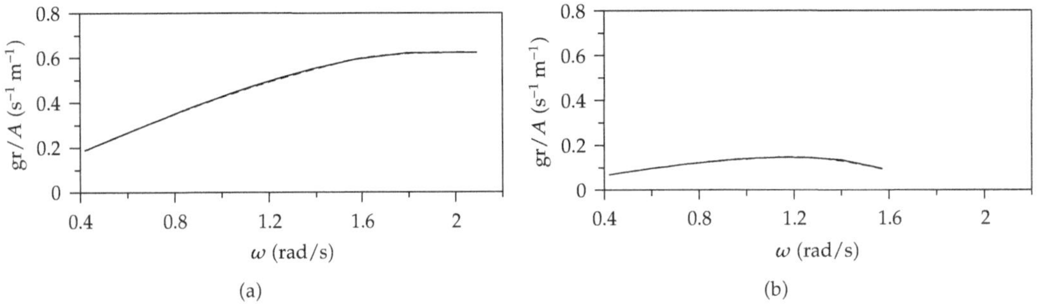

Figure 5: Theoretical determined internal wave growth rates as functions of surface wave frequency ω, $H = 4\,\text{m}$, $\gamma' = \gamma'' = 0.83$, (a) $h' = 0.1\,\text{m}$, and (b) $h = 0.3\,\text{m}$.

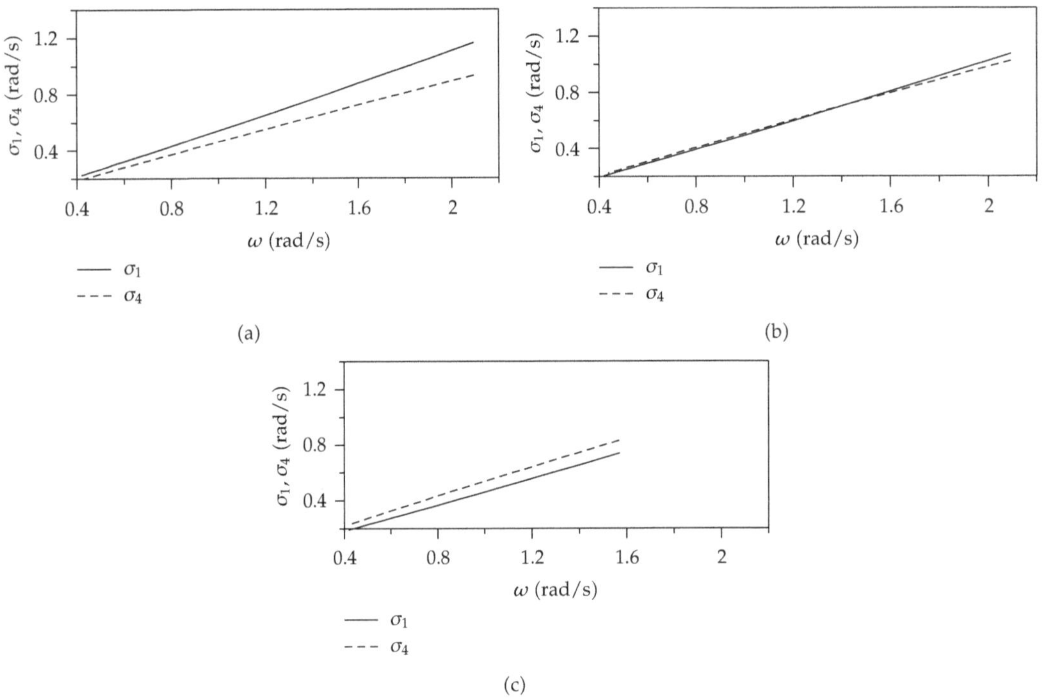

Figure 6: Two internal wave frequencies as functions of surface wave frequency ω, $H = 4\,\text{m}$, $h = 0.1\,\text{m}$ (a) $\gamma' = \gamma'' = 0.83$, (b) $\gamma' = 0.83$, $\gamma'' = 0.77$, and (c) $\gamma' = 0.83$, $\gamma'' = 0.71$.

"mud" waves each with very long wavelength will be simultaneously induced at the interface between two muddy sections. The wave numbers of these two long internal waves will be of the order $10^{-5} \sim 10^{-4}$ and have opposite value; $\lambda_2 = -\lambda_3 \approx 10^{-5} \sim 10^{-4}$ for the cases of different mud layer depth. Figure 6 shows that if the density of the right muddy section becomes increasingly larger than that of left side, then the right resonant internal wave frequency will also have increasingly higher value than the left. Meanwhile if the density difference between two muddy sections increases, the growth rate of the two resonant short internal waves will be suppressed. This result can be seen in Figure 7. The result means that it will

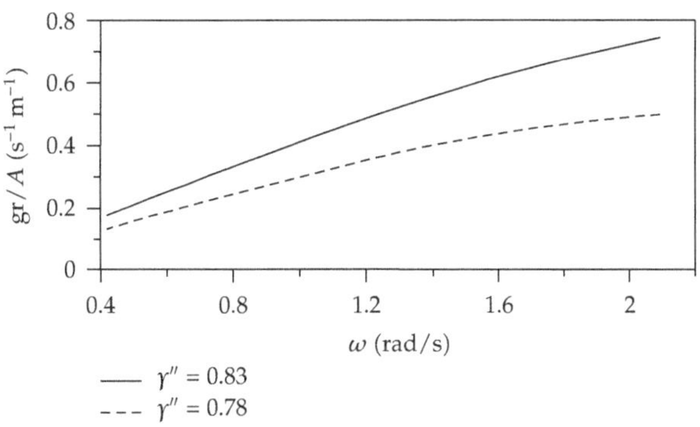

Figure 7: Growth rate of internal short waves for different density ratios of the right muddy section ($\gamma'' = 0.83, 0.78$), water depth $H = 4$ m, mud layer depth $h = 0.1$ m, and density ratio of the left muddy section $\gamma' = 0.83$.

be much difficult to excite the internal waves when surface water wave progressing over two muddy sections with the large density gap.

5. Conclusions

The resonant generation of internal waves on a sediment bed was presented as a new mechanism of sediment suspension. In this study, the corresponding theoretical analysis for equal density of two muddy sections is similar to that of the previous studies. A progressive surface wave simultaneously generates two opposite-traveling short internal waves. However, for a surface water wave progressing over two different muddy sections, the surface wave will also excite only two opposite-traveling interfacial short waves, forming the triad resonance with the surface wave. Meanwhile, two opposite-outgoing long internal waves will be also triggered at the interface between two muddy sections. Furthermore, it will be much difficult to excite the internal waves when surface water wave is progressing over two muddy sections with the large density gap.

References

[1] F. Wen, "Resonant generation of internal waves on the soft sea bed by a surface water wave," *Physics of Fluids*, vol. 7, no. 8, pp. 1915–1922, 1995.

[2] F. K. Ball, "Energy transfer between external and internal gravity waves," *Journal of Fluid Mechanics*, vol. 19, pp. 465–478, 1964.

[3] K. M. Watson, B. West, and B. I. Cohen, "Coupling of surface and internal gravity waves: a mode coupling model," *Journal of Fluid Mechanics*, vol. 77, pp. 185–208, 1976.

[4] D. F. Hill and M. A. Foda, "Subharmonic resonance of short internal standing waves by progressive surface waves," *Journal of Fluid Mechanics*, vol. 321, pp. 217–234, 1996.

[5] D. F. Hill and M. A. Foda, "Subharmonic resonance of oblique interfacial waves by a progressive surface wave," *The Royal Society of London Proceedings A*, vol. 454, no. 1972, pp. 1129–1144, 1998.

[6] M. Jamali, *Surface wave interaction with oblique internal waves [Ph.D. thesis]*, University of British Columbia, Vancouver, Canada, 1998.

[7] M. Jamali, G. A. Lawrence, and B. Seymour, "A note on the resonant interaction between a surface wave and two interfacial waves," *Journal of Fluid Mechanics*, vol. 491, pp. 1–9, 2003.

3

Peristaltic Transport of a Jeffrey Fluid with Variable Viscosity through a Porous Medium in an Asymmetric Channel

A. Afsar Khan,[1] R. Ellahi,[1,2] and K. Vafai[2]

[1] *Department of Mathematics & Statistics, FBAS, IIU, Islamabad, Pakistan*
[2] *Department of Mechanical Engineering, University of California Riverside, USA*

Correspondence should be addressed to R. Ellahi, rellahi@engr.ucr.edu

Academic Editor: Sanith Wijesinghe

The peristaltic flow of a Jeffrey fluid with variable viscosity through a porous medium in an asymmetric channel is investigated. The channel asymmetric is produced by choosing the peristaltic wave train on the wall of different amplitude and phase. The governing nonlinear partial differential equations for the Jeffrey fluid model are derived in Cartesian coordinates system. Analytic solutions for stream function, velocity, pressure gradient, and pressure rise are first developed by regular perturbation method, and then the role of pertinent parameters is illustrated graphically.

1. Introduction

Peristalsis is a mechanism to pump the fluid by means of moving contraction on the tubes or channel walls. This process has quite useful applications in many biological systems and industry. It occurs in swallowing food through the esophagus, chyme motion in the gastrointestinal tract, the vasomotion of small blood vessels such as venules, capillaries, and arterioles, urine transport from kidney to bladder, sanitary fluid transport of corrosive fluids, a toxic liquid transport in the nuclear industry, and so forth. In view of such physiological and industrial applications, the peristaltic flows has been studied with great interest by the various researchers for viscous and non-Newtonian fluids [1–9].

In most of the studies which deal with the peristaltic flows, the fluid viscosity is assumed to be constant. This assumption is not valid everywhere. In general the coefficients of viscosity for real fluids are functions of space coordinate, temperature, and pressure. For many liquids such as water, oils, and blood, the variation of viscosity due to space coordinate and temperature change is more dominant than other effects. Therefore, it is highly desirable to include the effect of variable viscosity instead of considering the viscosity of the fluid to be constant. Some important studies related to the variable viscosity are cited in [10–13].

A porous medium is the matter which contains a number of small holes distributed throughout the matter. Flows through a porous medium occur in filtration of fluids. Several investigations have been published by using generalized Darcy's law where the convective acceleration and viscous stress are taken into account [14–17].

Considering the importance of non-Newtonian fluid in peristalsis and keeping in mind the sensitivity of liquid viscosity, an attempt is made to study the peristaltic transport of Jeffrey having variable viscosity through a porous medium in a two-dimensional asymmetric channel under the assumption of long wave length and the low Reynolds number approximation. A regular perturbation method is used to solve the problem, and the solutions are expanded in a power series of viscosity parameter α. The obtained expressions are utilized to discuss the influences of various emerging parameters.

2. Mathematical Formulation

We consider an incompressible Jeffrey fluid in an asymmetric channel of width $d_1 + d_2$. A sinusoidal wave propagating with constant speed c on the channel walls induces the flow. The wall surfaces are chosen of the following forms:

$$
\begin{aligned}
H_1(X,t) &= a_1 + b_1 \cos\left[\frac{2\pi}{\lambda}(X - ct)\right], \quad \text{upper wall},\\
H_2(X,t) &= -a_2 - b_2 \cos\left[\frac{2\pi}{\lambda}(X - ct) + \phi\right], \quad \text{lower wall},
\end{aligned}
\tag{2.1}
$$

where b_1, b_2 are amplitude of the upper and lower waves, λ is the wave length, ϕ is the phase difference which varies in the range $0 \le \phi \le \pi$. Furthermore, a_1, a_2, b_1, b_2, and ϕ should satisfy the following condition

$$
b_1^2 + b_2^2 + 2b_1b_2\cos\phi \le (a_1 + a_2)^2. \tag{2.2}
$$

We assume that the flow becomes steady in the wave frame (x, y) moving with velocity c away from the fixed (laboratory) frame (X, Y). The transformation between these two frames is given by

$$
x = X - ct, \quad y = Y, \quad u = U - c, \quad v = V, \quad p(x) = P(X,t), \tag{2.3}
$$

where u and v are the velocity components in the wave frame (x, y), p and P are pressure in wave and fixed frame of reference, respectively. The governing equations in the wave frame of reference are the Brinkman extended Daray equations given by

$$\frac{\partial u}{\partial x} + \frac{\partial v}{\partial y} = 0, \tag{2.4}$$

$$\rho\left(u\frac{\partial u}{\partial x} + v\frac{\partial u}{\partial y}\right) = -\frac{\partial p}{\partial x} + \frac{1}{\varepsilon}\frac{\partial \tau_{xx}}{\partial x} + \frac{1}{\varepsilon}\frac{\partial \tau_{xy}}{\partial y} - \frac{\mu(y)}{k}(u+1), \tag{2.5}$$

$$\rho\left(u\frac{\partial v}{\partial x} + v\frac{\partial v}{\partial y}\right) = -\frac{\partial p}{\partial y} + \frac{\partial \tau_{yx}}{\partial x} + \frac{\partial \tau_{yy}}{\partial y} - \frac{\mu(y)}{k}v, \tag{2.6}$$

where

$$\begin{aligned}
\tau_{xx} &= \frac{2\mu(y)}{(1+\lambda_1)}\left[1 + \lambda_2\left(u\frac{\partial}{\partial x} + v\frac{\partial}{\partial y}\right)\right]\frac{\partial u}{\partial x}, \\
\tau_{xy} &= \frac{\mu(y)}{(1+\lambda_1)}\left[1 + \lambda_2\left(u\frac{\partial}{\partial x} + v\frac{\partial}{\partial y}\right)\right]\left(\frac{\partial u}{\partial y} + \frac{\partial v}{\partial x}\right), \\
\tau_{yy} &= \frac{2\mu(y)}{(1+\lambda_1)}\left[1 + \lambda_2\left(u\frac{\partial}{\partial x} + v\frac{\partial}{\partial y}\right)\right]\frac{\partial v}{\partial y},
\end{aligned} \tag{2.7}$$

where λ_1 is the ratio of relaxation to retardation times, λ_2 is the retardation time, ρ is the density, k is the permeability of the porous medium, and ε is the porosity of the porous medium.

Introducing the following nondimensional quantities:

$$\begin{aligned}
&\bar{x} = \frac{x}{\lambda}, \quad \bar{y} = \frac{y}{a_1}, \quad \bar{u} = \frac{u}{c}, \quad \bar{v} = \frac{v}{c\delta}, \quad h_1 = \frac{H_1}{a_1}, \quad h_2 = \frac{H_2}{a_1}, \quad \bar{\tau} = \frac{a_1\tau}{\mu_0 c} \\
&\bar{t} = \frac{ct}{\lambda}, \quad \mathrm{Da} = \frac{k}{a_1^2}, \quad \delta = \frac{a}{\lambda}, \quad \bar{p} = \frac{pa_1^2}{\mu_0 c\lambda}, \quad a = \frac{b_1}{a_1}, \quad b = \frac{b_2}{a_1}, \quad d = \frac{a_2}{a_1}.
\end{aligned} \tag{2.8}$$

With the help of (2.8), (2.4) to (2.6) after dropping the bars take the form

$$\frac{\partial u}{\partial x} + \frac{\partial v}{\partial y} = 0, \tag{2.9}$$

$$\mathrm{Re}\,\delta\left(u\frac{\partial u}{\partial x} + v\frac{\partial u}{\partial y}\right) = -\frac{\partial p}{\partial x} + \frac{\delta}{\varepsilon}\frac{\partial \tau_{xx}}{\partial x} + \frac{1}{\varepsilon}\frac{\partial \tau_{xy}}{\partial y} - \frac{\mu(y)}{\mathrm{Da}}(u+1), \tag{2.10}$$

$$\mathrm{Re}\,\delta^3\left(u\frac{\partial v}{\partial x} + v\frac{\partial v}{\partial y}\right) = -\frac{\partial p}{\partial y} + \frac{\delta^2}{\varepsilon}\frac{\partial \tau_{xy}}{\partial x} + \frac{\delta}{\varepsilon}\frac{\partial \tau_{yy}}{\partial y} - \frac{\delta^2\mu(y)}{\mathrm{Da}}v, \tag{2.11}$$

where Darcy's number is

$$\text{Da} = \frac{k}{a_1^2},$$

$$\begin{aligned}
\tau_{xx} &= \frac{2\delta\mu(y)}{(1+\lambda_1)}\left[1 + \frac{\lambda_2\delta c}{a_1}\left(u\frac{\partial}{\partial x} + v\frac{\partial}{\partial y}\right)\right]\frac{\partial u}{\partial x}, \\
\tau_{xy} &= \frac{\mu(y)}{(1+\lambda_1)}\left[1 + \frac{\lambda_2\delta c}{a_1}\left(u\frac{\partial}{\partial x} + v\frac{\partial}{\partial y}\right)\right]\left(\frac{\partial u}{\partial y} + \delta^2\frac{\partial v}{\partial x}\right), \\
\tau_{yy} &= \frac{2\mu(y)}{(1+\lambda_1)}\left[1 + \frac{\lambda_2\delta c}{a_1}\left(u\frac{\partial}{\partial x} + v\frac{\partial}{\partial y}\right)\right]\frac{\partial v}{\partial y}.
\end{aligned} \tag{2.12}$$

Using the longwave length and small Reynolds number approximation, (2.10) and (2.11) take the form

$$\frac{\partial p}{\partial x} = \frac{1}{\varepsilon}\frac{\partial}{\partial y}\left[\frac{\mu(y)}{(1+\lambda_1)}\frac{\partial u}{\partial y}\right] - \frac{\mu(y)}{\text{Da}}(u+1), \tag{2.13}$$

$$\frac{\partial p}{\partial y} = 0. \tag{2.14}$$

The corresponding boundary conditions are

$$u = -1, \quad \text{at } y = h_1, \tag{2.15a}$$

$$u = -1, \quad \text{at } y = h_2, \tag{2.15b}$$

where

$$h_1 = 1 + a\cos 2\pi x, \qquad h_2 = -d - b\cos(2\pi x + \phi). \tag{2.15c}$$

Equation (2.14) indicate that p is independent of y. Therefore, (2.10) can be written as

$$\frac{dp}{dx} = \frac{1}{\varepsilon}\frac{\partial}{\partial y}\left[\frac{\mu(y)}{(1+\lambda_1)}\frac{\partial u}{\partial y}\right] - \frac{\mu(y)}{\text{Da}}(u+1), \tag{2.16}$$

where $\mu(y)$ is the viscosity variation on peristaltic flow. For the present analysis, we assume viscosity variation in the dimensionless form [10]:

$$u(y) = e^{-\alpha y}, \quad u(y) = 1 - \alpha y + \frac{\alpha y^2}{2}, \quad \text{for } \alpha << 1. \tag{2.17}$$

The volume flow rate in the wave frame is given by

$$q = \int_{h_2}^{h_1} u \, dy. \tag{2.18}$$

The instantaneous flux $Q(x,t)$ in the laboratory frame is defined as

$$Q(x,t) = \int_{h_1}^{h_2} (u+1)dy = q + h_1 - h_2. \tag{2.19}$$

The average flux over one period $(T = \lambda/c)$ is given by

$$\overline{Q} = \frac{1}{T}\int_0^T Q dt = \frac{1}{T}\int_0^T (q + h_1 - h_2)dt = q + 1 + d. \tag{2.20}$$

3. Perturbation Solution

Equation (2.16) is a nonlinear differential equation so that it is not possible to obtain a closed form solution; so we seek perturbation solution. We expand u, p and q as

$$\begin{aligned} u &= u_0 + \alpha u_1 + \alpha^2 u_2 + o\left(\alpha^3\right), \\ p &= p_0 + \alpha p_1 + \alpha^2 p_2 + o\left(\alpha^3\right), \\ q &= q_0 + \alpha q_1 + \alpha^2 q_2 + o\left(\alpha^3\right). \end{aligned} \tag{3.1}$$

Substituting these equations into (2.15a), (2.15b), (2.15c), and (2.16), we have the following system of equations.

3.1. Zeroth-Order Equations α^0

$$\frac{\partial^2 u_0}{\partial y^2} - N^2 u_0 = \varepsilon(1+\lambda_1)\frac{dp_0}{dx} + N^2, \tag{3.2}$$

where

$$\begin{aligned} N &= \frac{\varepsilon(1+\lambda_1)}{\text{Da}}, \\ u_0 &= -1, \quad \text{at } y = h_1, h_2. \end{aligned} \tag{3.3}$$

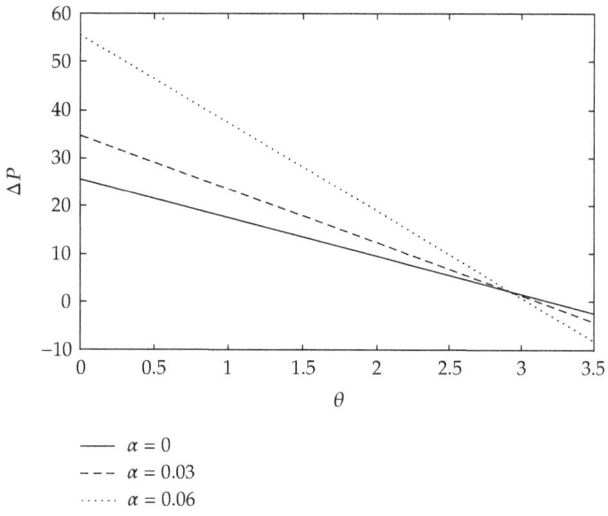

Figure 1: The pressure rise versus flow rate when $a = 0.2, b = 0.6,\ d = 0.8,\ \varepsilon = 0.3,\ \lambda_1 = 0.8,\ \mathrm{Da} = 0.6$, and $\phi = \pi/4$.

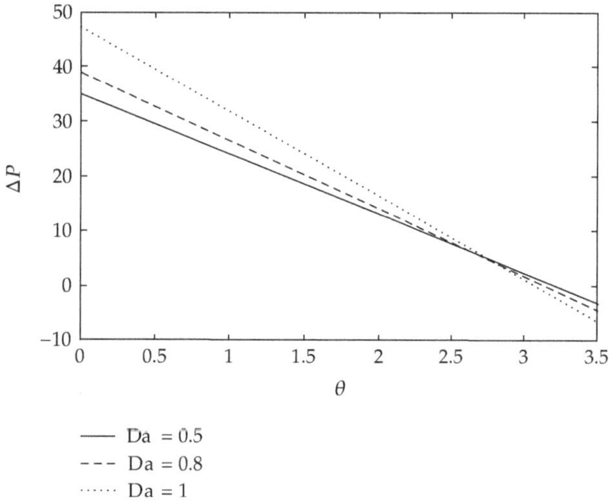

Figure 2: The pressure rise verses flow rate when $\alpha = 0.01,\ a = 0.2,\ b = 0.6,\ d = 0.8,\ \varepsilon = 0.3,\ \lambda_1 = 0.4$, and $\phi = \pi/4$.

3.2. *First-Order Equations α*

$$\frac{\partial^2 u_1}{\partial y^2} - \mathrm{N}^2 u_1 = \varepsilon(1+\lambda_1)\frac{dp_1}{dx} + \varepsilon y(1+\lambda_1)\frac{dp_0}{dx} + \frac{\partial u_0}{\partial y}, \tag{3.4}$$

$$u_1 = 0, \quad \text{at } y = h_1, h_2. \tag{3.5}$$

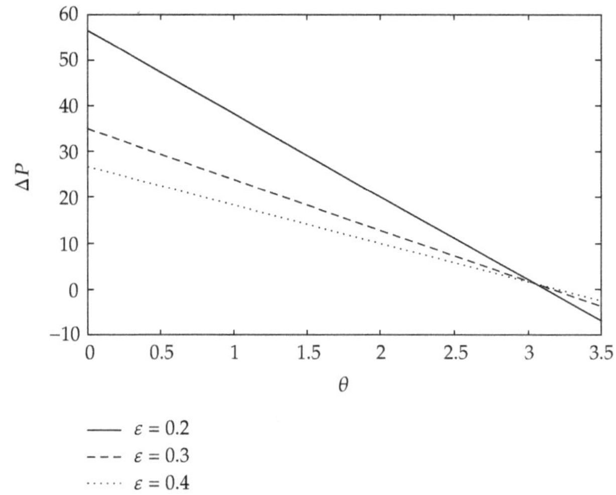

Figure 3: The pressure rise verses flow rate when $\alpha = 0.01$, $a = 0.2$, $b = 0.6$, $d = 0.8$, $\lambda_1 = 0.4$, $\text{Da} = 0.5$, and $\phi = \pi/4$.

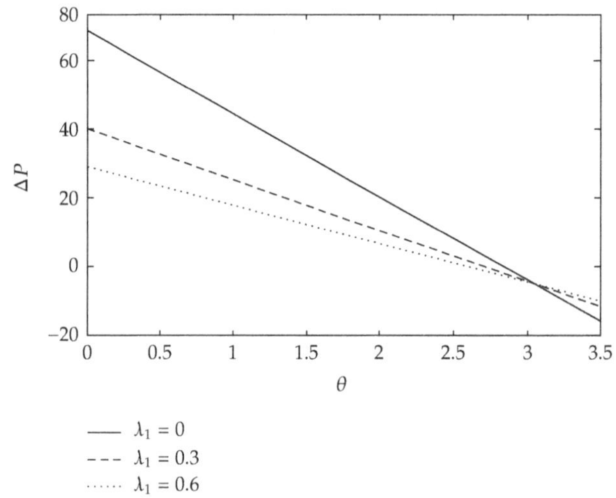

Figure 4: The pressure rise verses flow rate when $\alpha = 0.01$, $a = 0.2$, $b = 0.6$, $d = 0.8$, $\varepsilon = 0.3$, $\text{Da} = 0.8$, and $\phi = \pi/4$.

3.3. *Second-Order Equations* α^2

$$\frac{\partial^2 u_2}{\partial y^2} - N^2 u_2 = \varepsilon(1+\lambda_1)\frac{dp_2}{dx} + \varepsilon(1+\lambda_1)y\frac{dp_1}{dx} + \frac{y^2}{2}\varepsilon(1+\lambda_1)\frac{dp_0}{dx} + \frac{\partial u_1}{\partial y}, \tag{3.6}$$

$$u_2 = 0 \quad \text{at } y = h_1, h_2. \tag{3.7}$$

3.4. Zeroth-Order Solution

Solving (3.2) and (3.3), we get

$$u_0 = \frac{\varepsilon(1+\lambda_1)}{N^2}\frac{dp_0}{dx}\left[C_1 \cosh Ny + C_2 \sinh Ny - 1\right] - 1, \tag{3.8}$$

where

$$C_1 = \frac{(\sinh Nh_1 - \sinh Nh_2)}{\sinh N(h_1 - h_2)}, \qquad C_2 = \frac{(\cosh Nh_2 - \cosh Nh_1)}{\sinh N(h_1 - h_2)}, \tag{3.9}$$

and the volume flow rate q_0 is given by

$$q_0 = \int_{h_2}^{h_1} u_0 dy. \tag{3.10}$$

From (3.8), we have

$$\frac{dp_0}{dx} = (q_0 + h_1 - h_2)A, \tag{3.11}$$

where

$$A = \frac{N^3 \sinh N(h_1 - h_2)}{\varepsilon(1+\lambda_1)\left[2\cosh N(h_1 - h_2) - 2 - (h_1 - h_2)N \sinh N(h_1 - h_2)\right]}. \tag{3.12}$$

The dimensionless pressure rise at this order is

$$\Delta P_0 = \int_0^1 \frac{dp_0}{dx} dx. \tag{3.13}$$

3.5. First-Order Solution

Substituting zeroth order solution (3.8) into (3.4) and then solving the resulting system along with the corresponding boundary conditions, we arrive at

$$\begin{aligned} u_1 = {} & \frac{\varepsilon(1+\lambda_1)}{N^2}\frac{dp_1}{dx}\left[C_1 \cosh Ny + C_2 \sinh Ny - 1\right] \\ & + \frac{\varepsilon(1+\lambda_1)}{2N^2}\frac{dp_0}{dx}\Bigg[-2y + C_1 y \cosh Ny + C_2 y \sinh Ny \\ & \times \frac{\sinh Ny(h_1 \cosh Nh_2 - h_2 \cosh Nh_1)}{\sinh N(h_1 - h_2)} \\ & + \frac{\cosh Ny(h_2 \sinh Nh_1 - h_1 \sinh Nh_2)}{\sinh N(h_1 - h_2)}\Bigg], \end{aligned} \tag{3.14}$$

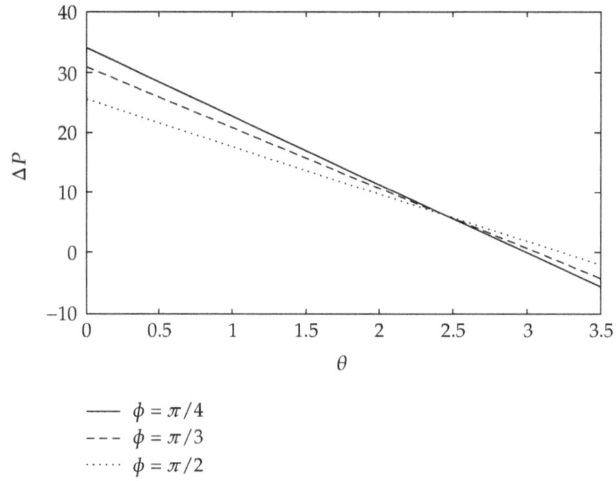

Figure 5: The pressure rise verses flow rate when $\alpha = 0.01$, $a = 0.4$, $b = 0.6$, $d = 0.8$, $\varepsilon = 0.4$, $\lambda_1 = 0.5$, and Da = 0.5.

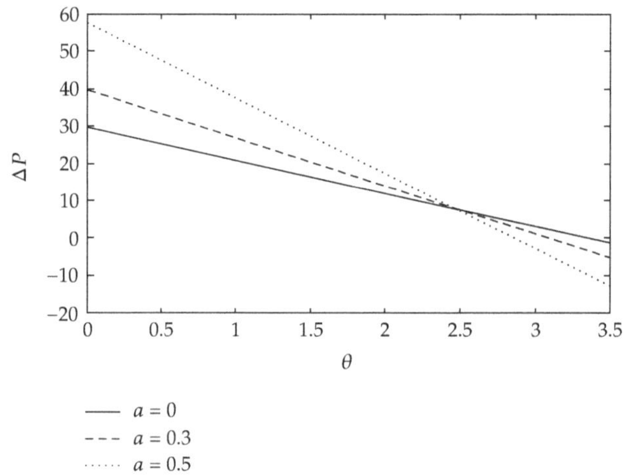

Figure 6: The pressure rise verses flow rate when $\alpha = 0.01$, $b = 0.6$, $d = 0.8$, $\varepsilon = 0.3$, $\lambda_1 = 0.4$, Da = 0.5, and $\phi = \pi/4$.

and the volume flow rate q_1 is given by

$$q_1 = \int_{h_2}^{h_1} u_1 dy. \tag{3.15}$$

From (3.14), we get

$$\frac{dp_1}{dx} = Aq_1 + \frac{A\varepsilon(1+\lambda_1)}{2N^3}\frac{dp_0}{dx}\left[N^2\left(h_1^2 - h_2^2\right) + \frac{(h_1+h_2)(1-\cosh N(h_1-h_2))}{\sinh N(h_1-h_2)}\right]. \tag{3.16}$$

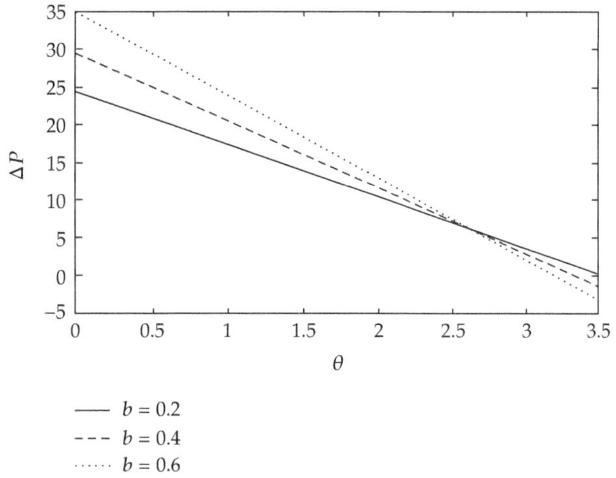

Figure 7: The pressure rise verses flow rate when $\alpha = 0.01$, $a = 0.2$, $d = 0.8$, $\varepsilon = 0.3$, $\lambda_1 = 0.4$, $\mathrm{Da} = 0.5$, and $\phi = \pi/4$.

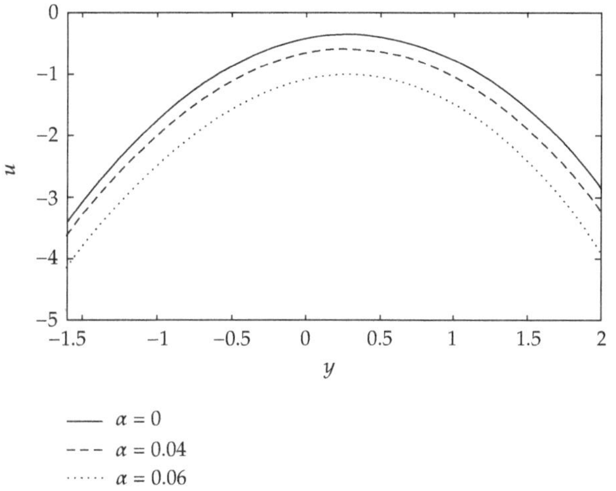

Figure 8: Axial velocity versus y at $a = 0.2$, $b = 0.6$, $d = 0.8$, $\varepsilon = 0.2$, $\lambda_1 = 1$, $\mathrm{Da} = 1$, $x = \pi/6$, $q = -1$, and $\phi = \pi/2$.

The dimensionless pressure rise at this order is

$$\Delta P_1 = \int_0^1 \frac{dp_1}{dx} dx. \tag{3.17}$$

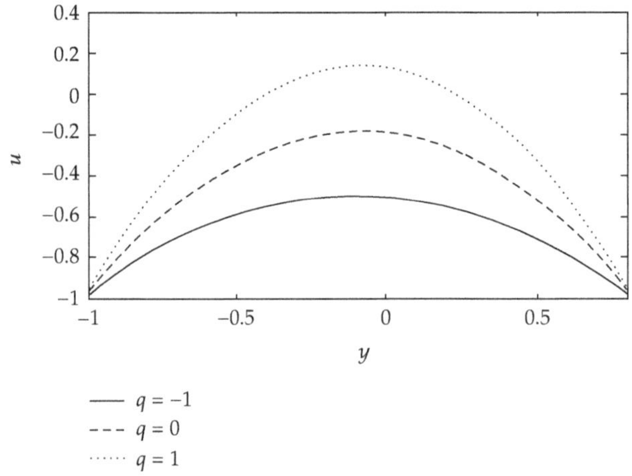

Figure 9: Axial velocity versus y at $\alpha = 0.05$, $a = 0.2$, $b = 0.6$, $d = 0.8$, $\varepsilon = 0.2$, $\lambda_1 = 1$, $\mathrm{Da} = 1$, $x = 0$, and $\phi = \pi/2$.

3.6. Second-Order Solution

Solving (3.6) by using (3.8) and (3.14) and the boundary condition (3.5), we obtain

$$
\begin{aligned}
u_2 = {} & \frac{\varepsilon(1+\lambda_1)}{N^2}\frac{dp_2}{dx}\left[C_1\cosh Ny + C_2\sinh Ny - 1\right] \\
& + \frac{\varepsilon(1+\lambda_1)}{2N^2}\frac{dp_1}{dx}\left[\frac{\sinh Ny(h_1\cosh Nh_2 - h_2\cosh Nh_1)}{\sinh N(h_1-h_2)} - 2y \right. \\
& \qquad + C_1 y\cosh Ny + C_2 y\sinh Ny \\
& \qquad \left. + \frac{\cosh Ny(h_2\sinh Nh_1 - h_1\sinh Nh_2)}{\sinh N(h_1-h_2)}\right] \\
& + \frac{dp_0}{dx}\frac{\varepsilon(1+\lambda_1)}{4N^2}\left[\frac{C_1(y\sinh Ny + Ny^2\cosh Ny) + C_2(y\cosh Ny + y^2N\sinh Ny)}{2N} \right. \\
& \qquad - y^2 + \frac{(1-\cosh N(h_1-h_2))(\sinh Ny(h_1\cosh Nh_2 + h_2\cosh Nh_1))}{2\sinh^2 N(h_1-h_2)} \\
& \qquad - \frac{(1-\cosh N(h_1-h_2))((h_1\sinh Nh_2 + h_2\sinh Nh_1)\cosh Ny)}{2\sinh^2 N(h_1-h_2)} \\
& \qquad - \frac{(h_1\cosh Nh_2 - h_2\cosh Nh_1)(C_1 y\cosh Ny + C_2 y\sinh Ny)}{(\cosh Nh_1 - \cosh Nh_2)} \\
& \qquad + \frac{(h_2^2\cosh Nh_1 - h_1^2\cosh Nh_2)\sinh Ny}{2\sinh N(h_1-h_2)}
\end{aligned}
$$

$$+\frac{(h_1^2\sinh Nh_1 - h_2^2\sinh Nh_2)\cosh Ny}{2\sinh N(h_1-h_2)}$$
$$\times\frac{8(1-C_1\cosh Ny - C_2\sinh Ny)}{N^2} + \frac{(h_1-h_2)y\cosh Ny}{(\cosh Nh_1 - \cosh Nh_2)}\Bigg],\tag{3.18}$$

and the volume flow rate q_2 is given by

$$q_2 = \int_{h_2}^{h_1} u_2 dy.\tag{3.19}$$

From (3.18), we have

$$\begin{aligned}\frac{dp_2}{dx} &= Aq_2 + \frac{A\varepsilon(1+\lambda_1)}{2N^3}\frac{dp_1}{dx}\left[\frac{2(h_1+h_2)(1-\cosh N(h_1-h_2))}{\sinh N(h_1-h_2)} + N\left(\left(h_1^2 - h_2^2\right)\right)\right]\\ &- \frac{A\varepsilon(1+\lambda_1)}{4N^3}\frac{dp_0}{dx}\left[\frac{8(h_1-h_2)}{N} - \frac{3(h_1-h_2)}{2N} - \left(\frac{h_1^3}{3} - \frac{h_2^3}{3}\right)\right]\\ &+ \frac{(h_1^2+h_2^2)(1-\cosh N(h_1-h_2))}{2N\sinh N(h_1-h_2)} + \frac{(h_1-h_2)(h_1\sin Nh_1 - h_2\sin Nh_2)}{(\cosh Nh_1 - \cosh Nh_2)}\\ &+ \frac{(h_1+h_2)(1-\cosh N(h_1-h_2))(h_1\cosh Nh_2 - h_2\cosh Nh_1)}{\sinh N(h_1-h_2)(\cosh Nh_1 - \cosh Nh_2)}\\ &+ \frac{(h_1^2N^2 + h_2^2N^2 + 2)(\cosh N(h_1-h_2) - 1)}{2N^2\sinh N(h_1-h_2)} + \frac{16(1-\cosh N(h_1-h_2))}{N^2\sinh N(h_1-h_2)}.\end{aligned}\tag{3.20}$$

The dimensionless pressure rise at this order is

$$\Delta P_2 = \int_0^2 \frac{dp_2}{dx}dx.\tag{3.21}$$

Summarizing the result obtained from (3.11), (3.16), and (3.20), we write

$$\Delta P = \Delta P_0 + \alpha\Delta P_1 + \alpha^2\Delta P_2.\tag{3.22}$$

Corresponding stream functions can be defined as

$$u = \frac{\partial\Psi}{\partial y}, \qquad v = -\delta\frac{\partial\Psi}{\partial x}.\tag{3.23}$$

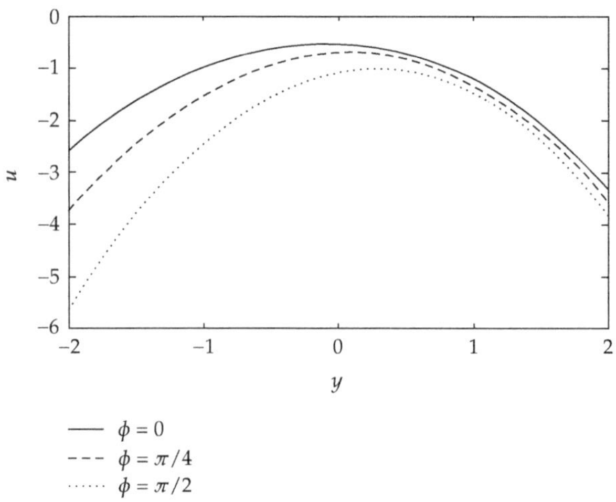

Figure 10: Axial velocity versus y at $\alpha = 0.05$, $a = 0.2$, $b = 0.6$, $d = 0.8$, $\varepsilon = 0.2$, $\lambda_1 = 1$, $\mathrm{Da} = 1$, $x = \pi/2$, and $q = -1$.

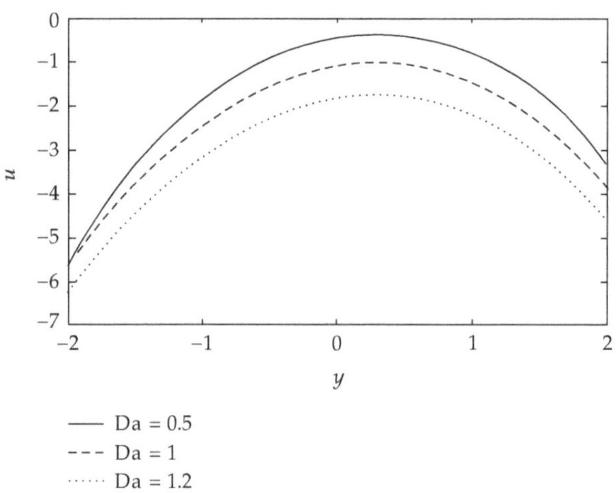

Figure 11: Axial velocity versus y at $\alpha = 0.05$, $a = 0.2$, $b = 0.6$, $d = 0.8$, $\varepsilon = 0.2$, $\lambda_1 = 1$, $q = -1$, $x = \pi/6$, and $\phi = \pi/2$.

4. Results and Discussion

We have used a regular perturbation series in term of the dimensional viscosity parameter α to obtain analytical solution of the field equations for peristaltic flow of Jeffrey fluid in an asymmetric channel. To study the behavior of solutions, numerical calculations for several values of viscosity parameter α, Daray number Da, porosity ε, amplitude ratio ϕ, Jeffrey fluid parameter λ_1, a and b have been calculated numerically using MATHEMATICA software.

Figure 1 shows the variation of ΔP with flow rate θ for different values of α. It is depicted that the time-average flux θ increase with increasing the viscosity parameter α. Figure 2 represents the variation of ΔP with the flow rate θ for different values of Da. We observe that an increase in the peristaltic pumping rate pressure rises. Figures 3 and 4 are graphs of pressure rise ΔP with the flow rate θ for values of ε and λ_1. It is observed that the

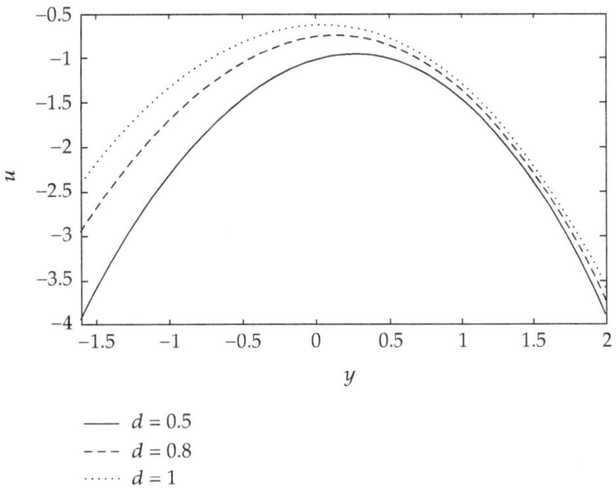

Figure 12: Axial velocity versus y at $\alpha = 0.05$, $a = 0.2$, $b = 0.6$, $q = -1$, $\varepsilon = 0.2$, $\lambda_1 = 1$, $\text{Da} = 1$, $x = \pi/6$, and $\phi = \pi/2$.

pumping rate decreases with increase of ε and λ_1. Figure 5 is the graph of the variation of ΔP versus the flow rate θ for different values of phase difference ϕ. It is observed that the pumping rate decreases with the increase of ϕ. Figures 6 and 7 plot the relation between pressure rise ΔP and flow rate θ for different values of a and b, respectively. Figure 8 represents the graph of axial velocity u versus y. It can be seen that an increase in α decreases the magnitude of axial velocity u. The effects of q on the axial velocity u are seen through Figure 9. It is noticed that an increase in q increase the magnitude of the axial velocity. Figures 10 and 11 illustrate the effect of phase difference ϕ and Daray's number Da on the axial velocity u. It is observed that the magnitude of axial velocity decreases with the increasing phase difference ϕ and Daray's number Da. In Figure 12 the axial velocity u is graphed versus y. We note that the magnitude of axial velocity increases as the channel width d increases. It is worth mentioning that in the absence of porosity parameter the solutions of [10] can be derived as special case of the present analysis. This provides the useful check. It may be remarked that the problem for this particular model was not solved earlier even by any traditional perturbation technique. The results presented in this paper will now be available for experimental verification.

Acknowledgments

R. Ellahi thanks to United State Education Foundation Pakistan and CIES USA for honoring him by the Fulbright Scholar Award for the year 2011-2012. R. Ellahi also grateful to the Higher Education Commission and PCST of Pakistan for awarding him with the awards of NRPU and Productive Scientist, respectively.

References

[1] M. Y. Malik, A. Hussain, and S. Nadeem, "Flow of a Jeffrey-six constant fluid between coaxial cylinders with heat transfer," *Communications in Theoretical Physics*, vol. 56, pp. 345–351, 2011.

[2] W. Tan and T. Masuoka, "Stokes' first problem for a second grade fluid in a porous half-space with heated boundary," *International Journal of Non-Linear Mechanics*, vol. 40, no. 4, pp. 515–522, 2005.
[3] M. Hameed and S. Nadeem, "Unsteady MHD flow of a non-Newtonian fluid on a porous plate," *Journal of Mathematical Analysis and Applications*, vol. 325, no. 1, pp. 724–733, 2007.
[4] F. M. Mahomed and T. Hayat, "Note on an exact solution for the pipe flow of a third-grade fluid," *Acta Mechanica*, vol. 190, no. 1–4, pp. 233–236, 2007.
[5] K. S. Mekheimer and Y. Abd elmaboud, "The influence of heat transfer and magnetic field on peristaltic transport of a Newtonian fluid in a vertical annulus: application of an endoscope," *Physics Letters A*, vol. 372, no. 10, pp. 1657–1665, 2008.
[6] K. S. Mekheimer, "Effect of the induced magnetic field on peristaltic flow of a couple stress fluid," *Physics Letters A*, vol. 372, no. 23, pp. 4271–4278, 2008.
[7] S. Srinivas and M. Kothandapani, "Peristaltic transport in an asymmetric channel with heat transfer—a note," *International Communications in Heat and Mass Transfer*, vol. 35, no. 4, pp. 514–522, 2008.
[8] M. Kothandapani and S. Srinivas, "Peristaltic transport of a Jeffrey fluid under the effect of magnetic field in an asymmetric channel," *International Journal of Non-Linear Mechanics*, vol. 43, no. 9, pp. 915–924, 2008.
[9] M. Dehghan and F. Shakeri, "The numerical solution of the second painlevé equation," *Numerical Methods for Partial Differential Equations*, vol. 25, no. 5, pp. 1238–1259, 2009.
[10] S. Nadeem and N. S. Akbar, "Peristaltic flow of a Jeffrey fluid with variable viscosity in an asymmetric channel," *Zeitschrift fur Naturforschung A* , vol. 64, no. 11, pp. 713–722, 2009.
[11] S. Nadeem and N. S. Akbar, "Effects of heat transfer on the peristaltic transport of MHD Newtonian fluid with variable viscosity: application of Adomian decomposition method," *Communications in Nonlinear Science and Numerical Simulation*, vol. 14, no. 11, pp. 3844–3855, 2009.
[12] A. Ebaid, "A new numerical solution for the MHD peristaltic flow of a bio-fluid with variable viscosity in a circular cylindrical tube via Adomian decomposition method," *Physics Letters A*, vol. 372, no. 32, pp. 5321–5328, 2008.
[13] A. El Hakeem, A. El Naby, A. E. M. El Misiery, and I. I. El Shamy, "Hydromagnetic flow of fluid with variable viscosity in a uniform tube with peristalsis," *Journal of Physics A*, vol. 36, no. 31, pp. 8535–8547, 2003.
[14] S. Srinivas and R. Gayathri, "Peristaltic transport of a Newtonian fluid in a vertical asymmetric channel with heat transfer and porous medium," *Applied Mathematics and Computation*, vol. 215, no. 1, pp. 185–196, 2009.
[15] S. Srinivas and M. Kothandapani, "The influence of heat and mass transfer on MHD peristaltic flow through a porous space with compliant walls," *Applied Mathematics and Computation*, vol. 213, no. 1, pp. 197–208, 2009.
[16] S. Nadeem, N. S. Akbar, and M. Y. Malik, "Numerical solutions of peristaltic flow of a newtonian fluid under the effects ofmagnetic field and heat transfer in a porous concentric tubes," *Zeitschrift fur Naturforschung A*, vol. 65, no. 5, pp. 369–380, 2010.
[17] E. F. Elshehawey, N. T. Eldabe, E. M. Elghazy, and A. Ebaid, "Peristaltic transport in an asymmetric channel through a porous medium," *Applied Mathematics and Computation*, vol. 182, no. 1, pp. 140–150, 2006.

4

Multiparameter Statistical Models from $N^2 \times N^2$ Braid Matrices: Explicit Eigenvalues of Transfer Matrices $\mathbf{T}^{(r)}$, Spin Chains, Factorizable Scatterings for All N

B. Abdesselam[1,2] and A. Chakrabarti[3]

[1] *Laboratoire de Physique Théorique, Université d'Oran Es-Sénia, 31100 Oran, Algeria*
[2] *Institut des Sciences et de la Technologie, Centre Universitaire d'Ain Témouchent, 46000 Ain Témouchent, Algeria*
[3] *Centre de Physique Théorique, Ecole Polytechnique, 91128 Palaiseau Cedex, France*

Correspondence should be addressed to B. Abdesselam, boucif@yahoo.fr

Academic Editor: Yao-Zhong Zhang

For a class of multiparameter statistical models based on $N^2 \times N^2$ braid matrices, the eigenvalues of the transfer matrix $\mathbf{T}^{(r)}$ are obtained explicitly for all (r, N). Our formalism yields them as solutions of sets of linear equations with simple constant coefficients. The role of zero-sum multiplets constituted in terms of roots of unity is pointed out, and their origin is traced to circular permutations of the indices in the tensor products of basis states induced by our class of $\mathbf{T}^{(r)}$ matrices. The role of free parameters, increasing as N^2 with N, is emphasized throughout. Spin chain Hamiltonians are constructed and studied for all N. Inverse Cayley transforms of the Yang-Baxter matrices corresponding to our braid matrices are obtained for all N. They provide potentials for factorizable S-matrices. Main results are summarized, and perspectives are indicated in the concluding remarks.

1. Introduction

Statistical models "exact" in the sense of Baxter [1] satisfy "star-triangle" relations leading to transfer matrices commuting for different values of the spectral parameter. Crucial in the study of such models is the spectrum of eigenvalues of these matrices. But even for the extensively studied 6-vertex and 8-vertex models based on 4×4 braid matrices (the braid property guaranteeing star-triangle relations), after the first few simple steps, one has to resort to numerical computations. (We cannot adequately discuss here the vast associated literature but refer to texts citing major sources [2, 3].) But the basic reason for such situation is that, starting with 2×2 blocks T_{ab} of the transfer matrix and constructing $2^r \times 2^r$ blocks via coproduct rules for order r ($r = 1,2,3,\ldots$), one faces increasingly complicated nonlinear systems of equations to be solved in constructing eigenstates and eigenvalues. Even when a systematic approach is available, such as the Bethe Ansatz for the 6-vertex case, it only means that the relevant nonlinear equations can be written down systematically. The task of solving them remains. For the 8-vertex case one explores analytical properties to extract informations (See [4], e.g.). Also the number of free parameters remains strictly limited for such models, including the multistate generalization of the 6-vertex one [2].

Our class of models exhibits the following properties.

(1) A systematic construction for all dimensions. One starts from $N^2 \times N^2$ braid matrices leading to $N^r \times N^r$ blocks of $\mathbf{T}^{(r)}$, the transfer matrix of order r for $N = 2,3,4,5,\ldots; r = 1,2,3,\ldots$ with no upper limit.

(2) The number of free parameters increases with N as N^2. This is a unique feature of our models.

(3) For all (N, r) one solves sets of linear equations with systematically obtained simple constant coefficients to construct eigenvalues of $\mathbf{T}^{(r)} = \sum_{a=1}^{N} T_{aa}^{(r)}$. This is a consequence of our starting point: braid matrices constructed on a basis of "nested sequence" of projectors [5].

The total dimension of the base space on which such equations have to be solved increases as N^r. This will be seen to break up into subspaces closed under the action of $\mathbf{T}^{(r)}$, thus reducing the work considerably. Evidently one cannot continue to display the results explicitly, as N^r increases. But it is possible to implement a general and particularly efficient approach. This will be illustrated for all N and $r = 1,2,3,4,5$. The generalization to $r > 5$ will be clearly visible. This is the central purpose of our paper. Certain other features will be explored with comments in conclusion.

2. Braid and Transfer Matrices

The roots of our class of braid matrices and the transfer matrices they generate are to be found in the nested sequence of projectors [5]. In [5] it was noted that the 4×4 projectors providing the basis of the 6-vertex and 8-vertex models (i.e., the braid matrices leading to those models) can be generalized to higher dimensions—all higher ones (N odd or even). Braid matrices for odd dimensions were exhaustively constructed on such a basis [6], and related statistical models were studied [7]. Then even dimensional ones were presented [8], and the corresponding braid matrices and statistical models were studied [9]. In previous works the eigenvalues and eigenfunctions of the transfer matrices were presented mostly for the lowest values of N, namely, the 4×4 and 9×9 braid matrices. We present below a general approach for all N.

Let us just mention that systematic study of the exotic bialgebras that arise from the 9×9 unitary braid matrices was constructed in [10]. The dual bialgebra of one of these exotic bialgebras was also presented.

We start now by recapitulating the construction of our nested sequence of projectors, leading to the remarkable properties of our solutions.

2.1. Even Dimensions

Let $N = 2n$ ($n = 1, 2, 3, \ldots$). Define ((ab) denoting a matrix with unity on row a and column b)

$$P_{ij}^{(\epsilon)} = \frac{1}{2}\left\{(ii)\otimes(jj) + \left(\bar{i}\bar{i}\right)\otimes\left(\bar{j}\bar{j}\right) + \epsilon\left[\left(i\bar{i}\right)\otimes\left(j\bar{j}\right) + \left(\bar{i}i\right)\otimes\left(\bar{j}j\right)\right]\right\}, \tag{2.1}$$

where $i, j \in \{1, \ldots, n\}$, $\epsilon = \pm$, $\bar{i} = 2n + 1 - i$, $\bar{j} = 2n + 1 - j$. Interchanging $j \rightleftharpoons \bar{j}$ on the right one obtains $P_{i\bar{j}}^{(\epsilon)}$. One thus obtains a complete basis of projectors satisfying (with $P_{ij}^{(\epsilon)} = P_{\bar{i}\bar{j}}^{(\epsilon)}$, $P_{i\bar{j}}^{(\epsilon)} = P_{\bar{i}j}^{(\epsilon)}$ by definition)

$$\begin{gathered} P_{ab}^{(\epsilon)} P_{cd}^{(\epsilon')} = \delta_{ac}\delta_{bd}\delta_{\epsilon\epsilon'} P_{ab}^{(\epsilon)}, \qquad a, b, c, d \in \{1, \ldots, N = 2n\}, \\ \sum_{\epsilon=\pm}\sum_{i,j=1}^{n}\left(P_{ij}^{(\epsilon)} + P_{i\bar{j}}^{(\epsilon)}\right) = I_{(2n)^2\times(2n)^2}. \end{gathered} \tag{2.2}$$

Let $m_{ij}^{(\epsilon)}$ be an arbitrary set of parameters satisfying the crucial constraint

$$m_{ij}^{(\epsilon)} = m_{\bar{i}\bar{j}}^{(\epsilon)}. \tag{2.3}$$

Define the $N^2 \times N^2$ matrix

$$\hat{R}(\theta) = \sum_{\epsilon}\sum_{i,j} e^{m_{ij}^{(\epsilon)}\theta}\left(P_{ij}^{(\epsilon)} + P_{i\bar{j}}^{(\epsilon)}\right). \tag{2.4}$$

The θ-dependence entering exclusively through the exponentials as coefficients as above. Such construction guarantees [7–9] the braid property

$$\hat{R}_{12}(\theta)\hat{R}_{23}(\theta+\theta')\hat{R}_{12}(\theta') = \hat{R}_{23}(\theta')\hat{R}_{12}(\theta+\theta')\hat{R}_{23}(\theta), \tag{2.5}$$

where, in standard notations, $\hat{R}_{12} = \hat{R} \otimes I$ and $\hat{R}_{23} = I \otimes \hat{R}$, I denoting the $N \times N$ identity matrix.

Define the permutation matrix $\mathbf{P} = \sum_{a,b}(ab)\otimes(ba)$. Then

$$R(\theta) = \mathbf{P}\hat{R}(\theta) \tag{2.6}$$

satisfies the Yang-Baxter equation. The monodromy matrix of order 1 ($r = 1$) is given by

$$T^{(1)}(\theta) = R(\theta). \tag{2.7}$$

The $N \times N$ blocks are (with $a, b \in \{1, \ldots, 2n\}$)

$$T_{ab}^{(1)}(\theta) = \frac{1}{2}\left(e^{m_{ba}^{(+)}\theta} + e^{m_{ba}^{(-)}\theta}\right)(ba) + \frac{1}{2}\left(e^{m_{ba}^{(+)}\theta} - e^{m_{ba}^{(-)}\theta}\right)\left(\overline{b}\overline{a}\right) \equiv f_{ba}^{(+)}(\theta)(ba) + f_{ba}^{(-)}(\theta)\left(\overline{b}\overline{a}\right), \quad (2.8)$$

with

$$m_{ba}^{(\epsilon)} = m_{b\overline{a}}^{(\epsilon)} = m_{\overline{b}a}^{(\epsilon)} = m_{\overline{b}\overline{a}}^{(\epsilon)}. \quad (2.9)$$

Higher orders ($r > 1$) are obtained via coproducts

$$T_{ab}^{(r)}(\theta) = \sum_{c_1, \ldots, c_{r-1}} T_{ac_1}^{(1)}(\theta) \otimes T_{c_1 c_2}^{(1)}(\theta) \otimes \cdots \otimes T_{c_{r-2} c_{r-1}}^{(1)}(\theta) \otimes T_{c_{r-1} b}^{(1)}(\theta). \quad (2.10)$$

For some essential purposes, it is worthwhile to express (2.8) as (with $i, j \in \{1, \ldots, n\}$)

$$\begin{aligned}
T_{ij}^{(1)}(\theta) &= \frac{1}{2}\sum_{\epsilon} e^{m_{ji}^{(\epsilon)}\theta}\left((ji) + \epsilon\left(\overline{j}\,\overline{i}\right)\right), \\
T_{\overline{i}\overline{j}}^{(1)}(\theta) &= \frac{1}{2}\sum_{\epsilon} e^{m_{ji}^{(\epsilon)}\theta}\left(\left(\overline{j}\,\overline{i}\right) + \epsilon(ji)\right), \\
T_{i\overline{j}}^{(1)}(\theta) &= \frac{1}{2}\sum_{\epsilon} e^{m_{ji}^{(\epsilon)}\theta}\left(\left(\overline{j}i\right) + \epsilon\left(j\overline{i}\right)\right), \\
T_{\overline{i}j}^{(1)}(\theta) &= \frac{1}{2}\sum_{\epsilon} e^{m_{ji}^{(\epsilon)}\theta}\left(\left(j\overline{i}\right) + \epsilon\left(\overline{j}i\right)\right).
\end{aligned} \quad (2.11)$$

The transfer matrix is the trace on (a, b):

$$\mathbf{T}^{(r)}(\theta) = \sum_{a} T_{aa}^{(r)}(\theta). \quad (2.12)$$

The foregoing construction assures the crucial commutativity

$$\left[\mathbf{T}^{(r)}(\theta), \mathbf{T}^{(r)}(\theta')\right] = 0. \quad (2.13)$$

This implies that the eigenstates of $\mathbf{T}^{(r)}(\theta)$ are θ-independent (sum of basis vectors with θ-independent, constant relative coefficients).

2.2. Odd Dimensions

From the extensive previous studies [6, 7], we select the essential features. For $N = 2n - 1$ ($n = 2, 3, \ldots$),

$$\overline{n} = N - n + 1 = n \quad (2.14)$$

with the same definition as in (2.1). This is the crucial new feature. For $(i,j) \neq n$, one has the same $P_{ij}^{(\epsilon)}$, $P_{i\bar{j}}^{(\epsilon)}$ as before. But now

$$P_{in}^{(\epsilon)} = \frac{1}{2}\left\{(ii) + \left(\bar{i}\bar{i}\right) + \epsilon\left[\left(i\bar{i}\right) + \left(\bar{i}i\right)\right]\right\} \otimes (nn),$$
$$P_{ni}^{(\epsilon)} = \frac{1}{2}(nn) \otimes \left\{(ii) + \left(\bar{i}\bar{i}\right) + \epsilon\left[\left(i\bar{i}\right) + \left(\bar{i}i\right)\right]\right\},$$
$$P_{nn}^{(\epsilon)} = (nn) \otimes (nn). \tag{2.15}$$

Normalizing the coefficient of $P_{nn}^{(\epsilon)}$ to unity, $\hat{R}(\theta)$ of (2.4) now becomes

$$\hat{R}(\theta) = P_{nn}^{(\epsilon)} + \sum_{i,\epsilon}\left(e^{m_{ni}^{(\epsilon)}\theta} P_{ni}^{(\epsilon)} + e^{m_{in}^{(\epsilon)}\theta} P_{in}^{(\epsilon)}\right) + \sum_{i,j\epsilon} e^{m_{ij}^{(\epsilon)}\theta}\left(P_{ij}^{(\epsilon)} + P_{i\bar{j}}^{(\epsilon)}\right) \tag{2.16}$$

($i,j \in \{1,\ldots,n-1\}$, $\bar{i},\bar{j} \in \{2n-1,\ldots,n+1\}$, $\epsilon = \pm$). There are evident parallel modifications concerning $T_{ab}^{(r)}(\theta)$. Reference [7] contains detailed discussions and results for $N = 3$, $r = 1,2,3,4$. We will not repeat them here but will reconsider them later in the context of the general methods to be implemented below.

2.3. *Free Parameters*

A unique feature of our constructions is the number of free parameters $m_{ij}^{(\epsilon)}$, increasing as N^2 with N. For our choice of normalizations (apart from a possible altered choice of an overall factor, irrelevant for (2.5)), the exact numbers are

$$\frac{1}{2}N^2 = 2n^2 \quad \text{for } N = 2n,$$
$$\frac{1}{2}(N+3)(N-1) = 2\left(n^2 - 1\right) \quad \text{for } N = 2n-1. \tag{2.17}$$

This is to be contrasted with the multistate generalization of the 6-vertex model where the parametrization remain fixed at the 6-vertex level ([2] and sources cited here).

One of our principal aims is to display the roles of our free parameters concerning the basic features of our models.

3. Explicit Eigenvalues of Transfer Matrices for All Dimensions N and Orders r

We present below a unified approach for all N and illustrate it in some detail for $r = 1,2,3,4,5$. For $r > 5$ the extension will be evident.

3.1. Even Dimensions

In Section 5 of [9] it was pointed out that the antidiagonal matrix

$$K = \sum_{a=1}^{2n} (a\overline{a}) = \sum_{i=1}^{n} \left(\left(i\overline{i} \right) + \left(\overline{i}i \right) \right) \tag{3.1}$$

relates the blocks of $T^{(1)}$ as follows:

$$KT_{ab}^{(1)} = T_{\overline{a}b}^{(1)}, \qquad T_{ab}^{(1)}K = T_{a\overline{b}}^{(1)}, \qquad KT_{ab}^{(1)}K = T_{\overline{a}\overline{b}}^{(1)}. \tag{3.2}$$

For $N = 2$ this led to an iterative construction of eigenstates and eigenvalues of the transfer matrix $\mathbf{T}^{(r)}$ for increasing r. This works efficiently only for $N = 2$ (the 4×4 $\widehat{R}(\theta)$-matrix).

For our present approach the crucial ingredient shows up immediately in the structure (arising from those of our projectors and the coproducts rules)

$$\begin{aligned}\mathbf{T}^{(r)} = \frac{1}{2^r} \sum_{a_1,a_2,\ldots,a_r} \sum_{\epsilon_{21},\epsilon_{32},\ldots,\epsilon_{1r}} & e^{(m_{a_2a_1}^{(\epsilon_{21})} + m_{a_3a_2}^{(\epsilon_{32})} + \cdots + m_{a_1a_r}^{(\epsilon_{1r})})\theta} \\ & \times \left[(a_2a_1) + \epsilon_{21}(\overline{a}_2\overline{a}_1) \right] \\ & \otimes \left[(a_3a_2) + \epsilon_{32}(\overline{a}_3\overline{a}_2) \right] \otimes \cdots \otimes \left[(a_1a_r) + \epsilon_{1r}(\overline{a}_1\overline{a}_r) \right],\end{aligned} \tag{3.3}$$

where each ϵ is independently $(\pm)$. From (3.3), one obtains immediately

$$\begin{aligned}\left(\mathbf{T}^{(r)}\right)^{-1} = \frac{1}{2^r} \sum_{a_1,a_2,\ldots,a_r} \sum_{\epsilon_{21},\epsilon_{32},\ldots,\epsilon_{1r}} & e^{-(m_{a_2a_1}^{(\epsilon_{21})} + m_{a_3a_2}^{(\epsilon_{32})} + \cdots + m_{a_1a_r}^{(\epsilon_{1r})})\theta} \\ & \times \left[(a_1a_2) + \epsilon_{21}(\overline{a}_1\overline{a}_2) \right] \\ & \otimes \left[(a_2a_3) + \epsilon_{32}(\overline{a}_2\overline{a}_3) \right] \otimes \cdots \otimes \left[(a_ra_1) + \epsilon_{1r}(\overline{a}_r\overline{a}_1) \right].\end{aligned} \tag{3.4}$$

Consider the basis state

$$|b_1\rangle \otimes |b_2\rangle \otimes \cdots \otimes |b_r\rangle \equiv |b_1b_2\cdots b_r\rangle. \tag{3.5}$$

(1) If any one (even a single one) of the indices $b_i \neq a_{i+1}$ or $\overline{a_{i+1}}$ (cyclic), then the coefficient of $e^{(m_{a_2a_1}^{(\epsilon_{21})} + \cdots + m_{a_1a_r}^{(\epsilon_{1r})})\theta}$, namely, $[(a_2a_1) + \epsilon_{21}(\overline{a}_2\overline{a}_1)], \otimes \cdots \otimes [(a_1a_r) + \epsilon_{1r}(\overline{a}_1\overline{a}_r)]$, will annihilate it. On the other hand (2.9) plays a crucial role. Gathering together all the terms with

$e^{(m_{b_2b_1}^{(\epsilon_{21})}+m_{b_3b_2}^{(\epsilon_{32})}+\cdots+m_{b_rb_{r-1}}^{(\epsilon_{r,r-1})}+m_{b_1b_r}^{(\epsilon_{1r})})\theta}$ as coefficient, one obtains the action of $\mathbf{T}^{(r)}$ on (3.5). All essential features for the general case can be read off the first few simples examples:

$$2^2\mathbf{T}^{(2)}|b_1b_2\rangle = \sum_{\epsilon_{21},\epsilon_{12}} e^{(m_{b_2b_1}^{(\epsilon_{21})}+m_{b_1b_2}^{(\epsilon_{12})})\theta}(1+\epsilon_{21}\epsilon_{12})\left[|b_2b_1\rangle + \epsilon_{21}\left|\overline{b}_2\overline{b}_1\right\rangle\right], \tag{3.6}$$

$$\begin{aligned}2^3\mathbf{T}^{(3)}|b_1b_2b_3\rangle = \sum_{\epsilon_{21},\epsilon_{32},\epsilon_{13}} & e^{(m_{b_2b_1}^{(\epsilon_{21})}+m_{b_3b_2}^{(\epsilon_{32})}+m_{b_1b_3}^{(\epsilon_{13})})\theta}(1+\epsilon_{21}\epsilon_{32}\epsilon_{13}) \\ & \times\left[|b_2b_3b_1\rangle + \epsilon_{21}\left|\overline{b}_2b_3\overline{b}_1\right\rangle + \epsilon_{32}\left|\overline{b}_2\overline{b}_3b_1\right\rangle + \epsilon_{13}\left|b_2\overline{b}_3\overline{b}_1\right\rangle\right],\end{aligned} \tag{3.7}$$

$$\begin{aligned}2^4\mathbf{T}^{(4)}|b_1b_2b_3b_4\rangle = \sum_{\epsilon_{21},\epsilon_{32},\epsilon_{43},\epsilon_{14}} & e^{(m_{b_2b_1}^{(\epsilon_{21})}+m_{b_3b_2}^{(\epsilon_{32})}+m_{b_4b_3}^{(\epsilon_{43})}+m_{b_1b_4}^{(\epsilon_{14})})\theta}(1+\epsilon_{21}\epsilon_{32}\epsilon_{43}\epsilon_{14}) \\ & \times\left[|b_2b_3b_4b_1\rangle + \epsilon_{21}\left|\overline{b}_2b_3b_4\overline{b}_1\right\rangle + \epsilon_{32}\left|\overline{b}_2\overline{b}_3b_4b_1\right\rangle\right. \\ & + \epsilon_{43}\left|b_2\overline{b}_3\overline{b}_4b_1\right\rangle + \epsilon_{14}\left|b_2b_3\overline{b}_4\overline{b}_1\right\rangle + \epsilon_{21}\epsilon_{32}\left|b_2\overline{b}_3b_4\overline{b}_1\right\rangle \\ & \left. + \epsilon_{21}\epsilon_{43}\left|\overline{b}_2\overline{b}_3\overline{b}_4\overline{b}_1\right\rangle + \epsilon_{21}\epsilon_{14}\left|\overline{b}_2b_3\overline{b}_4b_1\right\rangle\right].\end{aligned} \tag{3.8}$$

Note that the bars correspond to the indies of the ϵ's. One has thus $\epsilon_{ij}|\overline{b}_i\overline{b}_jb_kb_l\rangle$, $\epsilon_{ij}\epsilon_{ki}|\overline{\overline{b}}_i\overline{b}_j\overline{b}_kb_l\rangle = \epsilon_{ij}\epsilon_{ki}|b_i\overline{b}_j\overline{b}_kb_l\rangle$ since $\overline{\overline{b}}_i = b_i$. Consequently only even number of additional bars appear on the right. Starting with even or odd number of bars on the left leads to "even" and "odd" closed subspaces.

$$\begin{aligned}2^5\mathbf{T}^{(5)}|b_1b_2b_3b_4b_5\rangle = \sum_{\epsilon_{21},\epsilon_{32},\epsilon_{43},\epsilon_{14}} & e^{(m_{b_2b_1}^{(\epsilon_{21})}+m_{b_3b_2}^{(\epsilon_{32})}+m_{b_4b_3}^{(\epsilon_{43})}+m_{b_5ba_4}^{(\epsilon_{54})}+m_{b_1b_5}^{(\epsilon_{15})})\theta}(1+\epsilon_{21}\epsilon_{32}\epsilon_{43}\epsilon_{54}\epsilon_{15}) \\ & \times\left[|b_2b_3b_4b_5b_1\rangle + \epsilon_{21}\left|\overline{b}_2b_3b_4b_5\overline{b}_1\right\rangle + \epsilon_{32}\left|\overline{b}_2\overline{b}_3b_4b_5b_1\right\rangle\right. \\ & + \epsilon_{43}\left|b_2\overline{b}_3\overline{b}_4b_5b_1\right\rangle + \epsilon_{54}\left|b_2b_3\overline{b}_4\overline{b}_5b_1\right\rangle + \epsilon_{15}\left|b_2b_3b_4\overline{b}_5\overline{b}_1\right\rangle \\ & + \epsilon_{21}\epsilon_{32}\left|b_2\overline{b}_3b_4b_5\overline{b}_1\right\rangle + \epsilon_{21}\epsilon_{43}\left|\overline{b}_2\overline{b}_3\overline{b}_4b_5\overline{b}_1\right\rangle \\ & + \epsilon_{21}\epsilon_{54}\left|\overline{b}_2b_3\overline{b}_4\overline{b}_5\overline{b}_1\right\rangle + \epsilon_{21}\epsilon_{15}\left|\overline{b}_2b_3b_4\overline{b}_5b_1\right\rangle \\ & + \epsilon_{32}\epsilon_{43}\left|\overline{b}_2b_3\overline{b}_4b_5b_1\right\rangle + \epsilon_{32}\epsilon_{54}\left|\overline{b}_2\overline{b}_3\overline{b}_4\overline{b}_5b_1\right\rangle \\ & + \epsilon_{32}\epsilon_{15}\left|\overline{b}_2\overline{b}_3b_4\overline{b}_5\overline{b}_1\right\rangle + \epsilon_{43}\epsilon_{54}\left|b_2\overline{b}_3b_4\overline{b}_5b_1\right\rangle \\ & \left. + \epsilon_{43}\epsilon_{15}\left|b_2\overline{b}_3\overline{b}_4\overline{b}_5\overline{b}_1\right\rangle + \epsilon_{54}\epsilon_{15}\left|b_2b_3\overline{b}_4b_5\overline{b}_1\right\rangle\right].\end{aligned} \tag{3.9}$$

(2) It is important to note that the preceding features are independent of N—for a given r they are valid for all N. For $r > N$ all the indices $(b_1,\ldots,b_r)$ cannot be distinct. But

even for $r < N$, some or all of them can coincide. This general validity permits a unified treatment for all N.

(3) Note also the crucial cyclic permutation of the indices under the action of $\mathbf{T}^{(r)}$: $(b_1, b_2, b_3, \ldots, b_{r-1}, b_r) \rightarrow (b_2, b_3, b_4, \ldots, b_r, b_1)$. This is a consequence of the trace condition in (2.12) incorporated in (3.3). It will be seen later how this leads to an essential role of the rth roots of unity $e^{\mathrm{i}((k\cdot 2\pi)/r)}$ $(k = 0, 1, \ldots, r-1)$ in our construction of eigenstates and also of eigenvalues as factors accompanying exponentials of the type appearing in (3.6)–(3.9).

We continue the study of the particular cases $r = 1, 2, 3, 4, 5$ each one for any N. General features will be better understood afterwards.

(i) $r = 1$

$$\mathbf{T}^{(1)} = \frac{1}{2}\sum_{a,\epsilon} e^{m_{aa}^{(\epsilon)}\theta}[(aa) + \epsilon(\bar{a}\,\bar{a})] = \frac{1}{2}\sum_{i,\epsilon}(1+\epsilon)e^{m_{ii}^{(\epsilon)}\theta}\left[(ii) + \epsilon\left(\bar{i}\,\bar{i}\right)\right], \tag{3.10}$$

where $a \in \{1, \ldots, 2n\}$, $i \in \{1, \ldots, n\}$. This is already in diagonal form. Eigenstates and values are obtained trivially.

(ii) $r = 2$

We start with (3.6). Separating "even" and "odd" spaces, define (for $i \neq j$ when $(i, j, \bar{i}, \bar{j})$ are all distinct)

$$\begin{aligned} \mathbf{T}^{(2)}\left(c_1|ij\rangle + c_2\left|\bar{i}\,\bar{j}\right\rangle + c_3|ji\rangle + c_4\left|\bar{j}\,\bar{i}\right\rangle\right) &= v_e\left(c_1|ij\rangle + c_2\left|\bar{i}\,\bar{j}\right\rangle + c_3|ji\rangle + c_4\left|\bar{j}\,\bar{i}\right\rangle\right), \\ \mathbf{T}^{(2)}\left(d_1\left|i\bar{j}\right\rangle + d_2\left|\bar{i}j\right\rangle + d_3\left|j\bar{i}\right\rangle + d_4\left|\bar{j}i\right\rangle\right) &= v_o\left(d_1\left|i\bar{j}\right\rangle + d_2\left|\bar{i}j\right\rangle + d_3\left|j\bar{i}\right\rangle + d_4\left|\bar{j}i\right\rangle\right). \end{aligned} \tag{3.11}$$

The solutions are, for $(c_1, c_2, c_3, c_4) = (1,1,1,1), (1,1,-1,-1), (1,-1,1,-1), (1,-1,-1,1)$ respectively,

$$v_e = e^{(m_{ij}^{(+)} + m_{ji}^{(+)})\theta}, -e^{(m_{ij}^{(+)} + m_{ji}^{(+)})\theta}, e^{(m_{ij}^{(-)} + m_{ji}^{(-)})\theta}, -e^{(m_{ij}^{(-)} + m_{ji}^{(-)})\theta}, \tag{3.12}$$

and for $(d_1, d_2, d_3, d_4) = (1,1,1,1), (1,1,-1,-1), (1,-1,1,-1), (1,-1,-1,1)$, respectively,

$$v_o = e^{(m_{ij}^{(+)} + m_{ji}^{(+)})\theta}, -e^{(m_{ij}^{(+)} + m_{ji}^{(+)})\theta}, -e^{(m_{ij}^{(-)} + m_{ji}^{(-)})\theta}, e^{(m_{ij}^{(-)} + m_{ji}^{(-)})\theta}. \tag{3.13}$$

For $i = j$, with $\epsilon = \pm$

$$\begin{aligned} \mathbf{T}^{(2)}\left(|ii\rangle + \epsilon\left|\bar{i}\,\bar{i}\right\rangle\right) &= e^{2m_{ii}^{(\epsilon)}\theta}\left(|ii\rangle + \epsilon\left|\bar{i}\,\bar{i}\right\rangle\right), \\ \mathbf{T}^{(2)}\left(\left|i\bar{i}\right\rangle + \epsilon\left|\bar{i}i\right\rangle\right) &= \epsilon e^{2m_{ii}^{(\epsilon)}\theta}\left(\left|i\bar{i}\right\rangle + \epsilon\left|\bar{i}i\right\rangle\right). \end{aligned} \tag{3.14}$$

Summing over all i, consistently with the general rule, one obtains

$$\mathrm{Tr}\left(\mathbf{T}^{(2)}\right) = 2\sum_{i=1}^{n} e^{2m_{ii}^{(+)}\theta} \tag{3.15}$$

contributions coming only for $i = j$. For $r = 2$, the rth roots of unity are ± 1 and their role above is not visible very distinctly due to the presence of ± 1 also from ϵ. The role of the roots of unity will be more evident from $r = 3$ onwards. The simple exercise above gives all eigenvalues and eigenstates for $\mathbf{T}^{(2)}$ for all N $(= 2n)$. For $N = 2$, $r = 2$, writing out the 16×16 $\mathbf{T}^{(2)}$ in full and effectively diagonalizing it we have confirmed the results above.

(iii) r=3

Now the starting point is (3.7). Define, for the even subset,

$$\begin{aligned} V_\lambda = {} & \left(|i_1 i_2 i_3\rangle + \epsilon_{12}\left|\bar{i}_1 \bar{i}_2 i_3\right\rangle + \epsilon_{23}\left|i_1 \bar{i}_2 \bar{i}_3\right\rangle + \epsilon_{31}\left|\bar{i}_1 i_2 \bar{i}_3\right\rangle\right) \\ & + \lambda\left(|i_3 i_1 i_2\rangle + \epsilon_{12}\left|i_3 \bar{i}_1 \bar{i}_2\right\rangle + \epsilon_{23}\left|\bar{i}_3 i_1 \bar{i}_2\right\rangle + \epsilon_{31}\left|\bar{i}_3 \bar{i}_1 i_2\right\rangle\right) \\ & + \lambda^2\left(|i_2 i_3 i_1\rangle + \epsilon_{12}\left|\bar{i}_2 i_3 \bar{i}_1\right\rangle + \epsilon_{23}\left|\bar{i}_2 \bar{i}_3 i_1\right\rangle + \epsilon_{31}\left|i_2 \bar{i}_3 \bar{i}_1\right\rangle\right), \end{aligned} \tag{3.16}$$

where

$$\lambda = \left(1, e^{\mathrm{i}(2\pi/3)}, e^{\mathrm{i}(4\pi/3)}\right), \tag{3.17}$$

and hence $\lambda^3 = 1$, and also $1 + e^{\mathrm{i}(2\pi/3)} + e^{\mathrm{i}(4\pi/3)} = 0$. Corresponding to $(i_1, i_2, i_3) \rightleftharpoons (\bar{i}_1, \bar{i}_2, \bar{i}_3)$, define for the odd-subset,

$$\begin{aligned} \overline{V_\lambda} = {} & \left(\left|\bar{i}_1 \bar{i}_2 \bar{i}_3\right\rangle + \epsilon_{12}\left|i_1 i_2 \bar{i}_3\right\rangle + \epsilon_{23}\left|\bar{i}_1 i_2 i_3\right\rangle + \epsilon_{31}\left|i_1 \bar{i}_2 i_3\right\rangle\right) \\ & + \lambda\left(\left|\bar{i}_3 \bar{i}_1 \bar{i}_2\right\rangle + \epsilon_{12}\left|\bar{i}_3 i_1 i_2\right\rangle + \epsilon_{23}\left|i_3 \bar{i}_1 i_2\right\rangle + \epsilon_{31}\left|i_3 i_1 \bar{i}_2\right\rangle\right) \\ & + \lambda^2\left(\left|\bar{i}_2 \bar{i}_3 \bar{i}_1\right\rangle + \epsilon_{12}\left|i_2 \bar{i}_3 i_1\right\rangle + \epsilon_{23}\left|i_2 i_3 \bar{i}_1\right\rangle + \epsilon_{31}\left|\bar{i}_2 i_3 i_1\right\rangle\right). \end{aligned} \tag{3.18}$$

The 8 values of the set $(\epsilon_{12}, \epsilon_{23}, \epsilon_{31})$ are restricted (see (3.7)) by $(1 + \epsilon_{12}\epsilon_{23}\epsilon_{31}) = 2\ (\neq 0)$. The remaining 4 possibilities along with 3 for λ (see (3.17)) yield $4 \times 3 = 12$ solutions for V_λ and $\overline{V_\lambda}$ each for distinct indices $i_1 \neq i_2 \neq i_3 \neq i_1$. The eigenvalues are both V_λ and $\overline{V_\lambda}$:

$$e^{(m_{i_1 i_2}^{(\epsilon_{12})} + m_{i_2 i_3}^{(\epsilon_{23})} + m_{i_3 i_1}^{(\epsilon_{31})})\theta}\left(1, e^{\mathrm{i}(2\pi/3)}, e^{\mathrm{i}(4\pi/3)}\right). \tag{3.19}$$

For two of the three indices equal $(i_1 = i_2 \neq i_3, \ldots)$ the preceding results can be carried over without crucial change. But for $i_1 = i_2 = i_3 \equiv i$, say, there is a special situation due to the fact

$$1 + \lambda + \lambda^2 = (3, 0, 0), \tag{3.20}$$

for $\lambda = (1, e^{\mathrm{i}(2\pi/3)}, e^{\mathrm{i}(4\pi/3)})$, respectively. Now

$$\begin{aligned} V_\lambda = {} & \left(1 + \lambda + \lambda^2\right)|iii\rangle + \left(\epsilon_{12} + \epsilon_{31}\lambda + \epsilon_{23}\lambda^2\right)\left|\bar{i}\,\bar{i}i\right\rangle \\ & + \left(\epsilon_{23} + \epsilon_{12}\lambda + \epsilon_{31}\lambda^2\right)\left|i\bar{i}\,\bar{i}\right\rangle + \left(\epsilon_{31} + \epsilon_{23}\lambda + \epsilon_{12}\lambda^2\right)\left|\bar{i}i\bar{i}\right\rangle. \end{aligned} \tag{3.21}$$

The eigenfunctions and eigenvalues are (for mutually orthogonal basis states)

$$\begin{aligned} & \mathbf{T}^{(3)}\left(|iii\rangle + \left|\bar{i}\,\bar{i}i\right\rangle + \left|i\bar{i}\,\bar{i}\right\rangle + \left|\bar{i}i\bar{i}\right\rangle\right) \\ & \quad = e^{3m_{ii}^{(+)}\theta}\left(|iii\rangle + \left|\bar{i}\,\bar{i}i\right\rangle + \left|i\bar{i}\,\bar{i}\right\rangle + \left|\bar{i}i\bar{i}\right\rangle\right), \\ & \mathbf{T}^{(3)}\left(3|iii\rangle - \left|\bar{i}\,\bar{i}i\right\rangle - \left|i\bar{i}\,\bar{i}\right\rangle - \left|\bar{i}i\bar{i}\right\rangle\right) \\ & \quad = e^{(m_{ii}^{(+)} + 2m_{ii}^{(-)})\theta}\left(3|iii\rangle - \left|\bar{i}\,\bar{i}i\right\rangle - \left|i\bar{i}\,\bar{i}\right\rangle - \left|\bar{i}i\bar{i}\right\rangle\right), \\ & \mathbf{T}^{(3)}\left(\left|\bar{i}\,\bar{i}i\right\rangle + e^{\mathrm{i}(2\pi/3)}\left|i\bar{i}\,\bar{i}\right\rangle + e^{\mathrm{i}(4\pi/3)}\left|\bar{i}i\bar{i}\right\rangle\right) \\ & \quad = e^{\mathrm{i}(2\pi/3)}e^{(m_{ii}^{(+)} + 2m_{ii}^{(-)})\theta}\left(\left|\bar{i}\,\bar{i}i\right\rangle + e^{\mathrm{i}(2\pi/3)}\left|i\bar{i}\,\bar{i}\right\rangle + e^{\mathrm{i}(4\pi/3)}\left|\bar{i}i\bar{i}\right\rangle\right), \\ & \mathbf{T}^{(3)}\left(\left|\bar{i}\,\bar{i}i\right\rangle + e^{\mathrm{i}(4\pi/3)}\left|i\bar{i}\,\bar{i}\right\rangle + e^{\mathrm{i}(2\pi/3)}\left|\bar{i}i\bar{i}\right\rangle\right) \\ & \quad = e^{\mathrm{i}(4\pi/3)}e^{(m_{ii}^{(+)} + 2m_{ii}^{(-)})\theta}\left(\left|\bar{i}\,\bar{i}i\right\rangle + e^{\mathrm{i}(4\pi/3)}\left|i\bar{i}\,\bar{i}\right\rangle + e^{\mathrm{i}(2\pi/3)}\left|\bar{i}i\bar{i}\right\rangle\right). \end{aligned} \tag{3.22}$$

For $\overline{V_\lambda}$ one follows exactly similar steps and obtains the same set of eigenvalues (with $i \rightleftharpoons \bar{i}$ in the eigenstates). Combining all the solutions above, we obtain complete results for $r = 3$ and all even N.

(iv) $r = 4$

The key result (3.8) indicates the following construction (a direct generalization of (3.16)). Define, for the even subset,

$$\begin{aligned} V_{(b_1 b_2 b_3 b_4)} = {} & |b_1 b_2 b_3 b_4\rangle + \epsilon_{12}\left|\bar{b}_1\,\bar{b}_2 b_3 b_4\right\rangle + \epsilon_{23}\left|b_1 \bar{b}_2\,\bar{b}_3 b_4\right\rangle + \epsilon_{34}\left|b_1 b_2 \bar{b}_3 \bar{b}_4\right\rangle \\ & + \epsilon_{41}\left|\bar{b}_1 b_2 b_3 \bar{b}_4\right\rangle + \epsilon_{12}\epsilon_{23}\left|\bar{b}_1 b_2 \bar{b}_3 b_4\right\rangle + \epsilon_{12}\epsilon_{34}\left|\bar{b}_1\,\bar{b}_2\,\bar{b}_3 \bar{b}_4\right\rangle + \epsilon_{23}\epsilon_{34}\left|b_1 \bar{b}_2 b_3 \bar{b}_4\right\rangle, \end{aligned} \tag{3.23}$$

and similarly, implementing circular permutations as in (3.16), $V_{(b_4 b_1 b_2 b_3)}$, $V_{(b_3 b_4 b_1 b_2)}$, $V_{(b_2 b_3 b_4 b_1)}$, define

$$V_\lambda = V_{(b_1 b_2 b_3 b_4)} + \lambda V_{(b_4 b_1 b_2 b_3)} + \lambda^2 V_{(b_3 b_4 b_1 b_2)} + \lambda^3 V_{(b_2 b_3 b_4 b_1)}, \tag{3.24}$$

where

$$\lambda = \left(1, e^{\mathrm{i}(2\pi/4)}, e^{\mathrm{i}2\cdot(2\pi/4)}, e^{\mathrm{i}3\cdot(2\pi/4)}\right) = (1, \mathbf{i}, -1, -\mathbf{i}). \tag{3.25}$$

One obtains for distinct (i_1, i_2, i_3, i_4)

$$\mathbf{T}^{(4)} V_\lambda = \lambda e^{(m_{i_1 i_2}^{(\epsilon_{12})} + m_{i_2 i_3}^{(\epsilon_{23})} + m_{i_3 i_4}^{(\epsilon_{34})} + m_{i_4 i_1}^{(\epsilon_{41})})\theta} V_\lambda, \tag{3.26}$$

where for nonzero results $(1 + \epsilon_{12}\epsilon_{23}\epsilon_{34}\epsilon_{41}) = 2$. For the 8 possibilities remaining for the ϵ's and the 4 values of λ, one obtains the full set of 32 eigenstates and the corresponding eigenvalues. For the complementary "odd" subspace formed by

$$\begin{aligned} &\left\{ \left|b_1 b_2 b_3 \bar{b}_4\right\rangle, \left|b_1 b_2 \bar{b}_3 b_4\right\rangle, \left|b_1 \bar{b}_2 b_3 b_4\right\rangle, \left|\bar{b}_1 b_2 b_3 b_4\right\rangle, \right.\\ &\left. \left|\bar{b}_1 \bar{b}_2 \bar{b}_3 b_4\right\rangle, \left|\bar{b}_1 \bar{b}_2 b_3 \bar{b}_4\right\rangle, \left|\bar{b}_1 b_2 \bar{b}_3 \bar{b}_4\right\rangle, \left|b_1 \bar{b}_2 \bar{b}_3 \bar{b}_4\right\rangle \right\}, \end{aligned} \tag{3.27}$$

and the sets related through circular permutations of the indices, one obtains, entirely in evident analogy of (3.24),

$$\overline{V_\lambda} = \overline{V_{(b_1 b_2 b_3 b_4)}} + \lambda \overline{V_{(b_4 b_1 b_2 b_3)}} + \lambda^2 \overline{V_{(b_3 b_4 b_1 b_2)}} + \lambda^3 \overline{V_{(b_2 b_3 b_4 b_1)}}. \tag{3.28}$$

One obtains, for distinct indices, the same set of eigenvalues as in (3.26). For some equal indices, but not all, the preceding results can be carried over essentially. For all equal indices one obtains the following spectrum of eigenvalues (grouping together the even and the odd subspaces):

$$\begin{gathered} e^{4m_{ii}^{(+)}\theta}(1,1), \\ [3\text{-times}] : e^{2(m_{ii}^{(+)} + m_{ii}^{(-)})\theta}\left(1, e^{\mathrm{i}(2\pi/4)}, e^{\mathrm{i}2\cdot(2\pi/4)}, e^{\mathrm{i}3\cdot(2\pi/4)}\right), \\ e^{4m_{ii}^{(-)}\theta}(1,-1), \end{gathered} \tag{3.29}$$

(In fact these are the same as in (A.13) of [7], since the distinguishing index for odd dimensions signalled in (2.14) and (2.15) is not involved.) We will not present here the 16 orthogonal eigenstates corresponding to (3.29). But a new feature, as compared to (3.22), should be pointed out. For $r = 3$, considering together even and odd spaces, one has for identical indices the eigenvalues

$$\begin{gathered} e^{3m_{ii}^{(+)}\theta}(1,1), \\ [2\text{-times}] : e^{(m_{ii}^{(+)} + 2m_{ii}^{(-)})\theta}\left(1, e^{\mathrm{i}(2\pi/3)}, e^{\mathrm{i}2\cdot(2\pi/3)}\right). \end{gathered} \tag{3.30}$$

In fact, for each r one obtains, for $\epsilon = 1$, $e^{rm_{ii}^{(+)}\theta}(1,1)$ giving

$$\mathrm{Tr}\left(\mathbf{T}^{(r)}\right) = 2\sum_{i=1}^{n} e^{rm_{ii}^{(\epsilon)}\theta}. \tag{3.31}$$

All other multiplets have each one zero sum. Concerning the zero sum multiplets, comparing (3.30) and (3.29), we note that

(1) for $r = 3$, one has only triplets;

(2) for $r = 4$, one has quadruplets and also a doublet $(1,-1)$. This last enters due to factorizability of r $(4 = 2 \times 2)$.

Such a feature is worth signalling since it generalizes.

When r is a prime number, the zero-sum multiplets appear only as r-plets (with rth roots of unity). How they consistently cover the whole base space (along with the doublets giving the trace as in (3.31)) has been amply discussed in our previous papers [7, 9] under the heading "An encounter with a theorem of Fermat." When r has many factors the multiplicity of submultiplets corresponding to each one is difficult to formulate giving, say, an explicit general prescription.

(v) $r \geq 5$

Another problem concerning systematic, complete enumeration one encounters already for $r = 5$. Starting with (3.9) the generalizations of (3.23)–(3.26) are quite evident. We will not present them here. But as compared to (3.29) one has now for the 32-dimensional subspace (for $i_1 = i_2 = \cdots = i_5$)

$$\begin{aligned}
&e^{5m_{ii}^{(+)}\theta}(1,1),\\
&[n_2\text{-times}] : e^{(3m_{ii}^{(+)}+2m_{ii}^{(-)})\theta}\left(1, e^{\mathrm{i}(2\pi/5)}, e^{\mathrm{i}2\cdot(2\pi/5)}, e^{\mathrm{i}3\cdot(2\pi/5)}, e^{\mathrm{i}4\cdot(2\pi/5)}\right),\\
&[n_4\text{-times}] : e^{(m_{ii}^{(+)}+4m_{ii}^{(-)})\theta}\left(1, e^{\mathrm{i}(2\pi/5)}, e^{\mathrm{i}2\cdot(2\pi/5)}, e^{\mathrm{i}3\cdot(2\pi/5)}, e^{\mathrm{i}4\cdot(2\pi/5)}\right),
\end{aligned} \tag{3.32}$$

where $n_2 + n_4 = 6$. At this stage it is not difficult to present complete construction. But as r increases, for prime numbers, one has

$$\begin{aligned}
&e^{rm_{ii}^{(+)}\theta}(1,1),\\
&[n_2\text{-times}] : e^{((r-2)m_{ii}^{(+)}+2m_{ii}^{(-)})\theta}\left(1, e^{\mathrm{i}(2\pi/r)}, \ldots, e^{\mathrm{i}(r-1)\cdot(2\pi/r)}\right),\\
&[n_4\text{-times}] : e^{((r-4)m_{ii}^{(+)}+4m_{ii}^{(-)})\theta}\left(1, e^{\mathrm{i}(2\pi/r)}, \ldots, e^{\mathrm{i}(r-1)\cdot(2\pi/r)}\right),\\
&\quad\vdots \qquad\qquad\qquad\qquad \vdots\\
&[n_{r-1}\text{-times}] : e^{(m_{ii}^{(+)}+(r-1)m_{ii}^{(-)})\theta}\left(1, e^{\mathrm{i}(2\pi/r)}, \ldots, e^{\mathrm{i}(r-1)\cdot(2\pi/r)}\right),
\end{aligned} \tag{3.33}$$

the sum of the multiplicities of such r-plets satisfying

$$n_2 + n_4 + \cdots + n_{r-1} = \frac{1}{r}(2^r - 2). \tag{3.34}$$

(That the right-hand side is an integer for r a prime number is guaranteed by a theorem of Fermat, as has been discussed before.) A general explicit prescription for the sequence $(n_2, n_4, \ldots, n_{r-1})$ is beyond the scope of this paper.

The situation is as above for prime numbers r. For factorizable r the presence, in addition, of submultiplets has already been pointed out. For $r = 4$, only such submultiplets were $(1, -1)$. For $r = 6$, one can have $(1, -1)$, $(1, e^{i(2\pi/3)}, e^{i2\cdot(2\pi/3)})$ corresponding to the factors $6 = 2 \times 3$, respectively. For $r = 30 = 2 \times 3 \times 5$ one can have also submultiplets corresponding to 5. And so on.

But apart from the above-mentioned limitations concerning multiplicities of zero-sum ("roots of unity") multiplets and submultiplets we can claim to have elucidated the spectrum of eigenvalues for all (N, r). For the generic case with distinct indices for any (N, r), one has eigenvalues

$$\lambda_k e^{(m_{i_1 i_2}^{(\epsilon_{12})} + m_{i_2 i_3}^{(\epsilon_{23})} + \cdots + m_{i_{r-1} i_r}^{(\epsilon_{r-1,r})} + m_{i_1 i_r}^{(\epsilon_{1r})})\theta}, \tag{3.35}$$

where $(1 + \epsilon_{12}\epsilon_{23} \cdots \epsilon_{r-1,r}\epsilon_{1,r}) = 2$ and $\lambda_k = e^{ik\cdot(2\pi/r)}$ $(k = 0, 1, 2, \ldots, r-1)$. Here each $m_{i_k i_l}^{(\epsilon_{kl})}$ is a free parameter. We solve only sets of linear equations with quite simple constant coefficients to obtain the eigenstates and the eigenvalues.

3.2. *Odd Dimensions*

We refer back to (2.14)–(2.16). When the special index n $(n = \overline{n})$ for $N = 2n - 1$ is not present in the basis state $|b_1 b_2 \cdots b_r\rangle$ and hence in the subspace it generates via cyclic permutations as in the forgoing examples, the foregoing constructions can be taken over wholesale. When n is present, the modifications are not difficult to take into account. The trace is now $\mathrm{Tr}\,(\mathbf{T}^{(r)}) = 2\sum_{i=1}^{n-1} e^{rm_{ii}^{(\epsilon)}\theta} + 1$, since the central state $|n\rangle \otimes |n\rangle \otimes \cdots |n\rangle \otimes \equiv |nn \cdots n\rangle$ contributes with our normalization 1 to the trace. Various aspects have been studied in considerable detail in our previous papers [6, 7] on odd N. Here we just refer to them.

4. Spin Chain Hamiltonians

Spin chains corresponding to our braid matrices have already been studied in our previous papers [7, 8]. Here we formulate a unified approach for all dimensions $(N = 2n - 1, 2n)$.

The basic formula (see sources cited in [7, 8]) is

$$H = \sum_{k=1}^{r} I \otimes \cdots \otimes I \otimes \dot{\hat{R}}_{k,k+1}(0) \otimes I \otimes \cdots \otimes I, \tag{4.1}$$

where for circular boundary conditions , $k+1 = r+1 \approx 1$. For even N $(N = 2n)$, from (2.4),

$$\dot{\hat{R}}(0) = \frac{d}{d\theta}\hat{R}(\theta)\Big|_{\theta=0} = \sum_{\epsilon,i,j} m_{ij}^{(\epsilon)}\left(P_{ij}^{(\epsilon)} + P_{\bar{i}\bar{j}}^{(\epsilon)}\right) \tag{4.2}$$

the projectors being given by (2.1) and the remark below (2.1). For N odd $(N = 2n-1;$ $n = 2,3,\ldots)$ we introduce a modified overall normalization factor to start with. Such a factor is trivial concerning the braid equation, but not for the Hamiltonian (since a derivative is involved) if the factor is θ-dependent. Multiply (2.16) by $e^{m\theta}$ and redefine

$$\left(\left(m + m_{ij}^{(\epsilon)}\right), \left(m + m_{ni}^{(\epsilon)}\right), \left(m + m_{in}^{(\epsilon)}\right)\right) \longrightarrow \left(m_{ij}^{(\epsilon)}, m_{ni}^{(\epsilon)}, m_{in}^{(\epsilon)}\right) \tag{4.3}$$

since $(m_{ij}^{(\epsilon)},\ldots)$ are arbitrary to start with. Now for odd N (with the ranges of (i,j), $(\bar{i},\bar{j})$ of (2.16)),

$$\dot{\hat{R}}(0) = mP_{nn} + \sum_{\epsilon,i}\left(m_{ni}^{(\epsilon)}P_{ni}^{(\epsilon)} + m_{in}^{(\epsilon)}P_{in}^{(\epsilon)}\right) + \sum_{\epsilon,i,j} m_{ij}^{(\epsilon)}\left(P_{ij}^{(\epsilon)} + P_{\bar{i}\bar{j}}^{(\epsilon)}\right). \tag{4.4}$$

This extends the $m = 0$ case by including the presence of P_{nn}.

In H, $\dot{\hat{R}}(0)_{k,k+1}$ acts on the basis $|V\rangle_{(k)} \otimes |V\rangle_{(k+1)}$. One can denote, using standard ordering of spin components for each k,

$$|V\rangle_{(k)} = \left| \begin{matrix} |n-1/2\rangle_k \\ \vdots \\ |1/2\rangle_k \\ |-1/2\rangle_k \\ \vdots \\ |-n+1/2\rangle_k \end{matrix} \right\rangle, \qquad |V\rangle_{(k)} = \left| \begin{matrix} |n-1\rangle_k \\ \vdots \\ |1\rangle_k \\ |0\rangle_k \\ |-1\rangle_k \\ \vdots \\ |-n+1\rangle_k \end{matrix} \right\rangle, \tag{4.5}$$

for $N = 2n, 2n-1$, respectively. Without being restricted to spin, one can consider more generally any system with N orthogonal states. We will continue, however, to use the terminology of spin. At each site one can consider a superposition at each level such as $(\sum_l c_{il}^{(k)}|l\rangle_k)$. To keep the notation tractable, we will just denote

$$|V\rangle_{(k)} = \left| \begin{matrix} |1\rangle_k \\ |2\rangle_k \\ \vdots \\ |\bar{2}\rangle_k \\ |\bar{1}\rangle_k \end{matrix} \right\rangle, \tag{4.6}$$

and keep possible significances of $|i\rangle_k$ in mind. Each index k is acted upon twice by H, namely, by $\dot{\hat{R}}_{k-1,k}(0)$, $\dot{\hat{R}}_{k,k+1}(0)$, and, for closed chains, the index 1 is also thus involved in $\dot{\hat{R}}_{12}(0)$ and $\dot{\hat{R}}_{r1}(0)$. Some simple examples are

(i) $N = 2$

$$\dot{\hat{R}}(0) = \begin{vmatrix} \hat{a}_+ & 0 & 0 & \hat{a}_- \\ 0 & \hat{a}_+ & \hat{a}_- & 0 \\ 0 & \hat{a}_- & \hat{a}_+ & 0 \\ \hat{a}_- & 0 & 0 & \hat{a}_+ \end{vmatrix}, \tag{4.7}$$

where $\hat{a}_\pm = (1/2)(m_{11}^{(+)} \pm m_{11}^{(-)})$ and hence in evident notations

$$\dot{\hat{R}}(0)|V\rangle_{(k)} \otimes |V\rangle_{(k+1)} = \left| \begin{matrix} \hat{a}_+|11\rangle + \hat{a}_-\left|\bar{1}\bar{1}\right\rangle \\ \hat{a}_+\left|1\bar{1}\right\rangle + \hat{a}_-\left|\bar{1}1\right\rangle \\ \hat{a}_-\left|1\bar{1}\right\rangle + \hat{a}_+\left|\bar{1}1\right\rangle \\ \hat{a}_-|11\rangle + \hat{a}_+\left|\bar{1}\bar{1}\right\rangle \end{matrix} \right\rangle_{(k,k+1)}, \tag{4.8}$$

(ii) $N = 4$

In the notation of Section 7 of [9]

$$\hat{R}(\theta) = \begin{vmatrix} \hat{D}_{11} & 0 & 0 & \hat{A}_{1\bar{1}} \\ 0 & \hat{D}_{22} & \hat{A}_{2\bar{2}} & 0 \\ 0 & \hat{A}_{\bar{2}2} & \hat{D}_{\bar{2}\bar{2}} & 0 \\ \hat{A}_{\bar{1}1} & 0 & 0 & \hat{D}_{\bar{1}\bar{1}} \end{vmatrix}, \tag{4.9}$$

where

$$\hat{D}_{11} = \hat{D}_{\bar{1}\bar{1}} = \begin{pmatrix} \hat{a}_+ & 0 & 0 & 0 \\ 0 & \hat{b}_+ & 0 & 0 \\ 0 & 0 & \hat{b}_+ & 0 \\ 0 & 0 & 0 & \hat{a}_+ \end{pmatrix}, \qquad \hat{D}_{22} = \hat{D}_{\bar{2}\bar{2}} = \begin{pmatrix} \hat{c}_+ & 0 & 0 & 0 \\ 0 & \hat{d}_+ & 0 & 0 \\ 0 & 0 & \hat{d}_+ & 0 \\ 0 & 0 & 0 & \hat{c}_+ \end{pmatrix},$$

$$\hat{A}_{1\bar{1}} = \hat{A}_{\bar{1}1} = \begin{pmatrix} 0 & 0 & 0 & \hat{a}_- \\ 0 & 0 & \hat{b}_- & 0 \\ 0 & \hat{b}_- & 0 & 0 \\ \hat{a}_- & 0 & 0 & 0 \end{pmatrix}, \qquad \hat{A}_{2\bar{2}} = \hat{A}_{\bar{2}2} = \begin{pmatrix} 0 & 0 & 0 & \hat{c}_- \\ 0 & 0 & \hat{d}_- & 0 \\ 0 & \hat{d}_- & 0 & 0 \\ \hat{c}_- & 0 & 0 & 0 \end{pmatrix}, \tag{4.10}$$

$$\hat{a}_\pm = \frac{1}{2}\left(m_{11}^{(+)} \pm m_{11}^{(-)}\right), \qquad \hat{b}_\pm = \frac{1}{2}\left(m_{12}^{(+)} \pm m_{12}^{(-)}\right),$$

$$\hat{c}_\pm = \frac{1}{2}\left(m_{21}^{(+)} \pm m_{21}^{(-)}\right), \qquad \hat{d}_\pm = \frac{1}{2}\left(m_{22}^{(+)} \pm m_{22}^{(-)}\right).$$

Hence,

$$\hat{\dot{R}}(0)|V\rangle_{(k)} \otimes |V\rangle_{(k+1)} = \left| \begin{matrix} |1\rangle \otimes \left(\hat{D}_{11}|V\rangle\right) + \left|\bar{1}\right\rangle \otimes \left(\hat{A}_{1\bar{1}}|V\rangle\right) \\ |2\rangle \otimes \left(\hat{D}_{22}|V\rangle\right) + \left|\bar{2}\right\rangle \otimes \left(\hat{A}_{2\bar{2}}|V\rangle\right) \\ |2\rangle \otimes \left(\hat{A}_{2\bar{2}}|V\rangle\right) + \left|\bar{2}\right\rangle \otimes \left(\hat{D}_{22}|V\rangle\right) \\ |1\rangle \otimes \left(\hat{A}_{1\bar{1}}|V\rangle\right) + \left|\bar{1}\right\rangle \otimes \left(\hat{D}_{11}|V\rangle\right) \end{matrix} \right\rangle_{(k,k+1)}, \tag{4.11}$$

where for each k, $|V\rangle = \left| \begin{matrix} |1\rangle \\ |2\rangle \\ |\bar{2}\rangle \\ |\bar{1}\rangle \end{matrix} \right\rangle$

(iii) $N = 3$

Adapting the results of Section 1 of [7] and Section 11 of [8] to notations analogous to the cases above, one can write

$$\hat{R}(\theta) = \begin{vmatrix} D & 0 & A \\ 0 & C & 0 \\ A & 0 & D \end{vmatrix}, \tag{4.12}$$

where

$$D = \begin{pmatrix} \hat{a}_+ & 0 & 0 \\ 0 & \hat{b}_+ & 0 \\ 0 & 0 & \hat{a}_+ \end{pmatrix}, \qquad A = \begin{pmatrix} 0 & 0 & \hat{a}_- \\ 0 & \hat{b}_- & 0 \\ \hat{a}_- & 0 & 0 \end{pmatrix}, \qquad C = \begin{pmatrix} \hat{c}_+ & 0 & \hat{c}_- \\ 0 & m & 0 \\ \hat{c}_- & 0 & \hat{c}_+ \end{pmatrix}, \tag{4.13}$$

the central element m corresponding to mP_{nn} of (4.4) and

$$\hat{a}_\pm = \frac{1}{2}\left(m_{11}^{(+)} \pm m_{11}^{(-)}\right), \qquad \hat{b}_\pm = \frac{1}{2}\left(m_{12}^{(+)} \pm m_{12}^{(-)}\right), \qquad \hat{c}_\pm = \frac{1}{2}\left(m_{21}^{(+)} \pm m_{21}^{(-)}\right). \tag{4.14}$$

Denote the basis factors for each k as, $|V\rangle_{(k)} = \left| \begin{matrix} |1\rangle \\ |2\rangle \\ |\bar{1}\rangle \end{matrix} \right\rangle_{(k)}$ and

$$|V\rangle_{(k)} \otimes |V\rangle_{(k+1)} = \left| \begin{matrix} |1\rangle \otimes |V\rangle \\ |2\rangle \otimes |V\rangle \\ \left|\bar{1}\right\rangle \otimes |V\rangle \end{matrix} \right\rangle_{(k,k+1)}. \tag{4.15}$$

One obtains

$$\begin{aligned}
\dot{\hat{R}}(0)|V\rangle_{(k)} \otimes |V\rangle_{(k+1)} &= \begin{vmatrix} D & 0 & A \\ 0 & C & 0 \\ A & 0 & D \end{vmatrix} \left| \begin{matrix} |1\rangle \otimes |V\rangle \\ |2\rangle \otimes |V\rangle \\ |\bar{1}\rangle \otimes |V\rangle \end{matrix} \right\rangle_{(k,k+1)} \\
&= \left| \begin{matrix} |1\rangle \otimes (D|V\rangle) + |\bar{1}\rangle \otimes (A|V\rangle) \\ |2\rangle \otimes (C|V\rangle) \\ |1\rangle \otimes (A|V\rangle) + |\bar{1}\rangle \otimes (D|V\rangle) \end{matrix} \right\rangle_{(k,k+1)} .
\end{aligned} \tag{4.16}$$

For $N = 5, 6, \ldots$, one can generate such results systematically. They, considering the action of all the terms of (4.1), furnish the transition matrix elements and expectation values for possible states of the chain and permit a study of correlations.

One can also consider higher-order conserved quantities (Section 1.5 of [10]) given by

$$H_l = \frac{d^l}{d\theta^l} \log \mathbf{T}^{(r)}(\theta) \bigg|_{\theta=0} . \tag{4.17}$$

For $l = 2$, (4.1) is generalized by the appearance of factors of the type (apart from non-overlapping derivatives)

$$\dot{\hat{R}}(0)_{k-1,k} \otimes \dot{\hat{R}}(0)_{k,k+1}, \qquad \ddot{\hat{R}}(0)_{k,k+1}. \tag{4.18}$$

For $l > 2$ this generalizes in an evident fashion. A study of spin chains for the "exotic" SØ3 can be found in [11]. It is interesting to compare it with the 4×4 (for $N = 2$) case briefly presented above by exploring the latter case in comparable detail.

One may note that for our class of braid matrices, for even N as compared to (4.2), are

$$\frac{d^l}{d\theta^l} \hat{R}(\theta) \bigg|_{\theta=0} = \sum_{c} \sum_{i,j} \left(m_{ij}^{(\epsilon)} \right)^l \left(P_{ij}^{(\epsilon)} + P_{\bar{i}\bar{j}}^{(\epsilon)} \right), \tag{4.19}$$

and for odd N, as compared to (4.4),

$$\frac{d^l}{d\theta^l} \hat{R}(\theta) \bigg|_{\theta=0} = m^l P_{nn} + \sum_{\epsilon,i} \left(\left(m_{ni}^{(\epsilon)} \right)^l P_{ni}^{(\epsilon)} + \left(m_{in}^{(\epsilon)} \right)^l P_{in}^{(\epsilon)} \right) + \sum_{\epsilon,i,j} \left(m_{ij}^{(\epsilon)} \right)^l \left(P_{ij}^{(\epsilon)} + P_{\bar{i}\bar{j}}^{(\epsilon)} \right). \tag{4.20}$$

5. Potentials for Factorizable S-Matrices

Such potentials can be obtained as inverse Cayley transforms of the Yang-Baxter matrices of appropriate dimensions (see, for example, Section 3 of [12] and Section 1 of [2]). Starting with the Yang-Baxter matrix $R(\theta) = \mathbf{P}\hat{R}(\theta)$, the required potential $\mathbf{V}(0)$ is given by

$$-\mathbf{i}\mathbf{V}(\theta) = (R(\theta) - \lambda(\theta)I)^{-1}(R(\theta) + \lambda(\theta)I). \tag{5.1}$$

We have emphasized in our previous studies (Section 5 of [7], Section 8 of [8]) that $\lambda(\theta)$ cannot be arbitrary, a set of values must be excluded for the inverse (5.1) to be well defined. Here we generalize our previous results to all N. As compared to other well-known studies [13] of fields corresponding to factorizable scatterings, here our construction starts with the braid matrices (2.4) and (2.16).

Define

$$X(\theta) = (R(\theta) - \lambda(\theta)I)^{-1} \tag{5.2}$$

when

$$-\mathbf{i}\mathbf{V}(\theta) = I + 2\lambda(\theta)X(\theta). \tag{5.3}$$

We give below immediately the general solution and then explain the notations more precisely. The solution was obtained via formal series expansions. But the final closed form can be verified directly. The solution is

$$\begin{aligned} X(\theta) = -\frac{1}{2}\sum_{\epsilon=\pm}\sum_{a,b=1}^{N}\frac{1}{\lambda^2(\theta) - e^{(m_{ab}^{(\epsilon)}+m_{ba}^{(\epsilon)})\theta}}\Big\{\lambda(\theta)\Big[(aa)\otimes(bb) + \epsilon(a\bar{a})\otimes\big(b\bar{b}\big)\Big] \\ +e^{m_{ba}^{(\epsilon)}\theta}\Big[(ab)\otimes(ba) + \epsilon\big(a\bar{b}\big)\otimes(b\bar{a})\Big]\Big\}, \end{aligned} \tag{5.4}$$

where

$$\lambda(\theta) \neq \pm e^{(1/2)(m_{ab}^{(\epsilon)}+m_{ba}^{(\epsilon)})\theta}. \tag{5.5}$$

For $N = 2n$, (2.9) is implicit in this result. For $N = 2n - 1$, when $\bar{n} = n$, the conventions (2.16) (or (4.4)) are to be implemented when a or b or both (a, b) are n. From (5.3) and (5.4),

$$\begin{aligned} \mathbf{V}(\theta) = -\frac{\mathbf{i}}{2}\sum_{\epsilon=\pm}\sum_{a,b=1}^{N}\Bigg\{\frac{\lambda^2(\theta) + e^{(m_{ab}^{(\epsilon)}+m_{ba}^{(\epsilon)})\theta}}{\lambda^2(\theta) - e^{[m_{ab}^{(\epsilon)}+m_{ba}^{(\epsilon)}]\theta}}\Big[(aa)\otimes(bb) + \epsilon(a\bar{a})\otimes\big(b\bar{b}\big)\Big] \\ +\frac{e^{m_{ab}^{(\epsilon)}\theta}}{\lambda^2(\theta) - e^{(m_{ab}^{(\epsilon)}+m_{ba}^{(\epsilon)})\theta}}\Big[(ba)\otimes(ab) + \epsilon(b\bar{a})\otimes\big(a\bar{b}\big)\Big]\Bigg\} \\ \equiv \sum_{ab,cd}\mathbf{V}(\theta)_{(ab,cd)}(ab)\otimes(cd). \end{aligned} \tag{5.6}$$

Now one can write down the Lagrangians for scalar and spinor fields. For the spinor case, for example,

$$\mathcal{L} = \int dx\big[\mathbf{i}\bar{\psi}_a\gamma_\nu\partial_\nu\psi_a - g(\bar{\psi}_a\gamma_\nu\psi_c)\mathbf{V}_{ab,cd}(\bar{\psi}_b\gamma_\nu\psi_d)\big]. \tag{5.7}$$

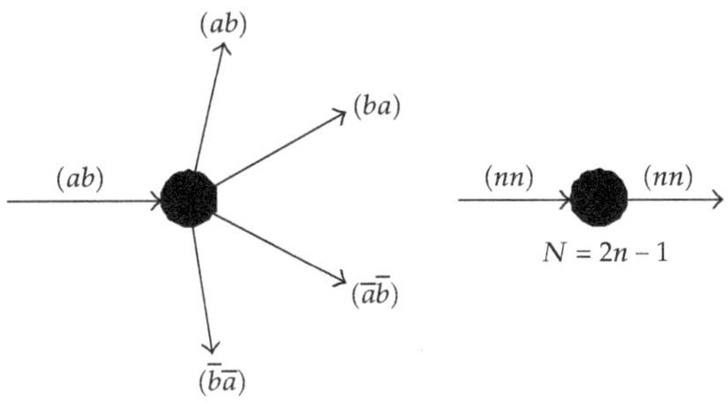

Figure 1: Scattering process.

The simpler scalar case can be written analogously. The scalar Lagrangian has an interaction term of the form $(\overline{\phi}_a\overline{\phi}_c)\mathbf{V}_{ab,cd}(\phi_b\phi_d)$. For $N = 2n$, one obtains the nonvanishing elements of $\mathbf{V}$ as

$$\begin{aligned}
\mathbf{V}_{bb,dd} &= -\frac{\mathrm{i}}{2}\sum_{\epsilon}\frac{\lambda^2(\theta) + e^{(m_{bd}^{(\epsilon)}+m_{db}^{(\epsilon)})\theta}}{\lambda^2(\theta) - e^{(m_{bd}^{(\epsilon)}+m_{db}^{(\epsilon)})\theta}},\\
\mathbf{V}_{\overline{b}b,\overline{d}d} &= -\frac{\mathrm{i}}{2}\sum_{\epsilon}\epsilon\frac{\lambda^2(\theta) + e^{(m_{bd}^{(\epsilon)}+m_{db}^{(\epsilon)})\theta}}{\lambda^2(\theta) - e^{(m_{bd}^{(\epsilon)}+m_{db}^{(\epsilon)})\theta}},\\
\mathbf{V}_{db,bd} &= -\frac{\mathrm{i}}{2}\sum_{\epsilon}\frac{e^{m_{bd}^{(\epsilon)}\theta}}{\lambda^2(\theta) - e^{(m_{bd}^{(\epsilon)}+m_{db}^{(\epsilon)})\theta}},\\
\mathbf{V}_{\overline{d}b,\overline{b}d} &= -\frac{\mathrm{i}}{2}\sum_{\epsilon}\epsilon\frac{e^{m_{bd}^{(\epsilon)}\theta}}{\lambda^2(\theta) - e^{(m_{bd}^{(\epsilon)}+m_{db}^{(\epsilon)})\theta}},
\end{aligned} \tag{5.8}$$

where $a, b \in \{1, \ldots, N\}$ and $(a, b) \neq (n, n)$ if $N = 2n - 1$. When N is odd $(N = 2n - 1)$, one has, for the special index n,

$$\mathbf{V}_{nn,nn} = -\mathrm{i}\left(\frac{\lambda^2(\theta) + 2}{\lambda^2(\theta) - 1}\right), \tag{5.9}$$

where $m_{nn}^{(\epsilon)}$ are taken to be zero. For our models, the scattering process can be schematically, presented as (see in Figure 1)

6. Remarks

We have obtained explicit eigenvalues of the transfer matrix $\mathbf{T}^{(r)}$ corresponding to our class of $N^2 \times N^2$ braid matrices for all (r, N). Starting with our nested sequence of projectors, we obtained a very specific structure of $\mathbf{T}^{(r)}$. Exploiting this structure fully, we obtained the eigenvalues and eigenstates. The zero-sum multiplets parametrized by sets of roots of unity arose from the circular permutations of the indices in the tensor products of the basis states under the action of $\mathbf{T}^{(r)}$. The same structure led to systematic constructions of multiparameter Hamiltonians of spin chains related to our class of braid matrices for all N. Finally we constructed the inverse Cayley transformation for the general case (any N) giving potentials compatible with factorizability of S-matrices. The contents of such matrices (generalizing Section 5 of [7]) will be further studied elsewhere. In the treatment of each aspect we emphasized the role of a remarkable feature of our formalism—the presence of free parameters whose number increase as N^2 with N. A single class of constraints ($\theta \geq 0$, $m_{ab}^{(+)} > m_{ab}^{(-)}$, $a, b \in \{1, \ldots, N\}$) assures nonnegative Boltzmann weights in the statistical models. But our constructions also furnish directly some important properties of such models. Thus $\operatorname{Tr}(\mathbf{T}^{(r)}) = 2\sum_{i=1}^{n} e^{rm_{ii}^{(+)}\theta}$ $(N = 2n)$ and $\operatorname{Tr}(\mathbf{T}^{(r)}) = 2\sum_{i=1}^{n-1} e^{rm_{ii}^{(+)}\theta} + 1$ $(N = 2n - 1)$ (for the normalization (2.16)). The eigenvalues can be ordered in magnitude by choosing the order of values of the parameters. Thus choosing $m_{11}^{(+)} > m_{22}^{(+)} > \cdots$, the largest eigenvalue of $\mathbf{T}^{(r)}$ is for $\theta > 0$, say $e^{rm_{11}^{(+)}\theta}$, the next largest is $e^{rm_{22}^{(+)}\theta}$ and so on. We intend to study elsewhere the properties of our models more thoroughly.

In previous papers [8, 14] we pointed out that for purely imaginary parameters ($im_{ab}^{(\pm)}$ with $m_{ab}^{(\pm)}$ real) our braid matrices are all unitary, providing an entire class with free parameters for all N. Here we have not repeated this discussion. But the fact that they generate parametrized entangled states is indeed of interest.

Unitary matrices provide valid transformations of a basis (of corresponding dimension) of quantum states. One may ask the following question. If such a matrix, apart from being unitary, also satisfies the braid equation what consequence might be implied? Link between quantum and topological entanglements has been discussed by several authors [15, 16] cited in our previous papers [8, 14]. For the braid property to be relevant, a triple tensor product $(V \otimes V \otimes V)$ of basis space is essential. We hope to explore elsewhere our multiparameter unitary matrices in such a context.

In a following paper we will present a quite different aspect of our multiparameter braid matrices. It will be shown how for imaginary parameters (when the matrices are unitary) they can be implemented to generate quantum entanglements. Topological and quantum entanglements will thus be brought together in the two papers.

Note Added

Professor J. H. H. Perk has kindly pointed out that a class of multiparameter generalization of the 6-vertex model is provided by $sl(m \mid n)$ ones [17, 18]. As the Perk-Schultz models they have been studied by many authors and have led to various important applications. (Relevant references can be easily found via ARXIV.) In this class the sources of parameters are multi-component rapidities. The study of eigenvectors and eigenvalues was pioneered in [18] the most recent followup being [19], where other references can be found. Such studies may be compared to our systematic explicit constructions for all (r, N).

References

[1] R. J. Baxter, *Exactly Solved Models in Statistical Mechanics*, Academic Press, London, UK, 1982.

[2] H. J. de Vega, "Yang-Baxter algebras, integrable theories and quantum groups," *International Journal of Modern Physics A*, vol. 4, no. 10, pp. 2371–2463, 1989.

[3] V. E. Korepin, N. M. Bogoliubov, and A. G. Izergin, *Quantum Inverse Scattering Method, Correlation Functions and Algebraic Bethe Ansatz*, Cambridge University Press, Cambridge, UK, 1993.

[4] M. Jimbo, T. Miwa, and A. Nakayashiki, "Difference equations for the correlation functions of the eight-vertex model," *Journal of Physics A*, vol. 26, no. 9, pp. 2199–2209, 1993.
[5] A. Chakrabarti, "Canonical factorization and diagonalization of Baxterized braid matrices: explicit constructions and applications," *Journal of Mathematical Physics*, vol. 44, no. 11, pp. 5320–5349, 2003.
[6] A. Chakrabarti, "A nested sequence of projectors and corresponding braid matrices $\widehat{R}(\theta)$: (1). I. Odd dimensions," *Journal of Mathematical Physics*, vol. 46, no. 6, p. 18, 2005.
[7] B. Abdesselam and A. Chakrabarti, "Nested sequence of projectors. II. Multiparameter multistate statistical models, Hamiltonians, S-matrices," *Journal of Mathematical Physics*, vol. 47, no. 5, p. 25, 2006.
[8] B. Abdesselam, A. Chakrabarti, V. K. Dobrev, and S. G. Mihov, "Higher dimensional multiparameter unitary and nonunitary braid matrices: even dimensions," *Journal of Mathematical Physics*, vol. 48, no. 10, p. 6, 2007.
[9] B. Abdesselam and A. Chakrabarti, "A new eight vertex model and higher dimensional, multiparameter generalizations," *Journal of Mathematical Physics*, vol. 49, no. 5, p. 21, 2008.
[10] B. Abdesselam, A. Chakrabarti, V. K. Dobrev, and S. G. Mihov, "Exotic bialgebras from 9 × 9 unitary braid matrices," *Physics of Atomic Nuclei*, vol. 74, no. 6, pp. 824–831, 2011, (851–857 Russian edition).
[11] J. B. Saleur and H. Zuber, "Integrable lattice models and quantum groups," in *proceedings of the Trieste Spring School on String Theory and Quantum Gravity*, 1990.
[12] D. Arnaudon, A. Chakrabarti, V. K. Dobrev, and S. G. Mihov, "Exotic bialgebra $S\phi 3$: representations, Baxterisation and applications," *Annales Henri Poincaré*, vol. 7, no. 7-8, pp. 1351–1373, 2006.
[13] A. B. Zamolodchikov and A. B. Zamolodchikov, "Factorized S-matrices in two dimensions as the exact solutions of certain relativistic quantum field theory models," *Annals of Physics*, vol. 120, no. 2, pp. 253–291, 1979.
[14] B. Abdesselam, A. Chakrabarti, V. K. Dobrev, and S. G. Mihov, "Higher dimensional unitary braid matrices: construction, associated structures, and entanglements," *Journal of Mathematical Physics*, vol. 48, no. 5, p. 21, 2007.
[15] S. J. Kauffman and L. H. Lomonaco Jr, "Braiding operators are universal quantum gates," *New Journal of Physics*, vol. 6, p. 34, 2004.
[16] Y. Zhang, L. H. Kauffman, and M.-L. Ge, "Yang-Baxterizations, universal quantum gates and Hamiltonians," *Quantum Information Processing*, vol. 4, no. 3, pp. 159–197, 2005.
[17] J. H. H. Perk and C. L. Schultz, "New families of commuting transfer matrices in Q-state vertex models," *Physics Letters A*, vol. 84, no. 8, pp. 407–410, 1981.
[18] H. Perk and J. H. H. Au-Yang, "Yang-Baxter equations," *Encyclopedia of Mathematical Physics*, vol. 5, pp. 465–473, 2006.
[19] C. L. Schultz, "Eigenvectors of the multicomponent generalization of the six-vertex model," *Physica A*, vol. 122, no. 1-2, pp. 71–88, 1983.

5

Characterization of Holomorphic Bisectional Curvature of *GCR*-Lightlike Submanifolds

Sangeet Kumar,[1] Rakesh Kumar,[2] and R. K. Nagaich[3]

[1] *Department of Applied Sciences, Rayat Institute of Engineering & Information Technology, Railmajra, SBS Nagar, Punjab 144533, India*
[2] *Department of Basic and Applied Sciences, University College of Engineering, Punjabi University, Patiala 147002, India*
[3] *Department of Mathematics, Punjabi University, Patiala 147002, India*

Correspondence should be addressed to Rakesh Kumar, dr_rk37c@yahoo.co.in

Academic Editor: M. Lakshmanan

We obtain the expressions for sectional curvature, holomorphic sectional curvature, and holomorphic bisectional curvature of a *GCR*-lightlike submanifold of an indefinite Kaehler manifold. We discuss the boundedness of holomorphic sectional curvature of *GCR*-lightlike submanifolds of an indefinite complex space form. We establish a condition for a *GCR*-lightlike submanifold of an indefinite complex space form to be null holomorphically flat. We also obtain some characterization theorems for holomorphic sectional and holomorphic bisectional curvature.

1. Introduction

The study of *CR* submanifolds of Kaehler manifolds was initiated by Bejancu [1], as a generalization of totally real and complex submanifolds and further developed by [2–7]. The *CR* structures on real hypresurfaces of complex manifolds have interesting applications to relativity. Penrose [8] discovered a correspondence, called Penrose correspondence, between points of a Minkowski space and projective lines of a certain real hypersurfaces in a complex projective space, which is an interesting means of passing from the geometry of a Minkowski space to the geometry of a *CR* manifold. Duggal [9, 10] studied the geometry of *CR* submanifolds with Lorentzian metric and obtained their interaction with relativity. The theory of lightlike submanifolds has interaction with some results on Killing horizon, electromagnetic, and radition fields and asymptotically flat spacetimes (for detail see chapters 7, 8, and 9 of [11]). Thus due to the significant applications of CR structures in relativity and growing importance of lightlike submanifolds in mathematical physics and relativity, Duggal and Bejancu [11] introduced the notion of CR-lightlike submanifolds of indefinite Kaehler manifolds which have direct relation with physically important asymptotically flat space time which further lead to Twistor theory of Penrose and Heaven theory of Newman. Moreover, they concluded that, contrary to the CR-non degenerate submanifolds, CR-lightlike submanifolds do not include invariant (complex) and totally real lightlike submanifolds. Therefore, Duggal and Sahin [12] introduced SCR-lightlike submanifolds of indefinite Kaehler manifold which contain complex and totally real subcases but there was no inclusion relation between CR and SCR cases. Later on, Duggal and Sahin [13] introduced GCR-lightlike submanifolds of indefinite Kaehler manifolds, which behaves as an umbrella of invariant (complex), screen real and CR-lightlike submanifolds and also studied the existence (or nonexistence) of this new class in an indefinite space form. R. Kumar et al. [14] studied geodesic GCR-lightlike submanifolds of indefinite Kaehler manifolds and obtained some characterization theorems for a GCR-lightlike submanifold to be a GCR-lightlike product.

Since sectional curvature offers a lot of information concerning the intrinsic geometry of Riemannian manifolds, therefore in this paper, we obtain the expressions for sectional curvature, holomorphic sectional curvature, and holomorphic bisectional curvature of a GCR-lightlike submanifold of an indefinite Kaehler manifold. In [15], Kulkarni showed that the boundedness of the sectional curvature on a semi-Riemannian manifold implies the constancy of the sectional curvature. In [16], Bonome et al. showed that the boundedness of the holomorphic sectional curvature on indefinite almost Hermitian manifolds leads to the space of pointwise constant holomorphic sectional curvature. Therefore in Section 4, we discuss the boundedness of holomorphic sectional curvature of GCR-lightlike submanifolds of an indefinite complex space form. In Section 5, we established a condition for a GCR-lightlike submanifold of an indefinite complex space form to be null holomorphically flat. We also obtain some characterization theorems on holomorphic sectional and holomorphic bisectional curvature.

2. Lightlike Submanifolds

Let $(\overline{M}, \overline{g})$ be a real $(m+n)$-dimensional semi-Riemannian manifold of constant index q such that $m, n \geq 1$, $1 \leq q \leq m+n-1$, (M, g) an m-dimensional submanifold of $\overline{M}$ and g the induced metric of $\overline{g}$ on M. If $\overline{g}$ is degenerate on the tangent bundle TM of M, then M is called a lightlike submanifold of $\overline{M}$ (see [11]). For a degenerate metric g on M, $TM^{\perp}$ is a degenerate n-dimensional subspace of $T_x\overline{M}$. Thus both T_xM and $T_xM^{\perp}$ are degenerate orthogonal subspaces but no longer complementary. In this case, there exists a subspace $\operatorname{Rad} T_xM = T_xM \cap T_xM^{\perp}$ which is known as radical (null) subspace. If the mapping $\operatorname{Rad} TM : x \in M \to \operatorname{Rad} T_xM$ defines a smooth distribution on M of rank $r > 0$, then the submanifold M of $\overline{M}$ is called an r-lightlike submanifold and $\operatorname{Rad} TM$ is called the radical distribution on M.

Screen distribution $S(TM)$ is a semi-Riemannian complementary distribution of $\operatorname{Rad}(TM)$ in TM, that is

$$TM = \operatorname{Rad} TM \perp S(TM), \tag{2.1}$$

and $S(TM^{\perp})$ is a complementary vector subbundle to $\operatorname{Rad} TM$ in $TM^{\perp}$. Let $\operatorname{tr}(TM)$ and $\operatorname{ltr}(TM)$ be complementary (but not orthogonal) vector bundles to TM in $T\overline{M}|_M$ and to $\operatorname{Rad} TM$ in $S(TM^{\perp})^{\perp}$, respectively. Then we have

$$\operatorname{tr}(TM) = \operatorname{ltr}(TM) \perp S\left(TM^{\perp}\right), \tag{2.2}$$

$$T\overline{M}|_M = TM \oplus \operatorname{tr}(TM) = (\operatorname{Rad} TM \oplus \operatorname{ltr}(TM)) \perp S(TM) \perp S\left(TM^{\perp}\right). \tag{2.3}$$

For a quasi-orthonormal fields of frames on TM, we have the following.

Theorem 2.1 (see [11]). *Let $(M, g, S(TM), S(TM^{\perp}))$ be an r-lightlike submanifold of a semi-Riemannian manifold $(\overline{M}, \overline{g})$. Then there exists a complementary vector bundle* $\operatorname{ltr}(TM)$ *of* $\operatorname{Rad} TM$ *in $S(TM^{\perp})^{\perp}$ and a basis of $\Gamma(\operatorname{ltr}(TM)|_u)$ consisting of smooth section $\{N_i\}$ of $S(TM^{\perp})^{\perp}|_u$, where u is a coordinate neighborhood of M such that*

$$\overline{g}(N_i, \xi_j) = \delta_{ij}, \quad \overline{g}(N_i, N_j) = 0, \quad \text{for any } i, j \in \{1, 2, \ldots, r\}, \tag{2.4}$$

where $\{\xi_1, \ldots, \xi_r\}$ is a lightlike basis of $\Gamma(\operatorname{Rad}(TM))$.

Let $\overline{\nabla}$ be the Levi-Civita connection on $\overline{M}$, then, according to the decomposition (2.3), the Gauss and Weingarten formulas are given by the following:

$$\overline{\nabla}_X Y = \nabla_X Y + h(X, Y), \qquad \overline{\nabla}_X U = -A_U X + \nabla_X^{\perp} U, \tag{2.5}$$

for any $X, Y \in \Gamma(TM)$ and $U \in \Gamma(\operatorname{tr}(TM))$, where $\{\nabla_X Y, A_U X\}$ and $\{h(X, Y), \nabla_X^{\perp} U\}$ belong to $\Gamma(TM)$ and $\Gamma(\operatorname{tr}(TM))$, respectively. Here ∇ is a torsion-free linear connection on M, h is a symmetric bilinear form on $\Gamma(TM)$ which is called second fundamental form, and A_U is a linear a operator on M and known as shape operator.

According to (2.2) considering the projection morphisms L and S of $\operatorname{tr}(TM)$ on $\operatorname{ltr}(TM)$ and $S(TM^{\perp})$, respectively, then Gauss and Weingarten formulas become

$$\overline{\nabla}_X Y = \nabla_X Y + h^l(X, Y) + h^s(X, Y), \qquad \overline{\nabla}_X U = -A_U X + D_X^l U + D_X^s U, \tag{2.6}$$

where we put $h^l(X, Y) = L(h(X, Y)), h^s(X, Y) = S(h(X, Y)), D_X^l U = L(\nabla_X^{\perp} U)$, $D_X^s U = S(\nabla_X^{\perp} U)$. As h^l and h^s are $\Gamma(\operatorname{ltr}(TM))$ valued and $\Gamma(S(TM^{\perp}))$ valued, respectively, therefore they are called the lightlike second fundamental form and the screen second fundamental form on M. In particular,

$$\overline{\nabla}_X N = -A_N X + \nabla_X^l N + D^s(X, N), \qquad \overline{\nabla}_X W = -A_W X + \nabla_X^s W + D^l(X, W), \tag{2.7}$$

where $X \in \Gamma(TM), N \in \Gamma(\operatorname{ltr}(TM))$, and $W \in \Gamma(S(TM^{\perp}))$. Using (2.6) and (2.7), we obtain

$$\overline{g}(h^s(X, Y), W) + \overline{g}\left(Y, D^l(X, W)\right) = g(A_W X, Y), \tag{2.8}$$

$$\overline{g}(D^s(X, N), W) = \overline{g}(A_W X, N), \tag{2.9}$$

for any $X \in \Gamma(TM)$, $W \in \Gamma(S(TM^{\perp}))$, and $N, N' \in \Gamma(\operatorname{ltr}(TM))$.

Let P be the projection morphism of TM on $S(TM)$, then, using (2.1), we can induce some new geometric objects on the screen distribution $S(TM)$ on M as follows:

$$\nabla_X PY = \nabla^*_X PY + h^*(X, PY), \qquad \nabla_X \xi = -A^*_\xi X + \nabla^{*t}_X \xi, \tag{2.10}$$

for any $X, Y \in \Gamma(TM)$ and $\xi \in \Gamma(\operatorname{Rad} TM)$, where $\{\nabla^*_X PY, A^*_\xi X\}$ and $\{h^*(X, PY), \nabla^{*t}_X \xi\}$ belong to $\Gamma(S(TM))$ and $\Gamma(\operatorname{Rad} TM)$, respectively. Using (2.6) and (2.10), we obtain

$$\overline{g}\left(h^l(X, PY), \xi\right) = g\left(A^*_\xi X, PY\right), \qquad \overline{g}(h^*(X, PY), N) = \overline{g}(A_N X, PY), \tag{2.11}$$

for any $X, Y \in \Gamma(TM), \xi \in \Gamma(\operatorname{Rad}(TM))$ and $N \in \Gamma(\operatorname{ltr}(TM))$.

Denote by $\overline{R}$ and R the curvature tensors of $\overline{\nabla}$ and ∇, respectively, then by straightforward calculations ([11]), we have

$$\begin{aligned}\overline{R}(X, Y)Z = {} & R(X, Y)Z + A_{h^l(X,Z)}Y - A_{h^l(Y,Z)}X + A_{h^s(X,Z)}Y \\ & - A_{h^s(Y,Z)}X + \left(\nabla_X h^l\right)(Y, Z) - \left(\nabla_Y h^l\right)(X, Z) \\ & + D^l(X, h^s(Y, Z)) - D^l(Y, h^s(X, Z)) + (\nabla_X h^s)(Y, Z) \\ & - (\nabla_Y h^s)(X, Z) + D^s\left(X, h^l(Y, Z)\right) - D^s\left(Y, h^l(X, Z)\right),\end{aligned} \tag{2.12}$$

where

$$(\nabla_X h^s)(Y, Z) = \nabla^s_X h^s(Y, Z) - h^s(\nabla_X Y, Z) - h^s(Y, \nabla_X Z) \tag{2.13}$$

$$\left(\nabla_X h^l\right)(Y, Z) = \nabla^l_X h^l(Y, Z) - h^l(\nabla_X Y, Z) - h^l(Y, \nabla_X Z). \tag{2.14}$$

Then Codazzi equation is given, respectively, by the following:

$$\begin{aligned}\left(\overline{R}(X, Y)Z\right)^{\perp} = {} & \left(\nabla_X h^l\right)(Y, Z) - \left(\nabla_Y h^l\right)(X, Z) \\ & + D^l(X, h^s(Y, Z)) - D^l(Y, h^s(X, Z)) + (\nabla_X h^s)(Y, Z) \\ & - (\nabla_Y h^s)(X, Z) + D^s\left(X, h^l(Y, Z)\right) - D^s\left(Y, h^l(X, Z)\right).\end{aligned} \tag{2.15}$$

Barros and Romero [17] defined indefinite Kaehler manifolds as follows.

Definition 2.2. Let $(\overline{M}, J, \overline{g})$ be an indefinite almost Hermitian manifold and $\overline{\nabla}$ the Levi-Civita connection on $\overline{M}$, with respect to an indefinite metric $\overline{g}$. Then $\overline{M}$ is called an indefinite Kaehler manifold if J is parallel, with respect to $\overline{\nabla}$, that is

$$J^2 = -I, \quad \left(\overline{\nabla}_X J\right)Y = 0, \quad \overline{g}(JX, JY) = \overline{g}(X, Y) \quad \forall X, Y \in \Gamma\left(T\overline{M}\right). \tag{2.16}$$

3. Generalized Cauchy-Riemann Lightlike Submanifolds

Definition 3.1. Let $(M, g, S(TM))$ be a real lightlike submanifold of an indefinite Kaehler manifold $(\overline{M}, \overline{g}, J)$, then M is called a generalized Cauchy-Riemann (GCR)-lightlike submanifold if the following conditions are satisfied.

(A) There exist two subbundles D_1 and D_2 of $\text{Rad}(TM)$ such that

$$\text{Rad}(TM) = D_1 \oplus D_2, \qquad J(D_1) = D_1, \qquad J(D_2) \subset S(TM). \tag{3.1}$$

(B) There exist two subbundles D_0 and D' of $S(TM)$ such that

$$S(TM) = \{JD_2 \oplus D'\} \perp D_0, \qquad J(D_0) = D_0, \qquad J(D') = L_1 \perp L_2, \tag{3.2}$$

where D_0 is a nondegenerate distribution on M, L_1 and L_2 are vector bundle of $\text{ltr}(TM)$ and $S(TM)^{\perp}$, respectively.

Then the tangent bundle TM of M is decomposed as $TM = D \perp D'$, where $D = \text{Rad}(TM) \oplus D_0 \oplus JD_2$. M is called a proper GCR-lightlike submanifold if $D_1 \neq \{0\}, D_2 \neq \{0\}, D_0 \neq \{0\}$, and $L_2 \neq \{0\}$. Let Q, P_1 and P_2 be the projections on D, $J(L_1) = M_1$, and $J(L_2) = M_2$, respectively. Then, for any $X \in \Gamma(TM)$, we have

$$X = QX + P_1X + P_2X, \tag{3.3}$$

applying J to (3.3), we obtain

$$JX = TX + wP_1X + wP_2X, \tag{3.4}$$

and we can write (3.4) as follwos:

$$JX = TX + wX, \tag{3.5}$$

where TX and wX are the tangential and transversal components of JX, respectively. Similarly,

$$JV = BV + CV, \tag{3.6}$$

for any $V \in \Gamma(\text{tr}(TM))$, where BV and CV are the sections of TM and $\text{tr}(TM)$, respectively. Applying J to (3.5) and (3.6), we get

$$T^2 = -I - Bw, \qquad C^2 = -I - wB. \tag{3.7}$$

Differentiating (3.4) and using (2.6), (2.7), and (3.6), we have

$$D^s(X, wP_1Y) = -\nabla^s_X wP_2Y + wP_2\nabla_X Y - h^s(X, TY) + Ch^s(X, Y), \tag{3.8}$$

$$D^l(X, wP_2Y) = -\nabla^l_X wP_1Y + wP_1\nabla_X Y - h^l(X, TY) + Ch^l(X, Y). \tag{3.9}$$

Using Kaehlerian property of $\overline{\nabla}$ with (2.7), we have the following lemmas.

Lemma 3.2. *Let M be a GCR-lightlike submanifold of an indefinite Kaehlerian manifold $\overline{M}$. Then one has*

$$(\nabla_X T)Y = A_{wY}X + Bh(X, Y), \qquad (\nabla^t_X w)Y = Ch(X, Y) - h(X, TY), \tag{3.10}$$

where $X, Y \in \Gamma(TM)$ and

$$(\nabla_X T)Y = \nabla_X TY - T\nabla_X Y, \qquad (\nabla^t_X w)Y = \nabla^t_X wY - w\nabla_X Y. \tag{3.11}$$

Lemma 3.3. *Let M be a GCR-lightlike submanifold of an indefinite Kaehlerian manifold $\overline{M}$. Then one has*

$$(\nabla_X B)V = A_{CV}X - TA_V X, \qquad (\nabla^t_X C)V = -wA_V X - h(X, BV), \tag{3.12}$$

where $X \in \Gamma(TM)$, $V \in \Gamma(\mathrm{tr}(TM))$, and

$$(\nabla_X B)V = \nabla_X BV - B\nabla^t_X V, \qquad (\nabla^t_X C)V = \nabla^t_X CV - C\nabla^t_X V. \tag{3.13}$$

4. Holomorphic Sectional Curvature of a *GCR*-Lightlike Submanifold

Let $\overline{M}$ be a complex space form of constant holomorphic curvature c. Then the curvature tensor $\overline{R}$ of $\overline{M}(c)$ is given by the following:

$$\begin{aligned}\overline{R}(X, Y)Z = \frac{c}{4}\{&\overline{g}(Y, Z)X - \overline{g}(X, Z)Y + \overline{g}(JY, Z)JX \\ &- \overline{g}(JX, Z)JY + 2\overline{g}(X, JY)JZ\},\end{aligned} \tag{4.1}$$

for X, Y, Z vector fields on $\overline{M}$. Using (4.1) and (2.12), we obtain

$$\begin{aligned} g(R(X,Y)Z,W) = {} & \frac{c}{4}\{g(Y,Z)g(X,W) - g(X,Z)g(Y,W) + g(JY,Z)g(JX,W) \\ & -g(JX,Z)g(JY,W) + 2g(X,JY)g(JZ,W)\} - g(A_{h^l(X,Z)}Y,W) \\ & + g(A_{h^l(Y,Z)}X,W) - g(A_{h^s(X,Z)}Y,W) + g(A_{h^s(Y,Z)}X,W) \\ & - \overline{g}\left(\left(\nabla_X h^l\right)(Y,Z),W\right) + \overline{g}\left(\left(\nabla_Y h^l\right)(X,Z),W\right) - \overline{g}\left(D^l(X,h^s(Y,Z)),W\right) \\ & + \overline{g}\left(D^l(Y,h^s(X,Z)),W\right). \end{aligned} \tag{4.2}$$

Using (2.8) in (4.2), we obtain

$$\begin{aligned} g(R(X,Y)Z,W) = {} & \frac{c}{4}\{g(Y,Z)g(X,W) - g(X,Z)g(Y,W) + g(JY,Z)g(JX,W) \\ & -g(JX,Z)g(JY,W) + 2g(X,JY)g(JZ,W)\} - g(A_{h^l(X,Z)}Y,W) \\ & + g(A_{h^l(Y,Z)}X,W) - \overline{g}(h^s(Y,W),h^s(X,Z)) + \overline{g}(h^s(X,W),h^s(Y,Z)) \\ & - \overline{g}\left(\left(\nabla_X h^l\right)(Y,Z),W\right) + \overline{g}\left(\left(\nabla_Y h^l\right)(X,Z),W\right). \end{aligned} \tag{4.3}$$

Then the sectional curvature $K_M(X,Y) = g(R(X,Y)Y,X)$ of M determined by orthonormal vectors X and Y of $\Gamma(D_0 \oplus M_2)$ and given by the following:

$$\begin{aligned} K_M(X,Y) = {} & \frac{c}{4}\left\{1 + 3g(X,JY)^2\right\} - g(A_{h^l(X,Y)}Y,X) + g(A_{h^l(Y,Y)}X,X) \\ & - \overline{g}(h^s(Y,X),h^s(X,Y)) + \overline{g}(h^s(X,X),h^s(Y,Y)). \end{aligned} \tag{4.4}$$

Corollary 4.1. *Let M be a GCR-lightlike submanifold of an indefinite Complex space form $\overline{M}(c)$. Then sectional curvature of M is given by $K_M(X,Y) = (c/4)\{1 + 3g(X,JY)^2\}$, if*

(i) *M_2 defines a totally geodesic foliation in $\overline{M}$,*

(ii) *D_0 defines a totally geodesic foliation in $\overline{M}$,*

(iii) *M is totally geodesic in $\overline{M}$.*

Definition 4.2. The holomorphic sectional curvature $H(X) = g(R(X,JX)JX,X)$ of M determined by a unit vector $X \in \Gamma(D_0)$ is the sectional curvature of a plane section $\{X, JX\}$.

Then using (2.11) and (4.4), for a unit vector field $X \in \Gamma(D_0)$, we get

$$\begin{aligned} H(X) = {} & c - \overline{g}\left(h^l(X,JX),h^*(JX,X)\right) + \overline{g}\left(h^l(JX,JX),h^*(X,X)\right) \\ & - \overline{g}(h^s(JX,X),h^s(X,JX)) + \overline{g}(h^s(X,X),h^s(JX,JX)). \end{aligned} \tag{4.5}$$

From (3.8), for any $X, Y \in \Gamma(D_0)$, we have

$$h^s(X, JY) = wP_2\nabla_X Y + Ch^s(X, Y), \tag{4.6}$$

and further using (3.7) and (4.6), we have

$$h^s(JX, JY) = wP_2\nabla_{JX} Y - wBh^s(X, Y) - h^s(X, Y). \tag{4.7}$$

Hence using (4.6) and (4.7) in (4.5), we obtain the expression for holomorphic sectional curvature as follows:

$$\begin{aligned} H(X) = c &- \overline{g}\left(h^l(X, JX), h^*(JX, X)\right) + \overline{g}\left(h^l(JX, JX), h^*(X, X)\right) - \|wP_2\nabla_X X\|^2 \\ &- \|Ch^s(X, X)\|^2 + \overline{g}(h^s(X, X), wP_2\nabla_{JX} X) + \|Bh^s(X, X)\|^2 - \|h^s(X, X)\|^2. \end{aligned} \tag{4.8}$$

Theorem 4.3. *Let M be a GCR-lightlike submanifold of an indefinite complex space form $\overline{M}(c)$. If M is totally geodesic in $\overline{M}(c)$, then $H(X) \leq c$, for any unit vector field $X \in \Gamma(D_0)$.*

Proof. Using the hypothesis in (4.8), we get $H(X) = c - \|wP_2\nabla_X X\|^2$. Hence the result follows. □

Theorem 4.4 (see [13]). *Let M be a GCR-lightlike submanifold of an indefinite Kaehler manifold $\overline{M}$, then the distribution D is integrable if and only if $h(X, JY) = h(Y, JX)$, for any $X, Y \in \Gamma(D)$.*

Theorem 4.5. *Let M be a GCR-lightlike submanifold of an indefinite complex space form $\overline{M}(c)$, and D_0 is integrable, then $H(X) \leq c$ for any unit vector field $X \in \Gamma(D_0)$.*

Proof. Since D_0 is integrable therefore using Theorem 4.4, we have $h(JX, JY) = -h(X, Y)$, for any unit vector field $X \in \Gamma(D_0)$. Therefore, from (4.5), we obtain

$$\begin{aligned} H(X) = c &- \overline{g}\left(h^l(X, JX), h^*(JX, X)\right) - \overline{g}\left(h^l(X, X), h^*(X, X)\right) \\ &- \|h^s(X, JX)\|^2 - \|h^s(X, X)\|^2 \leq c. \end{aligned} \tag{4.9}$$

□

Theorem 4.6. *A GCR-lightlike submanifold of an indefinite complex space form $\overline{M}(c)$ is D_0-totally geodesic if and only if*

(i) *D_0 is integrable,*

(ii) *$H(X) = c$, for any unit vector field $X \in \Gamma(D_0)$.*

Proof. Proof follows from (4.9). □

Theorem 4.7 (see [13]). *Let M be a GCR-lightlike submanifold of an indefinite Kaehler manifold $\overline{M}$. Then the distribution D defines a totally geodesic foliation in M if and only if $Bh(X,Y) = 0$, for any $X, Y \in \Gamma(D)$.*

Theorem 4.8. *Let M be a GCR-lightlike submanifold of an indefinite complex space form $\overline{M}(c)$, and D_0 defines a totally geodesic foliation in M, then $H(X) \leq c$, for any unit vector field $X \in \Gamma(D_0)$.*

Proof. Since D_0 defines a totally geodesic foliation in M, therefore by definition $\nabla_X X \in \Gamma(D_0)$, this implies that $h^*(X,X) = 0$. Also by using Theorem 4.7, we have $Bh(X,X) = 0$, for any $X \in \Gamma(D_0)$; hence, (4.8) becomes $H(X) = c - 2\|Ch^s(X,X)\|^2$ and the result follows. □

Definition 4.9. The horizontal distribution D is called parallel with respect to the induced connection ∇ on M if $\nabla_X Y \in D$ for any $X \in \Gamma(TM)$ and $Y \in \Gamma(D)$.

Theorem 4.10. *Let M be a GCR-lightlike submanifold of an indefinite complex space form $\overline{M}(c)$ and D_0 is parallel, with respect to ∇, then $H(X) \leq c$, for any unit vector field $X \in \Gamma(D_0)$.*

Proof. Since D_0 is parallel, with respect to the induced connection ∇ on M, therefore $\nabla^*_X Y \in \Gamma(D_0)$ and $h^*(X,Y) = 0$, for any $X \in \Gamma(TM)$ and $Y \in \Gamma(D_0)$. Hence, from (4.8), we obtain $H(X) = c + \|Bh^s(X,X)\|^2 - \|h^s(X,X)\|^2 - \|Ch^s(X,X)\|^2$, and then by using (3.6), we get $H(X) = c - 2\|Ch^s(X,X)\|^2$. Hence the result is complete. □

Lemma 4.11. *Let M be a GCR-lightlike submanifold of an indefinite Kaehler manifold $\overline{M}$. If the distribution D_0 defines a totally geodesic foliation in $\overline{M}$, then M is D_0-geodesic.*

Proof. By the definition of GCR-lightlike submanifold, M is D_0-geodesic if $\overline{g}(h^l(X,Y),\xi) = \overline{g}(h^s(X,Y),W) = 0$, for any $X, Y \in \Gamma(D_0)$, $\xi \in \Gamma(\text{Rad}(TM))$, and $W \in \Gamma(S(TM^{\perp}))$. Since D_0 defines a totally geodesic foliation in $\overline{M}$, therefore $\overline{g}(h^l(X,Y),\xi) = \overline{g}(\overline{\nabla}_X Y,\xi) = 0$ and $\overline{g}(h^s(X,Y),W) = \overline{g}(\overline{\nabla}_X Y,W) = 0$. Hence, the assertion follows. □

Theorem 4.12. *Let M be a GCR-lightlike submanifold of an indefinite complex space form $\overline{M}(c)$. If D_0 defines a totally geodesic foliation in $\overline{M}$, then $H(X) = c$, for any unit vector field $X \in \Gamma(D_0)$.*

Proof. The result follows directly using Lemma 4.11 and (4.8). □

5. Null Holomorphically Flat *GCR*-Lightlike Submanifold

Let $x \in \overline{M}$ and U be a null vector of $T_x\overline{M}$. A plane π of $T_x\overline{M}$ is called a null plane directed by U if it contains U, $\overline{g}_x(U,V) = 0$, for any $V \in \pi$, and there exists $V_0 \in \pi$ such that $\overline{g}_x(V_0,V_0) \neq 0$. Following Beem-Ehrlich [18], the null sectional curvature of π, with respect to U and $\overline{\nabla}$, as a real number, is defined as follows:

$$\overline{K}_U(\pi) = \frac{\overline{g}_x\left(\overline{R}(V,U)U,V\right)}{\overline{g}_x(V,V)}, \tag{5.1}$$

where V is an arbitrary non null vector in π. Clearly $\overline{K}_U(\pi)$ is independent of V but depends in a quadratic fashion on U.

Consider $u \in M$ and a null plane π of T_uM directed by $\xi_u \in \mathrm{Rad}(TM)$, then the null sectional curvature of π, with respect to ξ_u and ∇, as a real number is defined as

$$K_{\xi_u}(\pi) = \frac{g_u(R(V_u, \xi_u)\xi_u, V_u)}{g_u(V_u, V_u)}, \tag{5.2}$$

where V_u is an arbitrary non-null vector in π.

Let M be a GCR-lightlike submanifold of an indefinite complex space form $\overline{M}(c)$ then using (4.3), the null sectional curvature of π, with respect to ξ, is given by the following:

$$K_\xi(\pi) = \frac{1}{g(V,V)}\{g(A_{h^l(\xi,\xi)}V, V) - g(A_{h^l(V,\xi)}\xi, V) + \overline{g}(h^s(V,V), h^s(\xi,\xi)) - \overline{g}(h^s(\xi,V), h^s(V,\xi)) - \overline{g}\left(\left(\nabla_V h^l\right)(\xi,\xi), V\right) + \overline{g}\left(\left(\nabla_\xi h^l\right)(V,\xi), V\right)\}. \tag{5.3}$$

Then, using (2.11), we obtain

$$K_\xi(\pi) = \frac{1}{g(V,V)}\{g\left(h^*(V,V), h^l(\xi,\xi)\right) - g\left(h^*(\xi,V), h^l(V,\xi)\right) + \overline{g}(h^s(V,V), h^s(\xi,\xi)) - \overline{g}(h^s(\xi,V), h^s(V,\xi)) - \overline{g}\left(\left(\nabla_V h^l\right)(\xi,\xi), V\right) + \overline{g}\left(\left(\nabla_\xi h^l\right)(V,\xi), V\right)\}. \tag{5.4}$$

We know that a plane π is called holomorphic if it remains invariant under the action of the almost complex structure $\overline{J}$, that is, if $\pi = \{Z, \overline{J}Z\}$. The sectional curvature associated with the holomorphic plane is called the holomorphic sectional curvature, denoted by $\overline{H}(\pi)$ and given by $\overline{H}(\pi) = \overline{R}(Z, \overline{J}Z, Z, \overline{J}Z)/\overline{g}(Z,Z)^2$. The holomorphic plane $\pi = \{Z, \overline{J}Z\}$ is called null or degenerate if and only if Z is a null vector. A manifold $(\overline{M}, \overline{g}, \overline{J})$ is called null holomorphically flat if the curvature tensor $\overline{R}$ satisfies (see [19]).

$$\overline{R}\left(Z, \overline{J}Z, Z, \overline{J}Z\right) = 0, \tag{5.5}$$

for all null vectors Z. Put $\overline{g}(\overline{R}(X,Y)Z, W) = \overline{R}(X,Y,Z,W)$, then, from (5.4), we obtain

$$\begin{aligned} R\left(\xi, \overline{J}\xi, \xi, \overline{J}\xi\right) = {} & g\left(h^*(\xi,\xi), h^l\left(\overline{J}\xi, \overline{J}\xi\right)\right) - g\left(h^*\left(\overline{J}\xi, \xi\right), h^l\left(\xi, \overline{J}\xi\right)\right) \\ & + \overline{g}\left(h^s(\xi,\xi), h^s\left(\overline{J}\xi, \overline{J}\xi\right)\right) - \overline{g}\left(h^s\left(\overline{J}\xi, \xi\right), h^s\left(\xi, \overline{J}\xi\right)\right) \\ & - \overline{g}\left(\left(\nabla_\xi h^l\right)\left(\overline{J}\xi, \overline{J}\xi\right), \xi\right) + \overline{g}\left(\left(\nabla_{\overline{J}\xi} h^l\right)\left(\xi, \overline{J}\xi\right), \xi\right). \end{aligned} \tag{5.6}$$

Definition 5.1 (see [20]). A lightlike submanifold (M, g) of a semi-Riemannian manifold $(\overline{M}, \overline{g})$ is said to be a totally umbilical in $\overline{M}$ if there is a smooth transversal vector field

$H \in \Gamma(\mathrm{tr}(TM))$ on M, called the transversal curvature vector field of M, such that, for $X, Y \in \Gamma(TM)$,

$$h(X,Y) = H\overline{g}(X,Y). \tag{5.7}$$

Using (2.6), it is clear that M is a totally umbilical if and only if on each coordinate neighborhood u there exist smooth vector fields $H^l \in \Gamma(\mathrm{ltr}(TM))$ and $H^s \in \Gamma(S(TM^{\perp}))$ such that

$$h^l(X,Y) = H^l g(X,Y), \qquad h^s(X,Y) = H^s g(X,Y), \qquad D^l(X,W) = 0, \tag{5.8}$$

for $X, Y \in \Gamma(TM)$ and $W \in \Gamma(S(TM^{\perp}))$. A lightlike submanifold is said to be totally geodesic if $h(X,Y) = 0$, for any $X, Y \in \Gamma(TM)$. Therefore, in other words, a lightlike submanifold is totally geodesic if $H^l = 0$ and $H^s = 0$.

Let M be a totally umbilical lightlike submanifold, then, using above definition, we have $h(\overline{J}\xi, \overline{J}\xi) = Hg(\overline{J}\xi, \overline{J}\xi) = Hg(\xi,\xi) = 0$ and $h(\xi, \overline{J}\xi) = Hg(\xi, \overline{J}\xi) = 0$, for any $\xi \in \Gamma(\mathrm{Rad}(TM))$. Thus, from (5.6), we have the following theorem.

Theorem 5.2. *Let M be a GCR-lightlike submanifold of an indefinite complex space form $\overline{M}(c)$. If M is totally umbilical lightlike submanifold in $\overline{M}(c)$, then M is null holomorphically flat.*

Moreover, from (5.6), it is clear that the expression of $R(\xi, \overline{J}\xi, \xi, \overline{J}\xi)$ is expressed in terms of screen second fundamental forms of M, thus GCR-lightlike submanifold M of an indefinite complex space form $\overline{M}(c)$ is null holomorphically flat if M is totally geodesic.

6. Holomorphic Bisectional Curvature of a *GCR*-Lightlike Submanifold

Definition 6.1. The holomorphic bisectional for the pair of unit vector fields $\{X, Y\}$ on $\overline{M}$ is given by $\overline{H}(X,Y) = \overline{g}(\overline{R}(X, JX)JY, Y)$.

Theorem 6.2. *Let M be a mixed totally geodesic GCR-lightlike submanifold of an indefinite Kaehler manifold $\overline{M}$ with D_0 parallel distribution. Then $\overline{H}(X,Z) = 0$, for any unit vector fields $X \in \Gamma(D_0)$ and $Z \in \Gamma(M_2)$.*

Proof. Let $X, Y \in \Gamma(D_0)$ and $Z \in \Gamma(M_2)$ then, by using that hypothesis that the distribution D_0 is a parallel, with respect to ∇ on M, we have $g(T\nabla_X Z, Y) = -\overline{g}(\overline{\nabla}_X Z, TY) = g(Z, \nabla_X TY) = 0$. Hence, the non degeneracy of the distribution D_0 implies that

$$\nabla_X Z \in \Gamma(D'), \tag{6.1}$$

for any $Z \in \Gamma(M_2)$. Now replacing Y by JX, respectively, in (2.15) and then taking inner product with JZ, for any $X \in \Gamma(D_0)$ and $Z \in \Gamma(M_2)$, we get

$$\begin{aligned}\overline{H}(X,Z) &= \overline{g}(h^s(\nabla_X JX, Z), JZ) - \overline{g}(\nabla_X h^s(JX,Z), JZ) + \overline{g}(h^s(JX, \nabla_X Z), JZ)\\ &\quad + \overline{g}(\nabla_{JX} h^s(X,Z), JZ) - \overline{g}(h^s(\nabla_{JX} X, Z), JZ) - \overline{g}(h^s(X, \nabla_{JX} Z), JZ)\\ &\quad - \overline{g}\left(D^s\left(X, h^l(JX,Z)\right), JZ\right) + \overline{g}\left(D^s\left(JX, h^l(X,Z)\right), JZ\right).\end{aligned} \tag{6.2}$$

Hence by using that M is mixed totally geodesic with (6.1), the assertion follows. □

Theorem 6.3. *In order that an indefinite complex space form $\overline{M}(c)$ may admit a mixed totally geodesic GCR-lightlike submanifold M with parallel horizontal distribution D_0, it is necessary that $c = 0$.*

Proof. Let $X \in \Gamma(D_0)$ and $Z \in \Gamma(M_2)$ be unit vector fields, then (4.1) implies that $\overline{H}(X,Z) = -(c/2)g(X,X)g(Z,Z)$, then the non degeneracy of the distributions D_0 and M_2 with the Theorem 6.2, we obtain $c = 0$. Hence, the result follows. □

Lemma 6.4. *Let M be a GCR-lightlike submanifold of an indefinite Kaehler manifold $\overline{M}$. Then we have the following:*

(i) *if D' defines a totally geodesic foliation in $\overline{M}$ then $\overline{g}(h^s(D', D_0), JD') = 0$,*

(ii) *if D_0 is a parallel distribution, with respect to ∇, then $h(Z, JX) = Ch(Z,X)$ for any $X \in \Gamma(D_0), Z \in \Gamma(M_2)$.*

Proof. (i) Let D' define a totally geodesic foliation in $\overline{M}$ this implies that $\overline{\nabla}_X Y = \nabla_X Y \in \Gamma(D')$ and $h(X,Y) = 0$, for any $X, Y \in \Gamma(D')$. Therefore by using (3.10), we obtain $A_{wY}X = -Bh(X,Y) = 0$. Let $Z \in \Gamma(D_0)$, then by using (2.8), we get $0 = g(A_{wY}X, Z) = \overline{g}(h^s(X,Z), wY)$. Thus we have $\overline{g}(h^s(D', D_0), JD') = 0$.

(ii) Let D_0 is a parallel distribution with respect to the induced connection ∇, therefore $\nabla_X Y \in \Gamma(D_0)$, for any $Y \in \Gamma(D_0), X \in \Gamma(TM)$. Since $\overline{M}$ is Kaehler manifold, therefore for $Z \in \Gamma(M_2)$ and $X \in \Gamma(D_0)$, we have $\overline{\nabla}_Z JX = J\overline{\nabla}_Z X$. This implies that $\nabla_Z JX + h(Z, JX) = J\nabla_Z X + Bh(Z,X) + Ch(Z,X)$, then by equating transversal components on both sides, we get the result. □

Theorem 6.5. *Let M be a GCR-lightlike submanifold of an indefinite Kaehler manifold $\overline{M}$. If D_0 is parallel with respect to the induced connection ∇, and M_2 defines a totally geodesic foliation in $\overline{M}$, then*

$$\begin{aligned}\overline{H}(X,Z) &= \overline{g}(h(JX,Z), J\nabla_X Z) - \overline{g}(h(X,Z), J\nabla_{JX} Z) + g(A_{JZ}JX, \nabla_X Z)\\ &\quad - g(A_{JZ}X, \nabla_{JX} Z) + 2\|Ch^s(X,Z)\|^2,\end{aligned} \tag{6.3}$$

for any unit vector fields $X \in \Gamma(D_0)$ and $Z \in \Gamma(M_2)$.

Proof. Let $X \in \Gamma(D_0)$ and $Z \in \Gamma(M_2)$, then the equation of Codazzi (2.15) becomes

$$\begin{aligned}\overline{g}\left(\overline{R}(X,JX)Z,JZ\right) &= \overline{g}((\nabla_X h^s)(JX,Z),JZ) - \overline{g}((\nabla_{JX} h^s)(X,Z),JZ) \\ &\quad + \overline{g}\left(D^s\left(X,h^l(JX,Z)\right),JZ\right) - \overline{g}\left(D^s\left(JX,h^l(X,Z)\right),JZ\right).\end{aligned} \tag{6.4}$$

By using (2.13) and (6.1) with the Lemma 6.4 (i), we obtain

$$\begin{aligned}\overline{H}(X,Z) &= \overline{g}\left(\nabla^s_{JX} h^s(X,Z),JZ\right) - \overline{g}(\nabla^s_X h^s(JX,Z),JZ) \\ &\quad + g(A_{JZ}JX,\nabla_X Z) - g(A_{JZ}X,\nabla_{JX}Z) \\ &\quad - \overline{g}\left(D^s\left(X,h^l(JX,Z)\right),JZ\right) + \overline{g}\left(D^s\left(JX,h^l(X,Z)\right),JZ\right).\end{aligned} \tag{6.5}$$

Now by using (2.7) with the Lemma 6.4 (ii), we have

$$\begin{aligned}\overline{g}(\nabla^s_X h^s(JX,Z),JZ) &= \overline{g}\left(\overline{\nabla}_X h^s(JX,Z),JZ\right) \\ &= -\overline{g}(h^s(JX,Z),J\nabla_X Z) - \|Ch^s(X,Z)\|^2,\end{aligned} \tag{6.6}$$

and similarly

$$\overline{g}\left(\nabla^s_{JX} h^s(X,Z),JZ\right) = -\overline{g}(h^s(X,Z),J\nabla_{JX}Z) + \|Ch^s(X,Z)\|^2. \tag{6.7}$$

By using (2.7), we have

$$\overline{g}\left(D^s\left(X,h^l(JX,Z)\right),JZ\right) = -\overline{g}\left(h^l(JX,Z),J\nabla_X Z\right), \tag{6.8}$$

$$\overline{g}\left(D^s\left(JX,h^l(X,Z)\right),JZ\right) = -\overline{g}\left(h^l(X,Z),J\nabla_{JX}Z\right). \tag{6.9}$$

Hence by using (6.6)–(6.9) in (6.5), the result follows. □

Definition 6.6 (see [13]). A GCR-lightlike submanifold M of an indefinite Kaehler manifold $\overline{M}$ is called a GCR-lightlike product if both the distributions D and D' define totally geodesic foliations in M.

Theorem 6.7 (see [14]). *Let M be a GCR-lightlike submanifold of an indefinite Kaehler manifold $\overline{M}$. Then, M is a GCR-lightlike product if and only if $(\nabla_X T)Y = 0$, for any $X, Y \in \Gamma(D)$ or $X, Y \in \Gamma(D')$.*

Theorem 6.8. *Let M be a GCR-lightlike submanifold of an indefinite Kaehler manifold $\overline{M}$. If $(\nabla_X T)Y = 0$, for any $X, Y \in \Gamma(TM)$, then*

$$\overline{H}(X, Z) = 2\overline{g}(h^s(JX, Z), Jh^s(X, Z)), \tag{6.10}$$

for any unit vector fields $X \in \Gamma(D_0)$ and $Z \in \Gamma(M_2)$.

Proof. Let $(\nabla_X T)Y = 0$, then (3.10) implies that

$$A_{wY}X + Bh(X, Y) = 0, \tag{6.11}$$

for any $X, Y \in \Gamma(TM)$. Let $Y \in \Gamma(D), X \in \Gamma(TM)$, then (6.11) gives

$$Bh(X, Y) = 0. \tag{6.12}$$

Let $Z \in \Gamma(D')$ and $X \in \Gamma(D)$, then using (6.11) and (6.12), we obtain

$$A_{wZ}X = 0. \tag{6.13}$$

Particularly choosing $X \in \Gamma(D_0)$ and $Z \in \Gamma(M_2)$ in (2.15), we get

$$\begin{aligned}\overline{g}\left(\overline{R}(X, JX)Z, JZ\right) &= \overline{g}((\nabla_X h^s)(JX, Z), JZ) - \overline{g}((\nabla_{JX} h^s)(X, Z), JZ) \\ &+ \overline{g}\left(D^s\left(X, h^l(JX, Z)\right), JZ\right) - \overline{g}\left(D^s\left(JX, h^l(X, Z)\right), JZ\right).\end{aligned} \tag{6.14}$$

Since $(\nabla_X T)Y = 0$ therefore, by using Theorem 6.7, the distributions D and D' define totally geodesic foliations in M. Then D' defines totally geodesic foliation in M implies that for any $Z_1, Z_2 \in \Gamma(D')$, we have $\nabla_{Z_1} Z_2 \in \Gamma(D')$. Therefore, using (3.10) and (3.11), we get $A_{wZ_2}Z_1 + Bh(Z_1, Z_2) = 0$. By taking inner product with $X \in \Gamma(D_0)$ and using (2.8), we get

$$\overline{g}(h^s(Z_1, X), wZ_2) = 0. \tag{6.15}$$

Also, (3.11) implies that $T\nabla_X Z = 0$, that is, $\nabla_X Z \in \Gamma(D')$. Therefore using (2.8), (2.13), (6.13), and (6.15) in (6.14), we obtain

$$\begin{aligned}\overline{g}\left(\overline{R}(X, JX)Z, JZ\right) &= \overline{g}(\nabla_X^s h^s(JX, Z), JZ) - \overline{g}\left(\nabla_{JX}^s h^s(X, Z), JZ\right) \\ &+ \overline{g}\left(D^s\left(X, h^l(JX, Z)\right), JZ\right) - \overline{g}\left(D^s\left(JX, h^l(X, Z)\right), JZ\right).\end{aligned} \tag{6.16}$$

Now using (2.6), (2.7), (2.8), and (6.15), we have

$$\overline{g}(\nabla_X^s h^s(JX, Z), JZ) = \overline{g}(Jh^s(JX, Z), h^s(X, Z)). \tag{6.17}$$

Similarly,

$$\overline{g}\left(\nabla^s_{JX}h^s(X,Z),JZ\right) = -\overline{g}(Jh^s(JX,Z),h^s(X,Z)). \tag{6.18}$$

Also using (2.7), we have

$$\overline{g}\left(D^s\left(X,h^l(JX,Z)\right),JZ\right) = 0, \qquad \overline{g}\left(D^s\left(JX,h^l(X,Z)\right),JZ\right) = 0. \tag{6.19}$$

Hence, using (6.17)–(6.19) in (6.16), the result follows. □

Lemma 6.9. *Let M be a GCR-lightlike submanifold of an indefinite Kaehler manifold such that $(\nabla_X B)Y = 0$. Then $h^s(JX,Z) \in \Gamma(L_2^{\perp})$ for any $X \in \Gamma(D_0)$, $Z \in \Gamma(M_2)$.*

Proof. Since $(\nabla_X B)Y = 0$, therefore, from (3.12) we have $TA_W X = 0$, for any $X \in \Gamma(TM)$ and $W \in \Gamma(L_2)$, this implies that

$$A_W X \in \Gamma(D'), \tag{6.20}$$

for any $W \in \Gamma(L_2)$ and $X \in \Gamma(TM)$. Since $h^s(JX,Z) \in \Gamma(S(TM^{\perp}))$, therefore, to prove that $h^s(JX,Z) \in \Gamma(L_2^{\perp})$, it is sufficient to prove that $\overline{g}(h^s(JX,Z),W) = 0$, for any $W \in \Gamma(L_2)$. Let $X \in \Gamma(D_0)$ and $Z \in \Gamma(M_2)$ such that $W = wZ$, we have $g(\nabla_U Z,X) = \overline{g}(\overline{\nabla}_U Z,X) = \overline{g}(\overline{\nabla}_U JZ,JX) = -g(A_{JZ}U,JX) = g(JA_{JZ}U,X)$, then using (2.8) we obtain

$$\overline{g}(h^s(JX,Z),W) = -g(JA_W Z,X) = -g(\nabla_Z Z,X) = -g(T\nabla_Z Z,TX). \tag{6.21}$$

Since, from (3.10), we have $T\nabla_Z Z = -A_{wZ}Z - Bh(Z,Z)$, then using (6.20) in (6.21), the result follows. □

Theorem 6.10. *Let M be a GCR-lightlike submanifold of an indefinite Kaehler manifold such that $(\nabla_X B)Y = 0$, then*

$$\overline{H}(X,Z) = 2\overline{g}(Jh^s(X,Z),h^s(JX,Z)) - g(A_{JZ}\nabla_{JX}X,Z) + g(A_{JZ}\nabla_X JX,Z), \tag{6.22}$$

for any unit vector fields $X \in \Gamma(D_0)$ and $Z \in \Gamma(M_2)$.

Proof. Let $X \in \Gamma(D_0)$ and $Z \in \Gamma(M_2)$, then, from (2.15) and (2.13), we obtain

$$\begin{aligned}\overline{g}\left(\overline{R}(X,JX)Z,JZ\right) = {} & \overline{g}(\nabla^s_X h^s(JX,Z),JZ) - \overline{g}(h^s(\nabla_X JX,Z),JZ) \\ & - \overline{g}(h^s(JX,\nabla_X Z),JZ) - \overline{g}\left(\nabla^s_{JX}h^s(X,Z),JZ\right) \\ & + \overline{g}(h^s(\nabla_{JX}X,Z),JZ) + \overline{g}(h^s(X,\nabla_{JX}Z),JZ) \\ & + \overline{g}\left(D^s\left(X,h^l(JX,Z)\right),JZ\right) - \overline{g}\left(D^s\left(JX,h^l(X,Z)\right),JZ\right),\end{aligned} \tag{6.23}$$

using (6.20) in (2.9), we obtain

$$\overline{g}\left(D^s\left(X, h^l(JX,Z)\right), JZ\right) = 0, \qquad \overline{g}\left(D^s\left(JX, h^l(X,Z)\right), JZ\right) = 0. \tag{6.24}$$

Now consider

$$\overline{g}(\nabla^s_X h^s(JX,Z), JZ) = -\overline{g}\left(h^s(JX,Z), J\overline{\nabla}_X Z\right) = \overline{g}(h^s(X,Z), Jh^s(JX,Z)), \tag{6.25}$$

and similarly

$$\overline{g}\left(\nabla^s_{JX} h^s(X,Z), JZ\right) = -\overline{g}(h^s(X,Z), Jh^s(JX,Z)). \tag{6.26}$$

Also using (2.8), we have

$$\begin{aligned} \overline{g}(h^s(\nabla_X JX, Z), JZ) &= g(A_{JZ}\nabla_X JX, Z), \\ \overline{g}(h^s(\nabla_{JX} X, Z), JZ) &= g(A_{JZ}\nabla_{JX} X, Z). \end{aligned} \tag{6.27}$$

Using (6.20) in (2.8), we have

$$\overline{g}(h^s(JX, \nabla_X Z), JZ) = 0, \qquad \overline{g}(h^s(X, \nabla_{JX} Z), JZ) = 0. \tag{6.28}$$

Thus using (6.24)–(6.28) in (6.23), the result follows. □

Theorem 6.11. *Let M be a mixed foliate GCR-lightlike submanifold of an indefinite Kaehler manifold $\overline{M}$, and $S(TM)$ is parallel distribution, with respect to the induced connection ∇, then*

$$\overline{H}(X,Z) = 2\overline{g}(A_{JZ}JX, JA_{JZ}X), \tag{6.29}$$

for any unit vector fields $X \in \Gamma(D_0)$ and $Z \in \Gamma(M_2)$.

Proof. Since M is mixed foliate therefore for any $X \in \Gamma(D_0)$ and $Z \in \Gamma(M_2)$, Codazzi equation (2.15) and (2.13), imply that

$$\begin{aligned} \overline{H}(X,Z) = {}& \overline{g}(h^s(\nabla_X JX, Z), JZ) + \overline{g}(h^s(JX, \nabla_X Z), JZ) \\ & - \overline{g}(h^s(\nabla_{JX} X, Z), JZ) - \overline{g}(h^s(X, \nabla_{JX} Z), JZ). \end{aligned} \tag{6.30}$$

Since using (2.8) and the hypothesis, we have $g(A_{JZ}X, JW) = 0$ and $g(A_{JZ}X, J\xi) = \overline{g}(h^s(X, J\xi), JZ) = \overline{g}(\overline{\nabla}_X J\xi, JZ) = -g(\xi, \nabla_X Z) = 0$. Therefore, by the definition of a GCR-lightlike submanifold, we have

$$A_{JZ}X \in \Gamma(D). \tag{6.31}$$

Thus, using (2.7), (2.8), and (6.31) with the hypothesis, we obtain

$$\begin{aligned}\overline{g}(h^s(JX,\nabla_X Z),JZ) &= \overline{g}\left(A_{JZ}JX,\overline{\nabla}_X Z\right) = \overline{g}\left(JA_{JZ}JX,J\overline{\nabla}_X Z\right)\\ &= \overline{g}(A_{JZ}JX,JA_{JZ}X) - \overline{g}\left(A_{JZ}JX,JD^l(X,JZ)\right)\\ &= \overline{g}(A_{JZ}JX,JA_{JZ}X).\end{aligned} \tag{6.32}$$

Similarly, we obtain

$$\begin{aligned}\overline{g}(h^s(X,\nabla_{JX}Z),JZ) &= \overline{g}(A_{JZ}X,JA_{JZ}JX),\\ \overline{g}(h^s(\nabla_{JX}X,Z),JZ) &= g(A_{JZ}Z,\nabla_{JX}X) - \overline{g}\left(\nabla_{JX}X,D^l(JZ,Z)\right),\\ \overline{g}(h^s(\nabla_X JX,Z),JZ) &= g(A_{JZ}Z,\nabla_X JX) - \overline{g}\left(\nabla_X JX,D^l(JZ,Z)\right).\end{aligned} \tag{6.33}$$

Thus, (6.30) becomes

$$\begin{aligned}\overline{H}(X,Z) = 2\overline{g}(A_{JZ}JX,JA_{JZ}X) - g(A_{JZ}Z,\nabla_{JX}X) + \overline{g}\left(\nabla_{JX}X,D^l(JZ,Z)\right)\\ + g(A_{JZ}Z,\nabla_X JX) - \overline{g}\left(\nabla_X JX,D^l(JZ,Z)\right).\end{aligned} \tag{6.34}$$

Since the distribution D is integrable, therefore $\nabla_{JX}X - \nabla_X JX = [X,JX] = X' \in \Gamma(D)$, then (6.34) becomes $\overline{H}(X,Z) = 2\overline{g}(A_{JZ}JX,JA_{JZ}X) - g(A_{JZ}Z,X') + \overline{g}(X',D^l(JZ,Z))$. Hence, using the hypothesis and (2.8), the result follows. □

Acknowledgment

The authors would like to thank the anonymous referee for his/her comments that helped us to improve this paper.

References

[1] A. Bejancu, "CR submanifolds of a Kaehler manifold. I," *Proceedings of the American Mathematical Society*, vol. 69, no. 1, pp. 135–142, 1978.

[2] A. Bejancu, "CR submanifolds of a Kaehler manifold. II," *Transactions of the American Mathematical Society*, vol. 250, pp. 333–345, 1979.

[3] A. Bejancu, M. Kon, and K. Yano, "CR-submanifolds of a complex space form," *Journal of Differential Geometry*, vol. 16, no. 1, pp. 137–145, 1981.

[4] D. E. Blair and B.-Y. Chen, "On CR-submanifolds of Hermitian manifolds," *Israel Journal of Mathematics*, vol. 34, no. 4, pp. 353–363, 1979.

[5] B.-Y. Chen, "CR-submanifolds of a Kaehler manifold. I," *Journal of Differential Geometry*, vol. 16, no. 2, pp. 305–322, 1981.

[6] K. Yano and M. Kon, "Differential geometry of CR-submanifolds," *Geometriae Dedicata*, vol. 10, no. 1–4, pp. 369–391, 1981.

[7] K. Yano and M. Kon, *CR Submanifolds of Kaehlerian and Sasakian Manifolds*, vol. 30, Birkhäuser, Boston, Mass, USA, 1983.

[8] R. Penrose, "Physical space-time and nonrealizable CR-structures," *Proceedings of Symposia in Pure Mathematics*, vol. 39, pp. 401–422, 1983.

[9] K. L. Duggal, "CR-Structures and Lorentzian Geometry," *Acta Applicandae Mathematicae*, vol. 7, no. 3, pp. 211–223, 1986.

[10] K. L. Duggal, "Lorentzian geometry of CR submanifolds," *Acta Applicandae Mathematicae*, vol. 17, no. 2, pp. 171–193, 1989.

[11] K. L. Duggal and A. Bejancu, *Lightlike Submanifolds of Semi-Riemannian Manifolds and Applications*, vol. 364 of *Mathematics and its Applications*, Kluwer Academic Publishers, Dordrecht, The Netherlands, 1996.

[12] K. L. Duggal and B. Sahin, "Screen Cauchy Riemann lightlike submanifolds," *Acta Mathematica Hungarica*, vol. 106, no. 1-2, pp. 125–153, 2005.

[13] K. L. Duggal and B. Sahin, "Generalized Cauchy-Riemann lightlike submanifolds of Kaehler manifolds," *Acta Mathematica Hungarica*, vol. 112, no. 1-2, pp. 107–130, 2006.

[14] R. Kumar, S. Kumar, and R. K. Nagaich, "*GCR*-lightlike product of indefinite Kaehler manifolds," *ISRN Geometry*, vol. 2011, Article ID 531281, 13 pages, 2011.

[15] R. S. Kulkarni, "The values of sectional curvature in indefinite metrics," *Commentarii Mathematici Helvetici*, vol. 54, no. 1, pp. 173–176, 1979.

[16] A. Bonome, R. Castro, E. García-Río, and L. M. Hervella, "On the holomorphic sectional curvature on an indefinite Kähler manifold," *Comptes Rendus de l'Académie des Sciences I*, vol. 315, no. 11, pp. 1183–1187, 1992.

[17] M. Barros and A. Romero, "Indefinite Kähler manifolds," *Mathematische Annalen*, vol. 261, no. 1, pp. 55–62, 1982.

[18] J. K. Beem and P. E. Ehrlich, *Global Lorentzian Geometry*, vol. 67, Marcel Dekker, New York, NY, USA, 1981.

[19] A. Bonome, R. Castro, E. García-Río, L. M. Hervella, and Y. Matsushita, "Null holomorphically flat indefinite almost Hermitian manifolds," *Illinois Journal of Mathematics*, vol. 39, no. 4, pp. 635–660, 1995.

[20] K. L. Duggal and D. H. Jin, "Totally umbilical lightlike submanifolds," *Kodai Mathematical Journal*, vol. 26, no. 1, pp. 49–68, 2003.

6

Nonlinear Waveforms for Ion-Acoustic Waves in Weakly Relativistic Plasma of Warm Ion-Fluid and Isothermal Electrons

S. A. El-Wakil, Essam M. Abulwafa, E. K. El-Shewy, H. G. Abdelwahed, and Hamdi M. Abd-El-Hamid

Theoretical Physics Group, Physics Department, Faculty of Science, Mansoura University, Mansoura 35516, Egypt

Correspondence should be addressed to Essam M. Abulwafa, abulwafa@mans.edu.eg

Academic Editor: Wen Xiu Ma

The reductive perturbation method has been employed to derive the Korteweg-de Vries (KdV) equation for small- but finite-amplitude electrostatic ion-acoustic waves in weakly relativistic plasma consisting of warm ions and isothermal electrons. An algebraic method with computerized symbolic computation is applied in obtaining a series of exact solutions of the KdV equation. Numerical studies have been made using plasma parameters which reveal different solutions, that is, bell-shaped solitary pulses, rational pulses, and solutions with singularity at finite points, which called "blowup" solutions in addition to the propagation of an explosive pulses. The weakly relativistic effect is found to significantly change the basic properties (namely, the amplitude and the width) of the ion-acoustic waves. The result of the present investigation may be applicable to some plasma environments, such as ionosphere region.

1. Introduction

Nonlinear evolution equations are widely used as models to describe complex physical phenomena and have a significant role in several scientific and engineering fields [1, 2]. The propagation of solitary waves is important as it describes characteristic nature of the interaction of the waves and the plasmas. In the case where the velocity of particles is much smaller than that of light, ion-acoustic waves present the nonrelativistic behaviors, but in the case where the velocity of particles approaches that of light, the relativistic effect becomes dominant [3]. Actually high-speed and energetic streaming ions with the energy from 0.1 to 100 MeV are frequently observed in solar atmosphere and interplanetary space. Nevertheless,

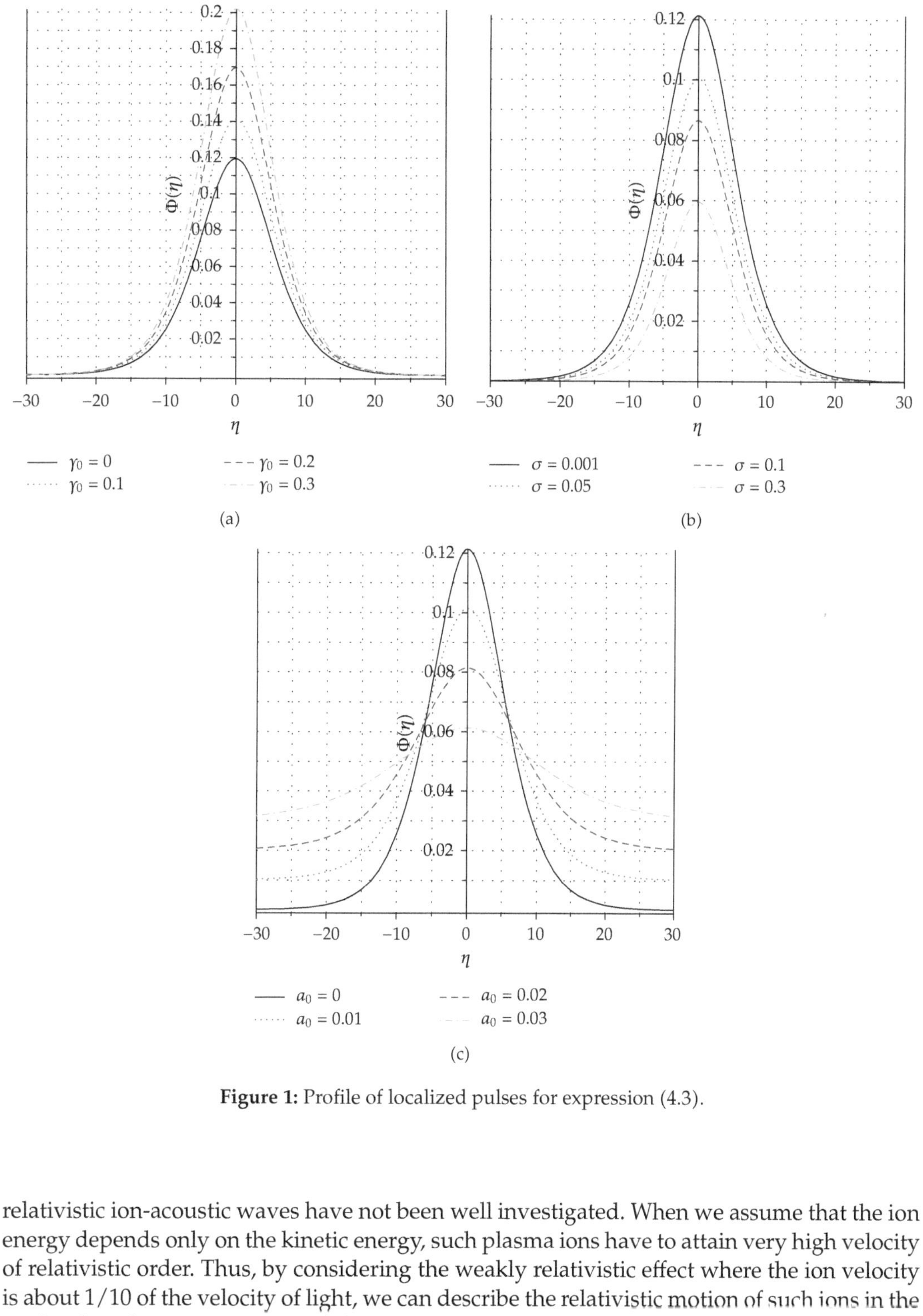

Figure 1: Profile of localized pulses for expression (4.3).

relativistic ion-acoustic waves have not been well investigated. When we assume that the ion energy depends only on the kinetic energy, such plasma ions have to attain very high velocity of relativistic order. Thus, by considering the weakly relativistic effect where the ion velocity is about 1/10 of the velocity of light, we can describe the relativistic motion of such ions in the study of nonlinear interaction of the waves and the plasmas [4]. It appears that the weakly relativistic and ion temperature effects play an important role in energetic ion-acoustic waves propagating in interplanetary space [5, 6].

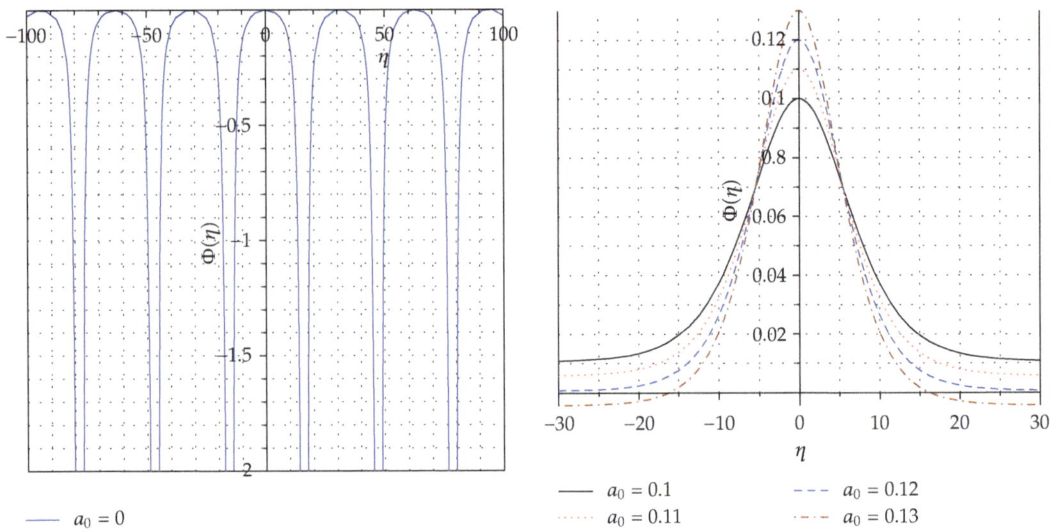

(a) (b)

Figure 2: Profile of pulses for expression (4.5) for $v = 0.04$, $\sigma = 0.001$, and $\gamma_0 = 0.01$.

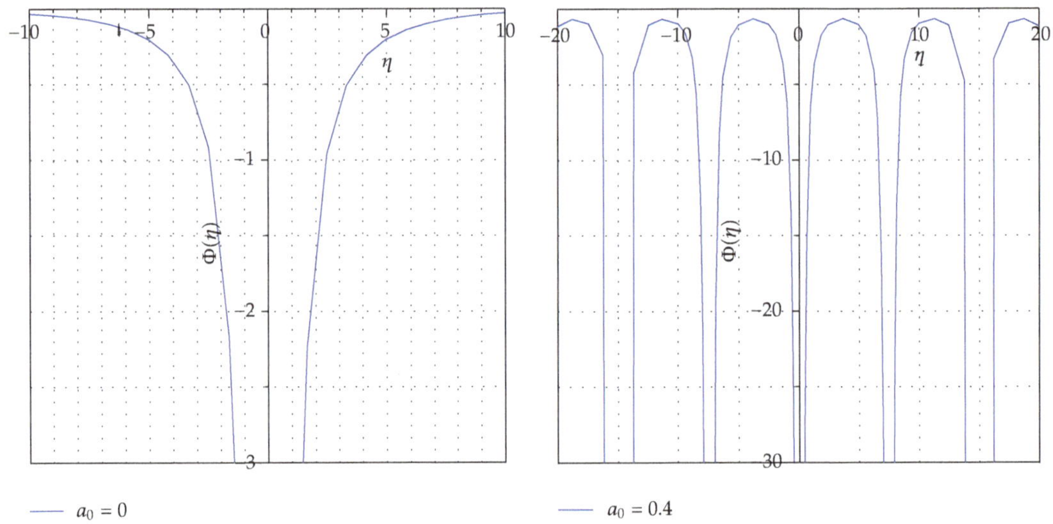

(a) (b)

Figure 3: The explosive solution for expression (4.7) for $v = 0.04$, $\sigma = 0.001$, and $\gamma_0 = 0.01$.

Washimi and Taniuti [7] were the first to use reductive perturbation method to study the propagation of a slow modulation of quasi-monochromatic waves through plasma. And then the attention has been focused by many authors [8–11].

The evolution of small-but finite-amplitude solitary waves, studied by means of the Korteweg-de Vries (KdV) equation, is of considerable interest in plasma dynamics.

Many powerful methods have been established and developed to study nonlinear evolution equations (NLEEs). These methods include the inverse scatting method [12],

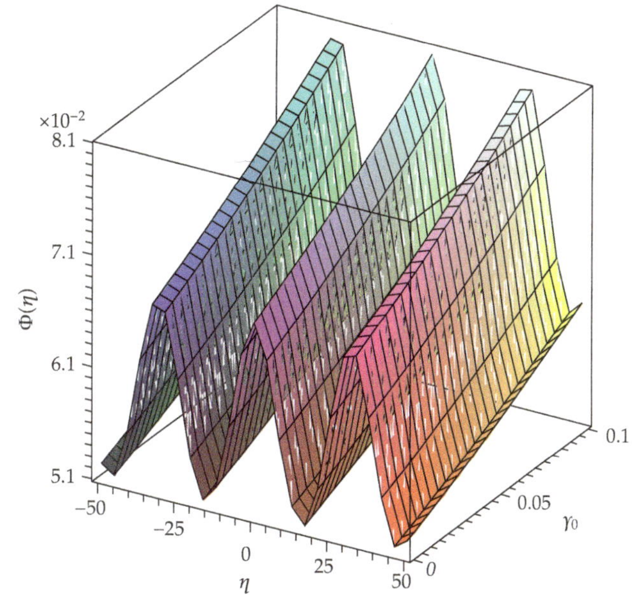

Figure 4: The asymptotical property of Jacobi elliptic doubly periodic wave for expression (4.10) with $v = 0.04$, $\sigma = 0.001$, $a_0 = 0$, and $m = 2$.

the tanh function method [13], extended tanh method [14–16], the Exp-function method [17], the extended F-expansion method [18], the Jacobi elliptic function expansion method [19], the homogeneous balance method [20], sech-function method [21], F-expansion method [22], and the multiple exp-function method [23]. There is no unified method that can be used to deal with all types of NLEEs. Fan [24] developed a new algebraic method with computerized symbolic computation, which greatly exceeds the applicability of the existing tanh, extended tanh methods, and Jacobi function expansion method in obtaining a series of exact solutions of nonlinear differential equations.

Recently, the ion-acoustic solitary wave in collisionless unmagnetized plasma consisting of warm ions fluid and isothermal electrons is studied using the time-fractional KdV equation by El-Wakil et al. [25]. They showed that the time fractional can be used to modulate the electrostatic potential wave.

The major topic of this work is to study the ion-acoustic solitary and other type waves in relativistic warm plasma. This paper is organized as follows. In Section 2, we present the basic set of fluid equations governing our plasma model. In Section 3, an algorithm describing the computerized symbolic computation method is presented. In Section 4, explicit solutions for KdV equation are obtained. Finally, some discussions and conclusions are given in Section 5.

2. Basic Equations and KdV Equation

Consider collisionless ionization-free unmagnetized plasma consisting of a mixture of warm ion-fluid and isothermal electrons. Assume that the ion flow velocity has a weak relativistic effect, and therefore there exist streaming ions in an equilibrium state when sufficiently

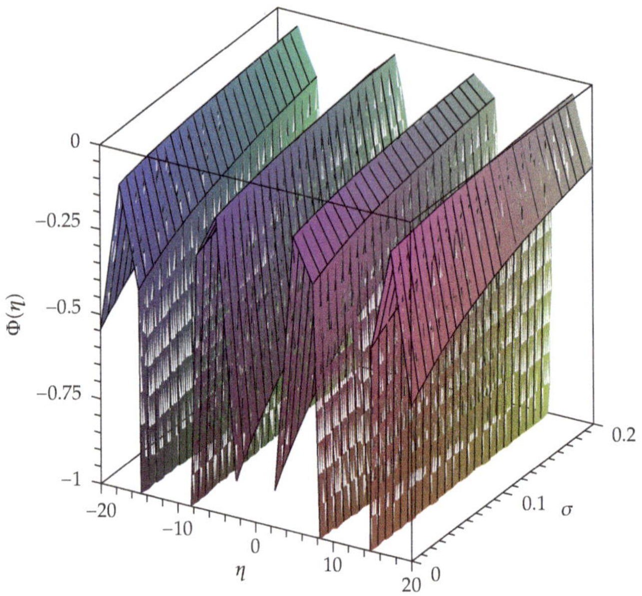

Figure 5: Weierstrass elliptic doubly periodic type solution (4.13) for $v = 0.04$, $\gamma_0 = 0.01$, $c_0 = -1$, $c_1 = 1$, and $c_3 = 0.1$.

small- but finite-amplitude waves propagate one-dimensionally. Such a system is governed by the following normalized equations [26]:

$$\frac{\partial}{\partial t}n(x,t) + \frac{\partial}{\partial x}[n(x,t)\ u(x,t)] = 0, \tag{2.1a}$$

$$\left[\frac{\partial}{\partial t} + u(x,t)\frac{\partial}{\partial x}\right][\gamma(x,t)\ u(x,t)] + \frac{\sigma}{n(x,t)}\frac{\partial}{\partial x}p(x,t) + \frac{\partial}{\partial x}\phi(x,t) = 0, \tag{2.1b}$$

$$\left[\frac{\partial}{\partial t} + u(x,t)\frac{\partial}{\partial x}\right]p(x,t) + 3p(x,t)\frac{\partial}{\partial x}[\gamma(x,t)\ u(x,t)] = 0, \tag{2.1c}$$

$$\frac{\partial^2}{\partial x^2}\phi(x,t) + n(x,t) - n_e(x,t) = 0. \tag{2.1d}$$

The electron temperature T_e is much larger than the ion temperature T_i and in this case for simplicity one can neglect the inertia of the electrons relative to that of the ions, that is, the high-frequency plasma oscillations are neglected. Since it is interested with the regime of density and velocity fluctuations near the ion plasma frequency, so the isothermal electrons density is given by

$$n_e(x,t) = \exp[\phi(x,t)]. \tag{2.1e}$$

For weakly relativistic effects, the relativistic factor $\gamma(x,t) = 1/\sqrt{1-u^2(x,t)/c^2}$ is approximated by

$$\gamma(x,t) \approx 1 + \frac{u^2(x,t)}{(2c^2)}, \tag{2.1f}$$

where c is the velocity of light.

In (2.1a)–(2.1f), $n(x,t)$ and $n_e(x,t)$ are the densities of ions and electrons, respectively, $u(x,t)$ is the ion flow velocity, $p(x,t)$ is the ion pressure, $\phi(x,t)$ is the electric potential, x is the space coordinate, and t is the time variable. $\sigma = T_i/T_e \ll 1$ is the ratio of the ion temperature to the electron temperature. All of these quantities are dimensionless and are normalized in terms of the following characteristic quantities: $n(x,t)$ and $n_e(x,t)$ by the unperturbed electron density n_0, $u(x,t)$ and c by the sound velocity $\sqrt{k_B T_e/m_i}$, $p(x,t)$ and $\phi(x,t)$ by $n_0 k_B T_i$ and $k_B T_e/e$, respectively; t and x by the inverse of the plasma frequency $\omega_{pi}^{-1} = 1/\sqrt{4\pi\, e^2 n_0/m_i}$ and the electron Debye length $\lambda_D = \sqrt{k_B T_e/(4\pi\, e^2 n_0)}$, respectively, k_B is Boltzmann's constant, and m_i is the mass of plasma ion.

According to the general method of reductive perturbation theory, the stretched variables are introduced as [7]

$$\tau = \varepsilon^{3/2} t, \qquad \xi = \varepsilon^{1/2}(x - \lambda t), \tag{2.2}$$

where λ is the phase velocity and ε represents the amplitude of the perturbation. All the physical quantities that appeared in (2.1a)–(2.1f) are expanded as power series in ε about the equilibrium values as

$$n(\xi,\tau) = 1 + \varepsilon n_1(\xi,\tau) + \varepsilon^2 n_2(\xi,\tau) + \cdots, \tag{2.3a}$$

$$u(\xi,\tau) = u_0 + \varepsilon u_1(\xi,\tau) + \varepsilon^2 u_2(\xi,\tau) + \cdots, \tag{2.3b}$$

$$p(\xi,\tau) = 1 + \varepsilon p_1(\xi,\tau) + \varepsilon^2 p_2(\xi,\tau) + \cdots, \tag{2.3c}$$

$$\phi(\xi,\tau) = \varepsilon \phi_1(\xi,\tau) + \varepsilon^2 \phi_2(\xi,\tau) + \cdots. \tag{2.3d}$$

The boundary conditions of this problem are imposed as $|\xi| \to \infty$, $n = n_e = p = 1$, $u = u_0$, and $\phi = 0$.

Substituting (2.2) and (2.3a)–(2.3d) into the system of (2.1a)–(2.1f) and equating the coefficients of like powers of ε, then from the lowest order

$$n_1(\xi,\tau) = \phi_1(\xi,\tau), \tag{2.4a}$$

$$u_1(\xi,\tau) = (\lambda - u_0)\phi_1(\xi,\tau), \tag{2.4b}$$

$$p_1(\xi,\tau) = 3\gamma_1\phi_1(\xi,\tau), \tag{2.4c}$$

with the transcendental equation of λ that is given as

$$1-\left(\lambda^2-3\sigma\right)\gamma_1=0 \Longrightarrow \lambda=\pm\sqrt{\frac{(3\sigma\gamma_1+1)}{\gamma_1}}, \tag{2.4d}$$

where

$$\gamma_1=1+\frac{3\gamma_0^2}{2}, \qquad \gamma_0=\frac{u_0}{c}. \tag{2.4e}$$

Using second-order equations in ε and eliminating the second-order perturbed quantities n_2, u_2, p_2, and ϕ_2, the following KdV equation for the first-order perturbed potential is obtained:

$$\frac{\partial}{\partial\tau}\phi_1(\xi,\tau)+A\phi_1(\xi,\tau)\frac{\partial}{\partial\xi}\phi_1(\xi,\tau)+B\,\frac{\partial^3}{\partial\xi^3}\phi_1(\xi,\tau)=0. \tag{2.5a}$$

Therefore, the nonlinear coefficient A and the dispersion coefficient B are represented by

$$A=B\left(g_1-\frac{g_2\gamma_2}{\sqrt{\gamma_1^3}}\right), \qquad B=\frac{1}{g_2\sqrt{\gamma_1}}, \tag{2.5b}$$

with

$$g_1=3\sigma\gamma_1(3\gamma_1+1)+2, \qquad g_2=2\sqrt{3\sigma\gamma_1+1}, \qquad \gamma_2=\frac{3\gamma_0}{(2c)}. \tag{2.5c}$$

3. Computerized Symbolic Computation Method

An algebraic method with computerized symbolic computation has been developed by Fan [24], which can be used to solve a given partial differential equation in $\phi_1(\xi,\tau)$ of the form

$$H\left(\phi_1,\frac{\partial\phi_1}{\partial\tau},\frac{\partial\phi_1}{\partial\xi},\frac{\partial^2\phi_1}{\partial\xi^2},\ldots\right)=0. \tag{3.1a}$$

This equation may be transformed into an ordinary differential equation of the form

$$H\left(\Phi,\frac{d\Phi}{d\eta},\frac{d^2\Phi}{d\eta^2},\ldots\right)=0 \tag{3.1b}$$

using a traveling frame of reference

$$\phi_1(\xi,\tau)=\Phi(\eta), \quad \eta=\xi-v\tau, \tag{3.1c}$$

where v is the traveling wave propagation velocity.

The computational technique used to solve the NLEE is described as follows [24].

Step 1. Reduce partial differential equation (3.1a) to the ordinary differential equation (3.1b) by considering the traveling wave transformation (3.1c).

Step 2. Expand the solution of (3.1b) in the form

$$\Phi(\eta) = \sum_{i=0}^{n} a_i \varphi^i(\eta), \tag{3.2a}$$

where the new variable $\varphi(\eta)$ is a solution of the following ordinary differential equation:

$$\frac{d\varphi(\eta)}{d\eta} = \epsilon \sqrt{\sum_{j=0}^{r} c_j \varphi^j(\eta)}, \quad \epsilon = \pm 1. \tag{3.2b}$$

Step 3. Substituting (3.2b) into (3.1b) and balancing the highest derivative term with the highest nonlinear term lead to a relation between n and r, from which the different possible values of n and r can be obtained. These values lead to the different series expansions of the solutions.

Step 4. Substituting the expansions (3.2a)–(3.2b) into (3.1b) and setting the coefficients of all powers of φ^i and $\varphi^i d\varphi/d\eta$ to zero will give a system of algebraic equations, from which the parameters a_i $(i = 0, 1, \ldots, n)$ and c_j $(j = 0, 1, \ldots, r)$ can be found explicitly.

Step 5. Substituting the parameters c_j $(j = 0, 1, \ldots, r)$ obtained in Step 4 into (3.2b) gives all the possible solutions $\varphi(\eta)$.

It is remarked here that the solutions of (3.1a) depend on the explicit solvability of (3.2b). The solutions of (3.2b) will get a series of fundamental solutions such as polynomial, exponential, soliton, rational, triangular periodic, Jacobi, and Weierstrass elliptic doubly periodic solutions.

4. Explicit Solutions for the KdV Equation

For KdV equation (2.5a), the traveling wave transformation (3.1c) leads to

$$-v\frac{d}{d\eta}\Phi(\eta) + A\Phi(\eta)\frac{d}{d\eta}\Phi(\eta) + B\frac{d^3}{d\eta^3}\Phi(\eta) = 0. \tag{4.1}$$

Balancing the highest derivative term with the highest nonlinear term leads to a relation between n and r as $r = n + 2$. Taking $n = 2$ gives $r = 4$ and leads to

$$\Phi(\eta) = a_0 + a_1\varphi(\eta) + a_2\varphi^2(\eta), \tag{4.2a}$$

$$\frac{d\varphi}{d\eta} = \in \sqrt{c_0 + c_1\varphi(\eta) + c_2\varphi^2(\eta) + c_3\varphi^3(\eta) + c_4\varphi^4(\eta)}. \tag{4.2b}$$

Substituting (4.2a)–(4.2b) into (4.1) and equating coefficients of all powers of φ^i and $\varphi^i d\varphi/d\eta$ to zero will get a system of algebraic equations, from which the parameters a_i $(i = 0, 1, 2)$ and c_j $(j = 0, 1, \ldots, 4)$ can be found explicitly. Substituting c_j into (4.2b) and using the symbolic software package *Maple* give explicit solutions of (4.2b).

Substituting the coefficients a_i and (4.2a) into (4.1) and using the symbolic software package *Maple*, we obtain, for KdV equation (4.1), the following solutions:

$$\Phi(\eta) = a_0 - 3\frac{(Aa_0 - v)}{A}\operatorname{sech}^2\left(\frac{1}{2}\sqrt{\frac{(v - Aa_0)}{B}}\eta\right), \qquad c_0 = c_1 = c_4 = 0, \qquad \frac{(v - Aa_0)}{B} > 0, \tag{4.3}$$

$$\Phi(\eta) = a_0 - 3\frac{(Aa_0 - v)}{A}\sec^2\left(\frac{1}{2}\sqrt{\frac{(Aa_0 - v)}{B}}\eta\right), \qquad c_0 = c_1 = c_4 = 0, \qquad \frac{(v - Aa_0)}{B} < 0, \tag{4.4}$$

$$\Phi(\eta) = a_0 + \frac{3}{2}\frac{(Aa_0 - v)}{A}\tan^2\left(\frac{1}{2}\sqrt{\frac{(v - Aa_0)}{2B}}\eta\right), \qquad c_0 \neq 0,$$
$$c_1 = c_3 = 0, \qquad \frac{(v - Aa_0)}{B} > 0, \tag{4.5}$$

$$\Phi(\eta) = a_0 - \frac{3}{2}\frac{(Aa_0 - v)}{A}\tanh^2\left(\frac{1}{2}\sqrt{\frac{(Aa_0 - v)}{2B}}\eta\right), \qquad c_0 \neq 0,$$
$$c_1 = c_3 = 0, \qquad \frac{(v - Aa_0)}{B} < 0, \tag{4.6}$$

$$\Phi(\eta) = a_0 + 3\frac{(Aa_0 - v)}{A}\operatorname{csch}^2\left(\frac{1}{2}\sqrt{\frac{(v - Aa_0)}{B}}\eta\right),$$
$$c_0 = c_1 = c_3 = 0, \qquad \frac{(v - Aa_0)}{B} > 0, \tag{4.7}$$

$$\Phi(\eta) = a_0 - 3\frac{(Aa_0 - v)}{A}\csc^2\left(\frac{1}{2}\sqrt{\frac{(Aa_0 - v)}{B}}\eta\right),$$
$$c_0 = c_1 = c_3 = 0, \qquad \frac{(v - Aa_0)}{B} < 0, \tag{4.8}$$

$$\Phi(\eta) = \frac{v}{A} - \frac{12B}{A\eta^2}, \qquad c_0 = c_1 = c_3 = 0, \qquad \frac{(v - Aa_0)}{B} = 0, \tag{4.9}$$

$$\Phi(\eta) = a_0 - 3\frac{(Aa_0 - v)}{A}\frac{m^2}{(2m^2 - 1)}\operatorname{cn}^2\left(\frac{1}{2}\sqrt{\frac{Aa_0 - v}{B(1 - 2m^2)}}\eta\right), \qquad c_0 \neq 0,$$
$$c_1 = c_3 = 0, \qquad \frac{v - a_0 A}{B} > 0, \tag{4.10}$$

$$\Phi(\eta) = a_0 + 3\frac{(Aa_0 - v)}{A}\frac{1}{(m^2 - 2)}\operatorname{dn}^2\left(\frac{1}{2}\sqrt{\frac{Aa_0 - v}{B\,(m^2 - 2)}}\eta\right), \qquad c_0 \neq 0,$$
$$c_1 = c_3 = 0, \qquad \frac{v - a_0 A}{B} > 0, \tag{4.11}$$

$$\Phi(\eta) = a_0 - 3\frac{(Aa_0 - v)}{A}\frac{m^2}{(m^2 + 1)}\operatorname{sn}^2\left(\frac{1}{2}\sqrt{\frac{Aa_0 - v}{B(m^2 + 1)}}\eta\right), \qquad c_0 \neq 0,$$
$$c_1 = c_3 = 0, \qquad \frac{v - a_0 A}{B} < 0, \tag{4.12}$$

$$\Phi(\eta) = \frac{v}{A} - 3\sqrt[3]{4c_3^2}\,\frac{B}{A}\wp\left(\sqrt[3]{\frac{c_3}{4}}\eta; -\sqrt[3]{\frac{4}{c_3}}c_1, -c_0\right), \qquad c_2 = c_4 = 0, \qquad c_3 > 0. \tag{4.13}$$

where m is the modulus of the Jacobi elliptic functions, a_0, c_0, c_1, and c_3 are arbitrary constants, and $\wp$ is the Weierstrass elliptic doubly periodic function.

5. Results and Discussion

Solutions (4.3) and (4.6) are hyperbolic wave solutions. In Figure 1, a profile of the bell-shaped solitary pulse is obtained for solution (4.3). Figures (1(a) and 1(b)) show that the soliton amplitude and width are sensitive to the relativistic factor γ_0 and the temperatures ratio σ. Also, the arbitrary value a_0 gives the same effect of adding a higher-order perturbation correction in increasing the amplitude and decreasing the width as shown in Figure 1(c).

Solutions (4.4) and (4.5) are triangular solutions that develop solitons with singularity at finite points, which are called "blowup" solutions [27] as in Figure 2(a). In Figure 2(b), for some values of a_0 the pulse does not vanish to infinity, so it has a localized form. Therefore, the arbitrary value a_0 plays a role in obtaining a pulse-shaped localized solution [28]. Solutions (4.7) and (4.8) lead to the propagation of an explosive pulses [29]. In Figure 3, the profiles of explosive (divergent) pulses are depicted for expression (4.7). For $a_0 = 0$, the solution gives a localized explosive pulse, as in Figure (2.3a) while a periodic solution is obtained for $a_0 = 0.4$, as shown in Figure (2.3b). The rational solution (4.9) may be helpful to explain certain physical phenomena. Because a rational solution is a disjoint union of manifolds, particle systems describing the motion of a pole of rational solutions for a KdV equation were analyzed [30]. Equations (4.10)–(4.12) are three Jacobi elliptic doubly periodic wave solutions. When $m \to 1$, solutions (4.10) and (4.11) reduce to (4.3) while (4.11) reduces to (4.6) [31]. In Figure 4, a profile of triangular periodic wave solution for expression (4.10) for $m = 2$ is shown. On the other hand, (4.13) gives the Weierstrass elliptic doubly periodic type solution as depicted in Figure 5.

In summary, it has been found that the amplitude and the width of the ion-acoustic waves as well as parametric regime where the solitons can exist are sensitive to the relativistic factor γ_0 and the ratio of the ion to the electron temperatures σ. Moreover, solutions for KdV equation have been obtained. It may be important to explain some physical phenomena in some plasma environments, such as ionosphere region.

References

[1] G. B. Whitham, *Linear and Nonlinear Waves*, John Wiley & Sons, New York, NY, USA, 1974, Pure and Applied Mathematics.

[2] R. Davidson, *Methods in Nonlinear Plasma Theory*, Academic Press, New York, NY, USA, 1972.

[3] G. C. Das and S. N. Paul, "Ion-acoustic solitary waves in relativistic plasmas," *Physics of Fluids*, vol. 28, no. 3, pp. 823–825, 1985.

[4] Y. Nejoh, "The effect of the ion temperature on the ion acoustic solitary waves in a collisionless relativistic plasma," *Journal of Plasma Physics*, vol. 37, no. 3, pp. 487–495, 1987.

[5] Y. Nejoh, "A two-dimensional ion acoustic solitary wave in a weakly relativistic plasma," *Journal of Plasma Physics*, vol. 38, no. 3, pp. 439–444, 1987.

[6] S. K. El-Labany, "Contribution of higher-order nonlinearity to nonlinear ion-acoustic waves in a weakly relativistic plasma," *Journal of Plasma Physics*, vol. 50, no. 3, pp. 495–504, 1993.

[7] H. Washimi and T. Taniuti, "Propagation of ion-acoustic solitary waves of small amplitude," *Physical Review Letters*, vol. 17, no. 19, pp. 996–998, 1966.

[8] T. Taniuti and C. C. Wei, "Reductive perturbation method in nonlinear wave propagation. I," *Journal of the Physical Society of Japan*, vol. 24, no. 4, pp. 941–946, 1968.

[9] T. Taniuti, "Reductive perturbation method and far fields of wave equations," *Progress of Theoretical Physics Supplement*, no. 55, pp. 1–35, 1974.

[10] V. I. Karpman, *Non-Linear Waves in Dispersive Media*, Pergamon Press, Oxford, UK, 1975.

[11] E. K. El-Shewy, "Effect of higher-order nonlinearity to nonlinear electron-acoustic solitary waves in an unmagnetized collisionless plasma," *Chaos, Solitons and Fractals*, vol. 26, no. 4, pp. 1073–1079, 2005.

[12] M. J. Ablowitz and P. A. Clarkson, *Solitons, Nonlinear Evolution Equations and Inverse Scattering*, vol. 149 of *London Mathematical Society Lecture Note Series*, Cambridge University Press, Cambridge, UK, 1991.

[13] E. J. Parkes and B. R. Duffy, "An automated tanh-function method for finding solitary wave solutions to non-linear evolution equations," *Computer Physics Communications*, vol. 98, no. 3, pp. 288–300, 1996.

[14] E. Fan, "Extended tanh-function method and its applications to nonlinear equations," *Physics Letters A*, vol. 277, no. 4-5, pp. 212–218, 2000.

[15] S. A. Elwakil, S. K. El-Labany, M. A. Zahran, and R. Sabry, "Modified extended tanh-function method for solving nonlinear partial differential equations," *Physics Letters A*, vol. 299, no. 2-3, pp. 179–188, 2002.

[16] E. Yusufoğlu and A. Bekir, "On the extended tanh method applications of nonlinear equations," *International Journal of Nonlinear Science*, vol. 4, no. 1, pp. 10–16, 2007.

[17] J.-H. He and M. A. Abdou, "New periodic solutions for nonlinear evolution equations using Exp-function method," *Chaos, Solitons and Fractals*, vol. 34, no. 5, pp. 1421–1429, 2007.

[18] M.-L. Wang and X.-Z. Li, "Extended *F*-expansion method and periodic wave solutions for the generalized Zakharov equations," *Physics Letters A*, vol. 343, no. 1–3, pp. 48–54, 2005.

[19] M. A. Abdou and A. Elhanbaly, "Construction of periodic and solitary wave solutions by the extended Jacobi elliptic function expansion method," *Communications in Nonlinear Science and Numerical Simulation*, vol. 12, no. 7, pp. 1229–1241, 2007.

[20] M.-L. Wang, "Solitary wave solutions for variant Boussinesq equations," *Physics Letters A*, vol. 199, no. 3-4, pp. 169–172, 1995.

[21] W.-X. Ma, "Travelling wave solutions to a seventh order generalized KdV equation," *Physics Letters A*, vol. 180, no. 3, pp. 221–224, 1993.

[22] Y. Zhou, M.-L. Wang, and Y.-M. Wang, "Periodic wave solutions to a coupled KdV equations with variable coefficients," *Physics Letters A*, vol. 308, no. 1, pp. 31–36, 2003.

[23] W.-X. Ma, T.-W. Huang, and Y. Zhang, "A multiple exp-function method for nonlinear differential equations and its application," *Physica Scripta*, vol. 82, no. 6, Article ID 065003, 8 pages, 2010.

[24] E. Fan, "A new algebraic method for finding the line soliton solutions and doubly periodic wave solution to a two-dimensional perturbed KdV equation," *Chaos, Solitons and Fractals*, vol. 15, no. 3, pp. 567–574, 2003.

[25] S. A. El-Wakil, E. M. Abulwafa, E. K. El-Shewy, and A. A. Mahmoud, "Ion-acoustic waves in plasma of warm ions and isothermal electrons using time-fractional KdV equation," *Chinese Physics B*, vol. 20, no. 4, Article ID 040508, 2011.

[26] S. K. El-Labany and A. M. El-Hanbaly, "Modulation of the non-linear ion acoustic waves in a plasma consisting of warm ions and isothermal electrons," *IL Nuovo Cimento D*, vol. 17, no. 6, pp. 547–556, 1995.

[27] W. M. Moslem, R. Sabry, U. M. Abdelsalam, I. Kourakis, and P. K. Shukla, "Solitary and blow-up electrostatic excitations in rotating magnetized electron-positron-ion plasmas," *New Journal of Physics*, vol. 11, no. 3, Article ID 033028, 2009.

[28] R. Sabry, W. M. Moslem, and P. K. Shukla, "Explosive and solitary excitations in a very dense magnetoplasma," *Physics Letters A*, vol. 372, no. 35, pp. 5691–5694, 2008.

[29] M. Adler and J. Moser, "On a class of polynomials connected with the Korteweg-de Vries equation," *Communications in Mathematical Physics*, vol. 61, no. 1, pp. 1–30, 1978.

[30] D. Baldwin, Ü. Göktaş, W. Hereman, L. Hong, R. S. Martino, and J. C. Miller, "Symbolic computation of exact solutions expressible in hyperbolic and elliptic functions for nonlinear PDEs," *Journal of Symbolic Computation*, vol. 37, no. 6, pp. 669–705, 2004.

[31] E. K. El-Shewy, H. G. Abdelwahed, and H. M. Abd-El-Hamid, "Computational solutions for the Korteweg—deVries equation in warm plasma," *Computational Methods in Science and Technology*, vol. 16, no. 1, pp. 13–18, 2010.

7

A Note on the Newman-Unti Group and the BMS Charge Algebra in Terms of Newman-Penrose Coefficients

Glenn Barnich[1, 2] and Pierre-Henry Lambert[1, 2]

[1] *Physique Théorique et Mathématique, Université Libre de Bruxelles, Campus Plaine, CP 231, 1050 Bruxelles, Belgium*
[2] *International Solvay Institutes, Campus Plaine, CP 231, 1050 Bruxelles, Belgium*

Correspondence should be addressed to Pierre-Henry Lambert, pilamber@ulb.ac.be

Academic Editor: Andrei D. Mironov

The symmetry algebra of asymptotically flat spacetimes at null infinity in four dimensions in the sense of Newman and Unti is revisited. As in the Bondi-Metzner-Sachs gauge, it is shown to be isomorphic to the direct sum of the abelian algebra of infinitesimal conformal rescalings with $\mathfrak{bms}_4$. The latter algebra is the semidirect sum of infinitesimal supertranslations with the conformal Killing vectors of the Riemann sphere. Infinitesimal local conformal transformations can then consistently be included. We work out the local conformal properties of the relevant Newman-Penrose coefficients, construct the surface charges, and derive their algebra.

1. Introduction

The definitions of asymptotically flat four dimensional spacetimes at null infinity by Bondi et al. [1, 2] (BMS) and Newman-Unti (NU) [3] in 1962 merely differ by the choice of the radial coordinate. Such a change of gauge should not affect the asymptotic symmetry algebra if, as we contend, this concept is to have a major physical significance.

The problem of comparing the symmetry algebra in both cases is that, besides the difference in gauge, the very definitions of these algebras are not the same. Indeed, NU allow the leading part of the metric induced on Scri to undergo a conformal rescaling. When this generalization is considered in the BMS setting, it turns out that the symmetry algebra is the direct sum of the BMS algebra $\mathfrak{bms}_4$ [4] with the abelian algebra of infinitesimal conformal rescalings [5, 6]. There are two novel and independent aspects in this computation.

(i) The first concerns the fact that the BMS algebra in 4 dimension involves the conformal Killing vectors of the unit, or equivalently, the Riemann sphere and can consistently accommodate infinitesimal local conformal transformations. The symmetry algebra $\mathfrak{bms}_4$ then involves two commuting copies of the noncentrally extended Virasoro algebra, called superrotations in [7], and simultaneously the supertranslations generators are expanded in Laurent series. The standard, globally well-defined symmetry algebra $\mathfrak{bms}_4^{\text{glob}}$ consists in restricting to the globally well-defined conformal Killing vectors of the sphere which correspond to infinitesimal Lorentz transformation, while the supertranslation generators are expanded into spherical harmonics.

This local versus global versions of the symmetry algebra are of course not related to the BMS gauge choice, but will also occur in alternative characterizations of the asymptotic symmetry algebra where the conformal Killing vectors of the sphere play a role. Examples of this are the geometrical approach of Geroch [8] based on Penrose's definition of null infinity [9] and also, as we will explicitly discuss in this paper, the asymptotic symmetries in the NU framework.

(ii) The second aspect is related to the modified Lie bracket that should be used when the vector fields parametrizing infinitesimal diffeomorphisms depend explicitly on the metric. Indeed, when using the modified Lie bracket, the spacetime vectors realize the asymptotic symmetry algebra everywhere in the bulk and furthermore, even on Scri, this bracket is needed to disentangle the algebra when conformal rescalings of the induced metric on Scri are allowed. Similarly, in the context of the AdS/CFT correspondence, this bracket allows one to realize the asymptotic symmetry algebra in the bulk and to disentangle the symmetry algebra at infinity when considering transformations that leave the Fefferman-Graham ansatz invariant only up to conformal rescaling of the boundary metric [10]. From a mathematical point of view, the modified Lie bracket is the natural bracket of the Lie algebroid that is associated to any theory with gauge invariance [11].

What we will do in this paper is to rederive from scratch the asymptotic symmetry algebra in the NU framework by focusing on metric aspects and on the two novel features discussed above. As expected, the symmetry algebra is again the direct sum of $\mathfrak{bms}_4$ with the abelian algebra of infinitesimal conformal rescalings of the metric on Scri and thus coincides, as it should, with the generalized symmetry algebra in the BMS approach. A related analysis of asymptotic symmetries in the NU context from the point of view of Scri and emphasizing global issues instead can be found in [12, 13].

Even though the results presented here are not really surprising in view of those in the BMS framework and the close relation between the NU and BMS approaches, the exercise of working out the details is justified because the NU framework is embedded in the context of the widely used Newman-Penrose formalism [14] so that explicit formulae in this context are directly relevant in many applications, see for example, the review article [15].

As a first application, we study the transformation properties of the Newman-Penrose coefficients parametrizing solution space in the NU approach. Our main focus is on the inhomogeneous terms in the transformation laws that contain the information on the central extensions of the theory. We then discuss the associated surface charges by following the analysis in the BMS gauge [16] and briefly compare with standard expressions that can be found in the literature. The algebra of these charges is derived and shown to involve field dependent central charges in the case of $\mathfrak{bms}_4$ which vanish for $\mathfrak{bms}_4^{\text{glob}}$.

2. NU Metric Ansatz for Asymptotically Flat Spacetimes

The metric ansatz of NU is based on a family of null hypersurfaces labelled by the first coordinate, $x^0 \equiv u = \text{const}$. The second coordinate $x^1 \equiv r$ is chosen as an affine parameter for the null geodesic generators l^μ of these hypersurfaces, so that $l^\mu = -\delta^\mu_r$. Up to a change of signature from $(+,-,-,-)$ to $(-,+,+,+)$, a renumbering of the indices, and the tetrad transformation that makes the conformal factor real, the line element considered in section 4 of NU [3] can be written as

$$ds^2 = W du^2 - 2dr\, du + g_{AB}\left(dx^A - V^A du\right)\left(dx^B - V^B du\right), \tag{2.1}$$

with associated inverse metric

$$g^{\mu\nu} = \begin{pmatrix} 0 & -1 & 0 \\ -1 & -W & -V^B \\ 0 & -V^A & g^{AB} \end{pmatrix}, \tag{2.2}$$

where

$$g_{AB}dx^A dx^B = r^2\overline{\gamma}_{AB}dx^A dx^B + rC_{AB}dx^A dx^B + o(r), \tag{2.3}$$

with $\overline{\gamma}_{AB}$ conformally flat. Below, we will use standard stereographic coordinates $\zeta = \cot(\theta/2)e^{i\phi}, \bar{\zeta}$, $\overline{\gamma}_{AB}dx^A dx^B = e^{2\widetilde{\varphi}}d\zeta\, d\bar{\zeta}$, $\widetilde{\varphi} = \widetilde{\varphi}(u,x)$. In particular, we use the notation $e^{2\widetilde{\varphi}}$ for the conformal factor. In Section 4, we will give the explicit dictionary that allows one to translate to the quantities originally used by NU.

In addition, the choice of origin for the affine parameter of the null geodesics is fixed through the requirement that the term proportional to r^{-2} in the expansion of the spin coefficient $-\rho = D_\rho l_\nu m^\rho \overline{m}^\nu$ is absent.

When expressed in terms of the metric, one finds

$$\rho = -\frac{1}{4}g^{AB}g_{AB,r} = -\frac{1}{4}\partial_r \ln|g| = -r^{-1} + \frac{1}{4}C^A_A r^{-2} + o\left(r^{-2}\right), \tag{2.4}$$

where $g = \det g_{\rho\nu}$ and the index has been raised with the inverse of $\overline{\gamma}_{AB}$. The requirement is thus equivalent to the condition

$$C^A_A = 0. \tag{2.5}$$

In the following we denote by $\overline{D}_A$ the covariant derivative with respect to $\overline{\gamma}_{AB}$ and by $\overline{\Delta}$ the associated Laplacian and by $\overline{R}$ the scalar curvature. In complex coordinates $\zeta, \bar{\zeta}$, $C_{\zeta\bar{\zeta}} = 0$ and we define for later convenience $C_{\zeta\zeta} = e^{2\widetilde{\varphi}}c$, $C_{\bar{\zeta}\bar{\zeta}} = e^{2\widetilde{\varphi}}\bar{c}$. Finally,

$$V^A = O\left(r^{-2}\right), \qquad W = -2r\partial_u\widetilde{\varphi} + \overline{\Delta}\widetilde{\varphi} + O\left(r^{-1}\right), \tag{2.6}$$

where $\overline{\Delta}\widetilde{\varphi} = 4e^{-2\widetilde{\varphi}}\partial\bar{\partial}\widetilde{\varphi}$ with $\partial = \partial_\zeta$, $\bar{\partial} = \partial_{\bar{\zeta}}$.

The more restrictive fall-off conditions in [3] are relevant for integrating the field equations but play no role in the discussion of the asymptotic symmetry algebra.

3. Asymptotic Symmetries in the NU Approach

The infinitesimal NU transformations can be defined as those infinitesimal transformations that leave the form (2.2) and the fall-off conditions (2.3)–(2.6) invariant, up to a rescaling of the conformal factor $\delta\widetilde{\varphi}(u,x^A) = \widetilde{\omega}(u,x^A)$. In other words, they satisfy

$$\mathcal{L}_\xi g^{uu} = 0, \qquad \mathcal{L}_\xi g^{uA} = 0, \qquad \mathcal{L}_\xi g^{ur} = 0, \tag{3.1}$$

$$\partial_r\left[\frac{1}{\sqrt{|g|}}\partial_\rho\left(\sqrt{|g|}\xi^\rho\right)\right] = o\left(r^{-2}\right), \tag{3.2}$$

$$\begin{aligned}\mathcal{L}_\xi g^{rA} &= O\left(r^{-2}\right), \qquad \mathcal{L}_\xi g^{AB} = -2\widetilde{\omega}g^{AB} + O\left(r^{-3}\right),\\ \mathcal{L}_\xi g^{rr} &= 2r\partial_u\widetilde{\omega} + 2\widetilde{\omega}\overline{\Delta}\widetilde{\varphi} - \overline{\Delta}\widetilde{\omega} + O\left(r^{-1}\right).\end{aligned} \tag{3.3}$$

Equations (3.1) are equivalent to

$$\partial_r\xi^\nu = g^{\nu\rho}\partial_\rho\xi^u \Longleftrightarrow \begin{cases}\partial_r\xi^u = 0,\\ \partial_r\xi^A = \partial_B\xi^u g^{BA},\\ \partial_r\xi^r = -\partial_u\xi^u - \partial_A\xi^u V^A,\end{cases} \tag{3.4}$$

and are explicitly solved by

$$\begin{aligned}\xi^u &= f,\\ \xi^A &= Y^A + I^A, \quad I^A = -\partial_B f\int_r^\infty dr' g^{AB},\\ \xi^r &= -r\partial_u f + Z + J, \quad J = \partial_A f\int_r^\infty dr' V^A,\end{aligned} \tag{3.5}$$

with $\partial_r f = 0 = \partial_r Y^A = \partial_r Z$. Equation (3.2) then implies

$$Z = \frac{1}{2}\overline{\Delta}f. \tag{3.6}$$

The first equation of (3.3) requires $\partial_u Y^A = 0$, the second that Y^A is a conformal Killing vector of $\overline{\gamma}_{AB}$, which amounts to

$$Y^\zeta \equiv Y = Y(\zeta), \qquad Y^{\bar{\zeta}} \equiv \overline{Y} = \overline{Y}\left(\bar{\zeta}\right), \tag{3.7}$$

in the coordinates $\zeta, \bar{\zeta}$, and also that

$$\partial_u f = f\partial_u\tilde{\varphi} + \frac{1}{2}\tilde{\psi}, \tag{3.8}$$

with $\psi = \overline{D}_A Y^A$, or more explicitly in $\zeta, \bar{\zeta}$ coordinates, $\psi = \partial Y + \overline{\partial Y} + 2Y\partial\tilde{\varphi} + 2\overline{Y}\bar{\partial}\tilde{\varphi}$, and $\tilde{\psi} = \psi - 2\tilde{\omega}$. Finally, the last equation of (3.3) implies

$$2(\partial_u Z + Z\partial_u\tilde{\varphi}) = Y^A\partial_A\overline{\Delta}\tilde{\varphi} + \psi\overline{\Delta}\tilde{\varphi} + 2\partial_A f\overline{\gamma}^{AB}\partial_B\partial_u\tilde{\varphi} + f\overline{\Delta}\partial_u\tilde{\varphi} - \overline{\Delta}\tilde{\omega}, \tag{3.9}$$

which is identically satisfied when taking the previous relations into account.

One approach is to consider that (3.8) fixes $\tilde{\omega}$ in terms of f and Y, $\tilde{\omega} = (1/2)\psi + f\partial_u\tilde{\varphi} - \partial_u f$. Consider Scri, the space $\mathcal{I}$ with coordinates $u, \zeta, \bar{\zeta}$ and metric

$$ds^2_{\mathcal{I}} = 0du^2 + e^{2\tilde{\varphi}}d\zeta\, d\bar{\zeta}. \tag{3.10}$$

The NU algebra is then defined as the commutator algebra of the vector fields

$$\bar{\xi} = f\frac{\partial}{\partial u} + Y^A\frac{\partial}{\partial x^A}, \tag{3.11}$$

with $f = f(u, x^A)$ arbitrary and $Y^A(x)$ conformal Killing vectors of a conformally flat metric in 2 dimensions, or equivalently, the algebra of conformal vector fields of the degenerate metric (3.10).

This is not the symmetry algebra of asymptotically flat spacetimes in the sense of NU however. Indeed, $\tilde{\varphi}$ is arbitrary, it can for instance be considered as the finite ambiguity related to Penrose's conformal approach [9, 17, 18] to null infinity. One can then interpret $\tilde{\varphi}$ as part of the background structure, or in other words, of the gauge fixing [8], and compute the asymptotic symmetries for a fixed choice of $\tilde{\varphi}$, that is, $\tilde{\omega} = 0$ in the formulae above, or ask the more general question of how the asymptotic symmetries depend on changes in $\tilde{\varphi}$ by an arbitrary infinitesimal amount $\tilde{\omega}$. In both cases, one has to consider (3.8) as a differential equation for f. As we now show, the symmetry algebra will then be isomorphic to the trivially extended $\mathfrak{bms}_4$ algebra by the abelian algebra of infinitesimal conformal rescalings, as it should, and as a consequence, the Poincaré algebra is embedded therein in a natural way. Furthermore, there is a natural realization of the asymptotic symmetry algebra on an asymptotically flat 4 dimensional bulk spacetime. Note also that, for $\tilde{\omega} = 0$, (3.8) has been interpreted from the point of view of Penrose's conformal approach to null infinity in [12] following [19] and related to the preservation of null angles, which is the standard way [9, 17, 20, 21] to recover the BMS algebra from geometrical data on Scri.

The general solution for (3.8) reads

$$f = e^{\tilde{\varphi}}\left[\tilde{T} + \frac{1}{2}\int_0^u du' e^{-\tilde{\varphi}}\tilde{\psi}\right], \quad \tilde{T} = \tilde{T}(\zeta, \bar{\zeta}), \tag{3.12}$$

and the general solution to (3.1)–(3.3) defining the asymptotic symmetries is given by ξ^ρ as in (3.5) where Z, Y^A, f satisfy (3.6), (3.7), (3.12) with $\widetilde{\omega}$ arbitrary. Asymptotic Killing vectors thus depend on Y^A, $\widetilde{T}$, $\widetilde{\omega}$ and the metric, $\xi = \xi[Y, \widetilde{T}, \widetilde{\omega}; g]$.

For such metric-dependent vector fields, consider on the one hand the suitably modified Lie bracket taking the metric dependence of the spacetime vectors into account,

$$[\xi_1, \xi_2]_M = [\xi_1, \xi_2] - \delta^g_{\xi_1}\xi_2 + \delta^g_{\xi_2}\xi_1, \tag{3.13}$$

where $\delta^g_{\xi_1}\xi_2$ denotes the variation in ξ_2 under the variation of the metric induced by ξ_1, $\delta^g_{\xi_1} g_{\mu\nu} = \mathcal{L}_{\xi_1} g_{\mu\nu}$.

Consider on the other hand the extended $\mathfrak{bms}_4$ algebra, that is, the semidirect sum of the algebra of conformal Killing vectors of the Riemann sphere with the abelian ideal of infinitesimal supertranslations, trivially extended by infinitesimal conformal rescalings of the conformally flat degenerate metric on Scri. More explicitly, the commutation relations are given by $[(Y_1, \widetilde{T}_1, \widetilde{\omega}_1), (Y_2, \widetilde{T}_2, \widetilde{\omega}_2)] = (\widehat{Y}, \widehat{\widetilde{T}}, \widehat{\widetilde{\omega}})$ where

$$\begin{aligned} \widehat{Y}^A &= Y_1^B \partial_B Y_2^A - Y_2^B \partial_B Y_1^A, \\ \widehat{\widetilde{T}} &= Y_1^A \partial_A \widetilde{T}_2 - Y_2^A \partial_A \widetilde{T}_1 + \frac{1}{2}\left(\widetilde{T}_1 \partial_A Y_2^A - \widetilde{T}_2 \partial_A Y_1^A\right), \\ \widehat{\widetilde{\omega}} &= 0. \end{aligned} \tag{3.14}$$

It thus follows the following.

Theorem 3.1. *The spacetime vectors $\xi[Y, \widetilde{T}, \widetilde{\omega}; g]$ realize the extended $\mathfrak{bms}_4$ algebra in the modified Lie bracket,*

$$\left[\xi\left[Y_1, \widetilde{T}_1, \widetilde{\omega}_1; g\right], \xi\left[Y_2, \widetilde{T}_2, \widetilde{\omega}_2; g\right]\right]_M = \xi\left[\widehat{Y}, \widehat{\widetilde{T}}, \widehat{\widetilde{\omega}}; g\right], \tag{3.15}$$

in the bulk of an asymptotically flat spacetime in the sense of Newman and Unti.

Note in particular that for two different choices of the conformal factor $\widetilde{\varphi}$ which is held fixed, $\widetilde{\omega} = 0$, the asymptotic symmetry algebras are isomorphic to $\mathfrak{bms}_4$, which is thus a gauge invariant statement.

Proof. The proof follows closely the one in [6] for the BMS gauge. In order to be self-contained we recall the different steps here. In a first stage, one shows that on $\mathcal{I}$, the vectors fields $\bar{\xi}[Y, \widetilde{T}, \widetilde{\omega}; \bar{\gamma}]$ given in (3.11) with f as in (3.12) realize the extended $\mathfrak{bms}_4$ algebra in terms of the modified Lie bracket. Indeed, this is obvious for the A components which do not depend on the metric so that the modified bracket reduces to the standard Lie bracket for these components. For the u component, taking into account that

$$\delta^g_{\xi_1} f_2 = \widetilde{\omega}_1 f_2 + \frac{1}{2} e^{\widetilde{\varphi}} \int_0^u du' e^{-\widetilde{\varphi}} \left[-\widetilde{\omega}_1 (\psi_2 - 2\widetilde{\omega}_2) + 2Y_2^A \partial_A \widetilde{\omega}_1\right], \tag{3.16}$$

we have $[\bar{\xi}_1, \bar{\xi}_2]^u_M|_{u=0} = e^{\tilde{\varphi}}|_{u=0}\hat{T}$. Direct computation then shows that $\partial_u([\bar{\xi}_1, \bar{\xi}_2]^u_M) = \hat{f}\partial_u\tilde{\varphi} + (1/2)\overline{D}_A\hat{Y}^A$ with $\hat{f}$ given by (3.12) with $\tilde{T}, Y, \tilde{\omega}$ replaced by their hatted counterparts, implying the result for the u component.

For the spacetime vectors, direct computation gives $[\xi_1, \xi_2]^u_M = [\bar{\xi}_1, \bar{\xi}_2]^u_M = \hat{f}$. Using the defining property (3.4), one then finds that $\partial_r([\xi_1, \xi_2]^\rho_M) = g^{\rho\nu}\partial_\nu\hat{f}$. For the A components the result then follows from the one on $\mathcal{I}$, $\lim_{r\to\infty}[\xi_1, \xi_2]^A_M = \hat{Y}^A$. This is due to the fact that I^A goes to zero at infinity, that the nonvanishing term at infinity does not involve the metric, and that the correction term in the bracket does not change the asymptotic behaviour. Finally, for the r component, we still need to check that the r independent component of $[\xi_1, \xi_2]^r_M$ is given by $(1/2)\overline{\Delta}\hat{f}$, which follows by direct computation. □

For completeness, let us also stress here that, if one focuses on local properties and expands the conformal Killing vectors $Y^A\partial_A$ and the infinitesimal supertranslations T in Laurent series,

$$l_n = -\zeta^{n+1}\frac{\partial}{\partial\zeta}, \qquad \bar{l}_n = -\bar{\zeta}^{n+1}\frac{\partial}{\partial\bar{\zeta}}, \qquad n \in \mathbb{Z},$$
$$\tilde{T}_{m,n} = \zeta^m\bar{\zeta}^n, \qquad m, n \in \mathbb{Z}, \tag{3.17}$$

the commutation relations for the complexified $\mathfrak{bms}_4$ algebra read

$$[l_m, l_n] = (m-n)l_{m+n}, \qquad \left[\bar{l}_m, \bar{l}_n\right] = (m-n)\bar{l}_{m+n}, \qquad \left[l_m, \bar{l}_n\right] = 0,$$
$$[l_l, T_{m,n}] = \left(\frac{l+1}{2} - m\right)T_{m+l,n}, \qquad \left[\bar{l}_l, T_{m,n}\right] = \left(\frac{l+1}{2} - n\right)T_{m,n+l}. \tag{3.18}$$

The $\mathfrak{bms}_4$ algebra contains as subalgebra the Poincaré algebra, which we identify with the algebra of exact Killing vectors of the Minkowski metric equipped with the standard Lie bracket. It is spanned by the generators

$$l_{-1}, l_0, l_1, \quad \bar{l}_{-1}, \bar{l}_0, \bar{l}_1, \quad \tilde{T}_{0,0}, \tilde{T}_{1,0}, \tilde{T}_{0,1}, \tilde{T}_{1,1}. \tag{3.19}$$

Nontrivial central extensions of the algebra (3.18) have been studied in [7]: the computation of $H^2(\mathfrak{bms}_4)$ reveals that there is only the standard ones for the Virasoro algebra extending the first two commutation relations.

4. Explicit Relation between the NU and the BMS Gauges

The definition of asymptotically flat spacetimes in the BMS approach [1, 2, 4] as reviewed in [5, 6] amounts to replacing $g_{uu} = 1/g^{uu} = -1$ by

$$g_{uu} = 1/g^{uu} = -e^{2\beta}, \qquad \beta = O\left(r^{-2}\right) \tag{4.1}$$

in (2.1) and (2.2) while imposing the additional requirement that

$$\det g_{AB} = r^4 \det \overline{\gamma}_{AB}. \tag{4.2}$$

Both definitions then differ just by a choice of radial coordinate. Indeed, replacing the radial coordinate by a function of the 4 coordinates preserves the zeros in (2.1) and (2.2) (see e.g., the discussion in [22]). Furthermore, for first nontrivial order in r, the determinant condition leads to the same restriction (2.5) as the choice of the origin of the affine parameter. It follows that the relation between the two radial coordinates does not involve constant terms and is of the form

$$r' = r + O\left(r^{-1}\right). \tag{4.3}$$

More explicitly, starting from the NU approach, BMS coordinates are obtained by defining the new radial coordinates as [23]

$$r_{\text{BMS}} = \left(\frac{\det g_{AB}}{\det \overline{\gamma}_{AB}}\right)^{1/4}. \tag{4.4}$$

Conversely, starting from the BMS approach with radial coordinate r, NU coordinates are obtained by changing the radial coordinate to

$$r_N = r - \int_r^{\infty} dr'\left(e^{2\beta} - 1\right). \tag{4.5}$$

These changes of coordinates only affect lower-order terms in the asymptotic expansion of the metric that plays no role in the definition of asymptotic symmetries and explains a posteriori why the asymptotic symmetry algebras in both approaches are isomorphic.

At this stage, the dynamics of the theory comes into play. The Einstein equations are solved order by order in r. In the first orders, there are integrations "constants" that appear as free data characterizing asymptotically flat solutions. We will now work out the explicit relation between these data in both approaches. The inverse metric in the BMS gauge (as discussed in [6]) is given by

$$g^{\mu\nu}_{\text{BMS}} = \begin{pmatrix} 0 & -e^{-2\beta} & 0 \\ -e^{-2\beta} & -e^{-2\beta}\frac{V}{r} & -e^{-2\beta}U^B \\ 0 & -e^{-2\beta}U^A & g^{AB} \end{pmatrix},$$
$$g_{AB} = r^2\overline{\gamma}_{AB} + rC_{AB} + \frac{1}{4}\overline{\gamma}_{AB}C^C_D C^D_C + O\left(r^{-1}\right). \tag{4.6}$$

For simplicity, we assume here that there is no trace-free part D_{AB} at order 0 and that the conformal factor is time-independent, $\partial_u\tilde{\varphi} = 0$, in which case the news tensor is simply $N_{AB} = \partial_u C_{AB}$ and $f = T + (1/2)u\tilde{\varphi}$ with $T = e^{\tilde{\varphi}}\tilde{T}$. Writing

$$C_{\zeta\zeta} = e^{2\tilde{\varphi}}c, \qquad C_{\bar{\zeta}\bar{\zeta}} = e^{2\tilde{\varphi}}\bar{c}, \qquad C_{\zeta\bar{\zeta}} = 0, \tag{4.7}$$

we have

$$\begin{aligned} \beta &= -\frac{1}{4}r^{-2}c\bar{c} + O\left(r^{-4}\right), \\ U^{\zeta} &= -\frac{2}{r^2}e^{-4\tilde{\varphi}}\eth\left(e^{2\tilde{\varphi}}\bar{c}\right) - \frac{2}{3r^3}\left[N^{\zeta} - 4e^{-4\tilde{\varphi}}\bar{c}\eth\left(e^{2\tilde{\varphi}}c\right)\right] + O\left(r^{-4}\right), \\ \frac{V}{r} &= 4e^{-2\tilde{\varphi}}\eth\bar{\eth}\tilde{\varphi} + r^{-1}2M + O\left(r^{-2}\right), \end{aligned} \tag{4.8}$$

which implies in particular that

$$r_N = r + \frac{c\bar{c}}{2r} + O\left(r^{-3}\right). \tag{4.9}$$

The only consequence of Einstein's equations on the angular momentum and mass aspects $N^{\zeta} = N^{\zeta}(u,\zeta,\bar{\zeta}),\ M = M(u,\zeta,\bar{\zeta})$ is the evolution equations

$$\partial_u M = -\frac{1}{8}N^A_B N^B_A + \frac{1}{8}\bar{\Delta}\bar{R} + \frac{1}{4}\bar{D}_A\bar{D}_C N^{CA}, \tag{4.10}$$

$$\begin{aligned} \partial_u N_A = {} & \partial_A M + \frac{1}{4}C^B_A\partial_B\bar{R} + \frac{1}{16}\partial_A\left[N^B_C C^C_B\right] - \frac{1}{4}\bar{D}_A C^C_B N^B_C \\ & - \frac{1}{4}\bar{D}_B\left[C^B_C N^C_A - N^B_C C^C_A\right] - \frac{1}{4}\bar{D}_B\left[\bar{D}^B\bar{D}_C C^C_A - \bar{D}_A\bar{D}_C C^{BC}\right]. \end{aligned} \tag{4.11}$$

Consider now the "eth" operators [24] defined here for a field η^s of spin weight s according to the conventions of [25] through

$$\eth\eta^s = P^{1-s}\bar{\eth}(P^s\eta^s), \quad \bar{\eth}\eta^s = P^{1+s}\eth(P^{-s}\eta^s), \quad P = \sqrt{2}e^{-\tilde{\varphi}}, \tag{4.12}$$

where $\eth, \bar{\eth}$ raise, respectively, lower the spin weight by one unit and satisfy

$$\left[\bar{\eth},\eth\right]\eta^s = \frac{s}{2}\bar{R}\eta^s. \tag{4.13}$$

The spin weights of the various quantities are summarized in Table 1. Note that the P used here differs from the one used in [3], which we will denote by P_N below. It also no longer denotes the particular function $(1/2)(1+\zeta\bar{\zeta})$, contrary to the notation used in [6, 16].

In order to compare with the notation used in [3], we use $\zeta = x^3 + ix^4$. With $x'^{\alpha} = u, r_N, x^3, x^4$ and $x^{\mu} = u, r, \zeta, \bar{\zeta}$, computing $g^{\alpha\beta}_N(x') = -((\partial x'^{\alpha}/\partial x^{\mu})g^{\mu\nu}_{BMS}(\partial x'^{\beta}/\partial x^{\nu}))(x(x'))$,

Table 1: Spin and conformal weights.

	σ^0	$\dot\sigma^0$	Ψ_4^0	Ψ_3^0	Ψ_2^0	Ψ_1^0	y	T
s	2	2	−2	−1	0	1	−1	0
w	−1	−2	−3	−3	−3	−3	1	1

where the overall minus sign takes the change of signature into account, then gives the following dictionary by comparing with [3]:

$$P_N = \frac{1}{\sqrt{2}}e^{-\tilde\varphi} = \frac{1}{2}P, \qquad \nabla = 2\bar\eth, \qquad \mu^0 = -P^2\eth\bar\eth\ln P = \frac{1}{2}\bar\Delta\tilde\varphi = -\frac{1}{4}\bar R,$$
$$\Psi_2^0 + \bar\Psi_2^0 = -2M - \partial_u(c\bar c), \quad \sigma^0 = \bar c\,, \quad \omega^0 = \bar\eth\sigma^0, \tag{4.14}$$
$$\Psi_1^0 = -PN_{\bar\zeta} - \sigma^0\eth\bar\sigma^0 - \frac{3}{4}\eth\left(\sigma^0\bar\sigma^0\right).$$

For convenience, let us also use

$$\Psi_3^0 = -\eth\dot{\bar\sigma}^0 - \frac{1}{4}\bar\eth\bar R, \qquad \Psi_4^0 = -\ddot{\bar\sigma}^0. \tag{4.15}$$

In these terms,

$$\dot\Psi_3^0 = \eth\Psi_4^0, \qquad \dot\Psi_2^0 = \eth\Psi_3^0 + \sigma^0\Psi_4^0, \qquad \dot\Psi_1^0 = \eth\Psi_2^0 + 2\sigma^0\Psi_3^0. \tag{4.16}$$

Indeed, the first equation holds by definition and the assumed time-independence of P. The evolution equation (4.10) is equivalent to the real part of the second equation. Taking into account the on-shell relation of the NU framework,

$$\Psi_2^0 - \bar\Psi_2^0 = \bar\eth^2\sigma^0 - \eth^2\bar\sigma^0 + \bar\sigma^0\dot\sigma^0 - \sigma^0\dot{\bar\sigma}^0, \tag{4.17}$$

we find

$$M = -\Psi_2^0 - \sigma^0\dot{\bar\sigma}^0 + \frac{1}{2}\bar\eth^2\sigma^0 - \frac{1}{2}\eth^2\bar\sigma^0, \tag{4.18}$$

in terms of which (4.10) is fully equivalent to the second equation of (4.16) and (4.11) is equivalent to the last equation of (4.16), in agreement with [3].

5. Transformation Laws of the NU Coefficients Characterizing Asymptotic Solutions

Let $\mathcal{Y} = P^{-1}\bar{Y}$ and $\bar{\mathcal{Y}} = P^{-1}Y$. The conformal Killing equations and the conformal factor then become

$$\eth\bar{\mathcal{Y}} = 0 = \bar{\eth}\mathcal{Y}, \qquad \varphi = \left(\eth\mathcal{Y} + \bar{\eth}\bar{\mathcal{Y}}\right). \tag{5.1}$$

It follows for instance that

$$\bar{\eth}\eth\mathcal{Y} = -\frac{\bar{R}}{2}\mathcal{Y}, \qquad \eth^2\varphi = \eth^3\mathcal{Y} - \frac{1}{2}\bar{\mathcal{Y}}\eth\bar{R}, \qquad \bar{\eth}\eth\varphi = -\frac{1}{2}\left[\eth\left(\bar{R}\mathcal{Y}\right) + \bar{\eth}\left(\bar{R}\bar{\mathcal{Y}}\right)\right]. \tag{5.2}$$

Using the notation $S = (Y, \tilde{T}, \tilde{\omega})$, we have $-\delta_S\bar{\gamma}_{AB} = 2\tilde{\omega}\bar{\gamma}_{AB}$ for the background metric and

$$\left[-\delta_S, \bar{\eth}\right]\eta^s = -\tilde{\omega}\bar{\eth}\eta^s + s\bar{\eth}\tilde{\omega}\eta^s, \qquad [-\delta_S, \eth]\eta^s = -\tilde{\omega}\eth\eta^s - s\eth\tilde{\omega}\eta^s. \tag{5.3}$$

To work out the transformation properties of the NU coefficients characterizing asymptotic solution space, one needs to evaluate the subleading terms in $\mathcal{L}_\xi g_N^{\alpha\beta}$ on-shell. This can also be done by translating the results from the BMS gauge, which yields

$$\begin{aligned}
-\delta_S\sigma^0 &= \left[f\partial_u + \mathcal{Y}\eth + \bar{\mathcal{Y}}\bar{\eth} + \frac{3}{2}\eth\mathcal{Y} - \frac{1}{2}\bar{\eth}\bar{\mathcal{Y}} - \tilde{\omega}\right]\sigma^0 - \eth^2 f,\\
-\delta_S\dot{\sigma}^0 &= \left[f\partial_u + \mathcal{Y}\eth + \bar{\mathcal{Y}}\bar{\eth} + 2\eth\mathcal{Y} - 2\tilde{\omega}\right]\dot{\sigma}^0 - \frac{1}{2}\eth^2\tilde{\varphi},\\
-\delta_S\Psi_4^0 &= \left[f\partial_u + \mathcal{Y}\eth + \bar{\mathcal{Y}}\bar{\eth} + \frac{1}{2}\eth\mathcal{Y} + \frac{5}{2}\bar{\eth}\bar{\mathcal{Y}} - 3\tilde{\omega}\right]\Psi_4^0,\\
-\delta_S\Psi_3^0 &= \left[f\partial_u + \mathcal{Y}\eth + \bar{\mathcal{Y}}\bar{\eth} + \eth\mathcal{Y} + 2\bar{\eth}\bar{\mathcal{Y}} - 3\tilde{\omega}\right]\Psi_3^0 + \eth f\Psi_4^0,\\
-\delta_S\Psi_2^0 &= \left[f\partial_u + \mathcal{Y}\eth + \bar{\mathcal{Y}}\bar{\eth} + \frac{3}{2}\eth\mathcal{Y} + \frac{3}{2}\bar{\eth}\bar{\mathcal{Y}} - 3\tilde{\omega}\right]\Psi_2^0 + 2\eth f\Psi_3^0,\\
-\delta_S\Psi_1^0 &= \left[f\partial_u + \mathcal{Y}\eth + \bar{\mathcal{Y}}\bar{\eth} + 2\eth\mathcal{Y} + \bar{\eth}\bar{\mathcal{Y}} - 3\tilde{\omega}\right]\Psi_1^0 + 3\eth f\Psi_2^0.
\end{aligned} \tag{5.4}$$

Following for instance the terminology in [26] Section 3, but now for general infinitesimal transformations $\zeta' = \zeta + \epsilon Y(\zeta)$, $\bar{\zeta}' = \bar{\zeta} + \epsilon\bar{Y}(\bar{\zeta})$ instead of those associated to linear fractional transformations on the sphere and also considering $\bar{\zeta}$ as the holomorphic coordinate instead of ζ, a field η has spin weight s and conformal weight w if it transforms as

$$-\delta_{Y,\bar{Y}}\eta = \left[Y^A\partial_A + \frac{s}{2}\left(\bar{\partial}\bar{Y} - \partial Y\right) - \frac{w}{2}\varphi\right]\eta. \tag{5.5}$$

Table 2: Rank and density weights.

	$P^{-1}\sigma^0$	$P^{-2}\dot{\sigma}^0$	$P^{-3}\Psi_4^0$	$P^{-3}\Psi_3^0$	$P^{-3}\Psi_2^0$	$P^{-3}\Psi_1^0$	$\overline{Y}$	$\widetilde{T}$
s	2	2	−2	−1	0	1	−1	0
n	−1/2	0	1/2	1	3/2	1	−1	−1/2

A tensor density of rank $s \geqslant 0$ and weight n transforms as

$$-\delta_{Y,\overline{Y}} A_{\bar{\zeta}\cdots\bar{\zeta}} = \left[Y^A \partial_A + s\bar{\partial}\overline{Y} + n\left(\partial Y + \bar{\partial}\overline{Y}\right)\right] A_{\bar{\zeta}\cdots\bar{\zeta}}, \tag{5.6}$$

while for rank $s \leqslant 0$ and weight n, we have

$$-\delta_{Y,\overline{Y}} A_{\zeta\cdots\zeta} = \left[Y^A \partial_A - s\partial Y + n\left(\partial Y + \bar{\partial}\overline{Y}\right)\right] A_{\zeta\cdots\zeta}. \tag{5.7}$$

It then follows that a tensor density of weights (s, n) defines a field of weights $(s, -(2n + |s|))$ and conversely, a field of weights (s, w) defines a tensor density of weights $(s, -(1/2)(w+|s|))$. For $s \geqslant 0$, this is done through $\eta = A_{\bar{\zeta}\cdots\bar{\zeta}} P^{2n+s}$ and $A_{\bar{\zeta}\cdots\bar{\zeta}} = P^{w}\eta$. For $s \leqslant 0$, we have $\eta = A_{\zeta\cdots\zeta} P^{2n-s}$ and $A_{\zeta\cdots\zeta} = P^{w}\eta$. Note that complex conjugation gives rise to opposite spin weight and rank but leaves the conformal and density weights unchanged. Alternatively, (5.5) can be written as

$$-\delta_{\mathcal{Y},\overline{\mathcal{Y}}}\eta = \left[\mathcal{Y}\eth + \overline{\mathcal{Y}}\bar{\eth} + \frac{s-w}{2}\eth\mathcal{Y} - \frac{s+w}{2}\bar{\eth}\overline{\mathcal{Y}}\right]\eta. \tag{5.8}$$

When focusing on $T = 0 = \tilde{\omega}$ at the surface $u = 0$ and on the homogeneous part of the transformations, this gives the weights summarized in Tables 1 and 2. These tables are extended to the Lie algebra elements, which are passive in all our computations, by writing $[Y, \widetilde{T}] = -\delta_{Y,\overline{Y}}\widetilde{T}$ and $[Y, Y']^A = -\delta_{Y,\overline{Y}} Y'^A$.

6. Surface Charge Algebra

In this section, $\tilde{\omega} = 0$ so that $f = T + (1/2)u\varphi$ and we use the notation $s = (\mathcal{Y}, \overline{\mathcal{Y}}, T)$ for elements of the symmetry algebra, which is given in these terms by $[s_1, s_2] = \hat{s}$ where

$$\begin{gathered}\hat{\mathcal{Y}} = \mathcal{Y}_1\eth\mathcal{Y}_2 - (1 \longleftrightarrow 2), \qquad \hat{\overline{\mathcal{Y}}} = \overline{\mathcal{Y}}_1\bar{\eth}\overline{\mathcal{Y}}_2 \quad (1 \longleftrightarrow 2), \\ \hat{T} = \left(\mathcal{Y}_1\eth + \overline{\mathcal{Y}}_1\bar{\eth}\right)T_2 - \frac{1}{2}\varphi_1 T_2 - (1 \longleftrightarrow 2).\end{gathered} \tag{6.1}$$

The translation of the charges, the nonintegrable piece due to the news, and the central charges computed in [16] is given here

$$Q_s[\mathcal{X}] = -\frac{1}{8\pi G}\int d^2\Omega^{\varphi}\left[\left(f\left(\Psi_2^0 + \sigma^0\dot{\overline{\sigma}}^0\right) + \mathcal{Y}\left(\Psi_1^0 + \sigma^0\eth\overline{\sigma}^0 + \frac{1}{2}\eth\left(\sigma^0\overline{\sigma}^0\right)\right)\right) + \text{c.c.}\right],$$

$$\Theta_s[\delta\mathcal{X},\mathcal{X}] = \frac{1}{8\pi G}\int d^2\Omega^{\varphi} f\left[\dot{\overline{\sigma}}^0\delta\sigma^0 + \text{c.c.}\right], \tag{6.2}$$

$$K_{s_1,s_2}[\mathcal{X}] = \frac{1}{8\pi G}\int d^2\Omega^{\varphi}\left[\left(\frac{1}{4}f_1\eth f_2\overline{\eth}R + \frac{1}{2}\overline{\sigma}^0 f_1\eth^2\varphi_2 - (1\longleftrightarrow 2)\right) + \text{c.c.}\right].$$

Note that one could also write the charges $Q_s[\mathcal{X}]$ by allowing for the additional terms $((1/2)\eth^2\overline{\sigma}^0 - (1/2)\overline{\eth}^2\sigma^0)$ in the first parenthesis since these terms were cancelled with the corresponding terms in the complex conjugate expression. Note also that not Ψ_2^0 but only $\Psi_2^0 + \overline{\Psi}_2^0$ is free data on-shell because of the relation (4.17).

We recognize all the ingredients of the surface charges described in [27], which in turn have been related there to previous expressions in the literature and, in particular, to the twistorial approach of Penrose [28]. More precisely, up to conventions, $Q_{0,0,T}$ agrees with Geroch's linear supermomentum [8] $Q_{gn} + \overline{Q}_{gn}$, as given in (A1.12) of [27]. The angular (super-)momentum that we get is

$$Q_{\mathcal{Y},0,0} = -\frac{1}{8\pi G}\int d^2\Omega^{\varphi}\mathcal{Y}\left[\Psi_1^0 + \sigma^0\eth\overline{\sigma}^0 + \frac{1}{2}\eth\left(\sigma^0\overline{\sigma}^0\right) - \frac{u}{2}\eth\left(\Psi_2^0 + \overline{\Psi}_2^0 + \partial_u\left(\sigma^0\overline{\sigma}^0\right)\right)\right]. \tag{6.3}$$

It differs from Q_{η_c} given in (4) of [27] by the explicitly u-dependent term of the second line. It thus has a similar structure to Penrose's angular momentum as described in (11), (12), and (17a) of [27] in the sense that it also differs by a specific amount of linear supermomentum, but the amount is different and explicitly u-dependent,

$$Q_{\mathcal{Y},0,0} = Q^{u=0}_{\mathcal{Y},0,0} + \frac{1}{2}uQ_{0,0,\eth\mathcal{Y}}. \tag{6.4}$$

The main result derived in [16] states that

(i) if one is allowed to integrate by parts,

$$\int d^2\Omega^{\varphi}\eth\eta^{-1} = 0 = \int d^2\Omega^{\varphi}\,\overline{\eth}\eta^1, \tag{6.5}$$

where $d^2\Omega^{\varphi} = (2d\zeta\wedge d\overline{\zeta})/iP^2$,

(ii) if one defines the "Dirac bracket" through

$$\{Q_{s_1}, Q_{s_2}\}^*[\mathcal{X}] = -\delta_{s_2}Q_{s_1}[\mathcal{X}] + \Theta_{s_2}[-\delta_{s_1}\mathcal{X},\mathcal{X}], \tag{6.6}$$

then the charges define a representation of the $\mathfrak{bms}_4$ algebra, up to a field-dependent central extension,

$$\{Q_{s_1}, Q_{s_2}\}^* = Q_{[s_1,s_2]} + K_{s_1,s_2}, \tag{6.7}$$

where K_{s_1,s_2} satisfies the generalized cocycle condition

$$K_{[s_1,s_2],s_3} - \delta_{s_3} K_{s_1,s_2} + \text{cyclic}(1,2,3) = 0. \tag{6.8}$$

The representation theorem contained in (6.7) and (6.8) can be verified directly in the present context by starting from (6.2), (4.17) and using the properties (4.13), (6.5) of $\eth$, the evolution equations (4.16), the conformal Killing equations (5.1), the $\mathfrak{bms}_4$ algebra (6.1), and the transformation laws (5.4).

Several remarks are in order as follows.

(i) Integrations by parts are justified for regular functions on the sphere and thus for $\mathfrak{bms}_4^{\text{glob}}$ and regular solutions. In the case of Laurent series more care is needed, see for example, [29]. We will address this question elsewhere.

(ii) For the globally well-defined $\mathfrak{bms}_4^{\text{glob}}$ algebra on the sphere, the central charge K_{s_1,s_2} vanishes.

(iii) The nonconservation of the charges follows by taking $s_2 = (0,0,1)$ and $s_1 = s$. Indeed, since $(d/du)Q_s = (\partial/\partial u)Q_s - \delta_{(0,0,1)}Q_s$, the equality of the right hand sides of (6.6) and (6.7) gives

$$\frac{d}{du}Q_s = -\frac{1}{8\pi G}\int d^2\Omega^{\varphi}\left[\dot{\bar{\sigma}}^0\left(-\delta_s\sigma^0\right) + \frac{1}{4}\eth f \bar{\eth}\bar{R} + \frac{1}{2}\bar{\sigma}^0\eth^2\psi + \text{c.c.}\right]. \tag{6.9}$$

For $s = (0,0,1)$, this gives the standard Bondi-Sachs mass loss formula,

$$\frac{d}{du}Q_{0,0,1} = -\frac{1}{8\pi G}\int d^2\Omega^{\varphi}\left[\dot{\bar{\sigma}}^0\dot{\sigma}^0 + \text{c.c.}\right]. \tag{6.10}$$

It also follows that the standard $\mathfrak{bms}_4^{\text{glob}}$ charges are all conserved on the sphere in the absence of news.

To the best of our knowledge, except for the previous analysis in the BMS gauge, the above representation result does not exist elsewhere in the literature. A more detailed discussion of its implications, a detailed comparison with results in the literature as well as a self-contained derivation of the $\mathfrak{bms}_4$ transformation laws in the context of the Newman-Penrose formalism will be given elsewhere.

Acknowledgments

The authors thank Cédric Troessaert for useful discussions. This work is supported in part by the Fund for Scientific Research-FNRS (Belgium), by the Belgian Federal Science Policy

Office through the Interuniversity Attraction Pole P6/11, by IISN-Belgium, by "Communauté française de Belgique—Actions de Recherche Concertées", and by Fondecyt Projects no. 1085322 and no. 1090753.

References

[1] H. Bondi, M. G. van der Burg, and A. W. Metzner, "Gravitational waves in general relativity. VII. Waves from axi-symmetric isolated systems," *Proceedings of the Royal Society A*, vol. 269, no. 1336, pp. 21–52, 1962.

[2] R. K. Sachs, "Gravitational waves in general relativity. 8. Waves in asymptotically flat space-times," *Proceedings of the Royal Society A*, vol. 270, pp. 103–126, 1962.

[3] E. T. Newman and T. W. J. Unti, "Behavior of asymptotically flat empty spaces," *Journal of Mathematical Physics*, vol. 3, no. 5, pp. 891–901, 1962.

[4] R. K. Sachs, "Asymptotic symmetries in gravitational theory," *Physical Review*, vol. 128, pp. 2851–2864, 1962.

[5] G. Barnich and C. Troessaert, "Symmetries of asymptotically flat four-dimensional spacetimes at null infinity revisited," *Physical Review Letters*, vol. 105, no. 11, Article ID 111103, 2010.

[6] G. Barnich and C. Troessaert, "Aspects of the BMS/CFT correspondence," *Journal of High Energy Physics*, vol. 1005, article 062, 2010.

[7] G. Barnich and C. Troessaert, "Supertranslations call for superrotations," *Proceedings of Science CNCFG*, 010, 2010, http://arxiv.org/abs/1102.4632.

[8] R. Geroch, "Asymptotic structure of space-time," in *Symposium on the Asymptotic Structure of Space-Time*, P. Esposito and L. Witten, Eds., pp. 1–105, Plenum, New York, NY, USA, 1977.

[9] R. Penrose, "Asymptotic properties of fields and space-times," *Physical Review Letters*, vol. 10, no. 2, pp. 66–68, 1963.

[10] C. Imbimbo, A. Schwimmer, S. Theisen, and S. Yankielowicz, "Diffeomorphisms and holographic anomalies," *Classical and Quantum Gravity*, vol. 17, no. 5, pp. 1129–1138, 2000.

[11] G. Barnich, "A note on gauge systems from the point of view of Lie algebroids," *AIP Conference Proceedings*, vol. 1307, pp. 7–18, 2010.

[12] J. Foster, "Conformal structure of i^+ and asymptotic symmetry. I. Definitions and local theory," *Journal of Physics A*, vol. 11, no. 1, pp. 93–102, 1978.

[13] J. Foster, "Asymptotic symmetry and the global structure of future null infinity," *International Journal of Theoretical Physics*, vol. 26, pp. 1107–1124, 1987.

[14] E. Newman and R. Penrose, "An approach to gravitational radiation by a method of spin coefficients," *Journal of Mathematical Physics*, vol. 3, no. 3, pp. 566–578, 1962.

[15] E. P. Newman and K. P. Tod, "Asymptotically flat space-times," in *General Relativity and Gravitation. 100 Years after the Birth of Albert Einstein. Volume 2*, pp. 1–36, Plenum Press, 1980.

[16] G. Barnich and C. Troessaert, "BMS charge algebra," *Journal of High Energy Physics*, vol. 1112, article 105, 2011.

[17] R. Penrose, "Conformal treatment of infinity," in *Relativity, Groups and Topology: Les Houches 1963*, B. D. C. DeWitt, Ed., pp. 563–584, Gordon and Breach, 1964.

[18] R. Penrose, "Zero rest mass fields including gravitation: asymptotic behavior," *Proceedings of the Royal Society A*, vol. 284, pp. 159–203, 1965.

[19] L. A. Tamburino and J. H. Winicour, "Gravitational fields in finite and conformal Bondi frames," *Physical Review*, vol. 150, no. 4, pp. 1039–1053, 1966.

[20] R. Penrose, "Relativistic symmetry groups," in *Group Theory in Non-Linear Problems*, A. O. Barut, Ed., pp. 1–58, Reidel Publishing Company, Dodrecht, The Netherlands, 1974.

[21] B. Schmidt, M. Walker, and P. Sommers, "A characterization of the Bondi-Metzner-Sachs group," *General Relativity and Gravitation*, vol. 6, no. 5, pp. 489–497, 1975.

[22] S. J. Fletcher and A. W. C. Lun, "The Kerr spacetime in generalized Bondi-Sachs coordinates," *Classical and Quantum Gravity*, vol. 20, no. 19, pp. 4153–4167, 2003.

[23] J. A. V. Kroon, "A comment on the outgoing radiation condition for the gravitational field and the peeling theorem," *General Relativity and Gravitation*, vol. 31, no. 8, pp. 1219–1224, 1999.

[24] E. T. Newman and R. Penrose, "Note on the Bondi-Metzner-Sachs group," *Journal of Mathematical Physics*, vol. 7, no. 5, pp. 863–870, 1966.

[25] R. Penrose and W. Rindler, *Spinors and Space-Time, Volume 2: Spinor and Twistor Methods in Space-Time Geometry*, Cambridge University Press, 1986.
[26] A. Held, E. T. Newman, and R. Posadas, "The Lorentz group and the sphere," *Journal of Mathematical Physics*, vol. 11, no. 11, pp. 3145–3154, 1970.
[27] T. Dray and M. Streubel, "Angular momentum at null infinity," *Classical and Quantum Gravity*, vol. 1, no. 1, pp. 15–26, 1984.
[28] R. Penrose, "Quasi-local mass and angular momentum in general relativity," *Proceedings of the Royal Society A*, vol. 381, no. 1780, pp. 53–63, 1982.
[29] E. H. Saidi and M. Zakkari, "Harmonic distributions, Diff(S2), and Virasoro algebra," Tech. Rep. IC/90/257, ICTP, 1990.

8

A New Approach to Black Hole Quasinormal Modes: A Review of the Asymptotic Iteration Method

H. T. Cho,[1] A. S. Cornell,[2] Jason Doukas,[3] T.-R. Huang,[1] and Wade Naylor[4]

[1] *Department of Physics, Tamkang University, Tamsui, New Taipei City 25137, Taiwan*
[2] *National Institute for Theoretical Physics, School of Physics, University of the Witwatersrand, Wits 2050, South Africa*
[3] *Yukawa Institute for Theoretical Physics, Kyoto University, Kyoto 606-8502, Japan*
[4] *International College and Department of Physics, Osaka University, Toyonaka, Osaka 560-0043, Japan*

Correspondence should be addressed to A. S. Cornell, alan.cornell@wits.ac.za

Academic Editor: Ricardo Weder

We discuss how to obtain black hole quasinormal modes (QNMs) using the asymptotic iteration method (AIM), initially developed to solve second-order ordinary differential equations. We introduce the standard version of this method and present an improvement more suitable for numerical implementation. We demonstrate that the AIM can be used to find radial QNMs for Schwarzschild, Reissner-Nordström (RN), and Kerr black holes in a unified way. We discuss some advantages of the AIM over the continued fractions method (CFM). This paper presents for the first time the spin 0, 1/2 and 2 QNMs of a Kerr black hole and the gravitational and electromagnetic QNMs of the RN black hole calculated via the AIM and confirms results previously obtained using the CFM. We also present some new results comparing the AIM to the WKB method. Finally we emphasize that the AIM is well suited to higher-dimensional generalizations and we give an example of doubly rotating black holes.

1. Introduction

The study of quasinormal modes (QNMs) of black holes is an old and well-established subject, where the various frequencies are indicative of both the parameters of the black hole and the type of emissions possible. Initially the calculation of these frequencies was done in a purely numerical way, which requires selecting a value for the complex frequency, integrating the differential equation, and checking whether the boundary conditions are satisfied. Note that in the following we will use the definition that QNMs are defined as solutions of the perturbed field equations with boundary conditions:

$$\psi(x) \longrightarrow \begin{cases} e^{-i\omega x}, & x \longrightarrow -\infty, \\ e^{i\omega x}, & x \longrightarrow \infty, \end{cases} \tag{1.1}$$

for an $e^{-i\omega t}$ time dependence (which corresponds to ingoing waves at the horizon and outgoing waves at infinity). Also note the boundary condition as $x \rightarrow \infty$ does not apply to asymptotically anti-de-Sitter spacetimes, where instead something like a Dirichlet boundary condition is imposed, for example, see [1]. Since those conditions are not satisfied in general, the complex frequency plane must be surveyed for discrete values that lead to QNMs. This technique is time consuming and cumbersome, making it difficult to systematically survey the QNMs for a wide range of parameter values. Following early work by Vishveshwara [2], Chandrasekhar and Detweiler [3] pioneered this method for studying QNMs.

In order to improve on this, a few semianalytic analyses were also attempted. In one approach, employed by Ferrari and Mashhoon et al. [4], the potential barrier in the effective one-dimensional Schrödinger equation is replaced by a parameterized analytic potential barrier function for which simple exact solutions are known. The overall shape approximates that of the true black hole barrier, and the parameters of the barrier function are adjusted to fit the height and curvature of the true barrier at the peak. The resulting estimates for the QNM frequencies have been applied to the Schwarzschild, Reissner-Nordström, and Kerr black holes, with agreement within a few percent with the numerical results of Chandrasekhar and Detweiler in the Schwarzschild case [3], and with Gunter [5] in the Reissner-Nordström case. However, as this method relies upon a specialized barrier function, there is no systematic way to estimate the errors or to improve the accuracy.

The method by Leaver [6], which is a hybrid of the analytic and the numerical, successfully generates QNM frequencies by making use of an analytic infinite-series representation of the solutions, together with a numerical solution of an equation for the QNM frequencies which involves, typically by applying a Frobenius series solution approach, the use of continued fractions. This technique is known as the continued fraction method (CFM).

Historically, another commonly applied technique is the WKB approximation [7–9]. Even though it is based on an approximation, this approach is powerful as the WKB approximation is known in many cases to be more accurate and can be carried to higher orders, either as a means to improve accuracy or as a means to estimate the errors explicitly. Also it allows a more systematic study of QNMs than has been possible using outright numerical methods. The WKB approximation has since been extended to sixth order [10].

However, all of these approaches have their limitations, where in recent years a new method has been developed which can be more efficient in some cases, called the asymptotic iteration method (AIM). Previously this method was used to solve eigenvalue problems [11, 12] as a semi analytic technique for solving second-order homogeneous linear differential equations. It has also been successfully shown by some of the current authors that the AIM is an efficient and accurate technique for calculating QNMs [13].

As such, we will review the AIM as applied to a variety of black hole spacetimes, making (where possible) comparisons with the results calculated by the WKB method and the CFM á la Leaver [6]. Therefore, the structure of this paper will be as following: In Section 2 we will review the AIM and the improved method of Ciftci et al. [11, 12] (also see [14]), along with a discussion of how the QNM boundary conditions are ensured. Applications to simple concrete examples, such as the harmonic oscillator and the Poschl-Teller potential, are also provided. In Section 3 the case of Schwarzschild (A)dS black holes will be discussed, developing the integer and half-spin equations. In Section 4 a review of the QNMs of the Reissner-Nordström black holes will be made, with several frequencies calculated in the AIM and compared with previous results. Section 5 will review the application of the AIM to Kerr black holes for spin 0, 1/2, 2 fields. Section 6 will discuss the spin-zero QNMs for doubly rotating black holes. We then summarize and conclude in Section 7.

2. The Asymptotic Iteration Method

2.1. The Method

To begin we will now review the idea behind the AIM, where we first consider the homogeneous linear second-order differential equation for the function $\chi(x)$:

$$\chi'' = \lambda_0(x)\chi' + s_0(x)\chi, \tag{2.1}$$

where $\lambda_0(x)$ and $s_0(x)$ are functions in $C_\infty(a,b)$. In order to find a general solution to this equation, we rely on the symmetric structure of the right-hand side of (2.1) [11, 12]. If we differentiate (2.1) with respect to x, we find that

$$\chi''' = \lambda_1(x)\chi' + s_1(x)\chi, \tag{2.2}$$

where

$$\lambda_1 = \lambda_0' + s_0 + (\lambda_0)^2, \qquad s_1 = s_0' + s_0\lambda_0. \tag{2.3}$$

Taking the second derivative of (2.1) we get

$$\chi'''' = \lambda_2(x)\chi' + s_2(x)\chi, \tag{2.4}$$

where

$$\lambda_2 = \lambda_1' + s_1 + \lambda_0\lambda_1, \qquad s_1 = s_0' + s_0\lambda_0. \tag{2.5}$$

Iteratively, for the $(n+1)$th and the $(n+2)$th derivatives, $n = 1, 2, \ldots$, we have

$$\chi^{(n+1)} = \lambda_{n-1}(x)\chi' + s_{n-1}(x)\chi, \tag{2.6}$$

and thus bringing us to the crucial observation in the AIM that differentiating the pervious equation n times with respect to x leaves a symmetric form for the right-hand side:

$$\chi^{(n+2)} = \lambda_n(x)\chi' + s_n(x)\chi, \tag{2.7}$$

where

$$\lambda_n(x) = \lambda_{n-1}'(x) + s_{n-1}(x) + \lambda_0(x)\lambda_{n-1}(x), \qquad s_n(x) = s_{n-1}'(x) + s_0(x)\lambda_{n-1}(x). \tag{2.8}$$

For sufficiently large n the asymptotic aspect of the "method" is introduced, that is

$$\frac{s_n(x)}{\lambda_n(x)} = \frac{s_{n-1}(x)}{\lambda_{n-1}(x)} \equiv \beta(x), \tag{2.9}$$

where the QNMs are obtained from the "quantization condition":

$$\delta_n = s_n\lambda_{n-1} - s_{n-1}\lambda_n = 0, \tag{2.10}$$

which is equivalent to imposing a termination to the number of iterations [14]. From the ratio of the $(n+1)$th and the $(n+2)$th derivatives, we have

$$\frac{d}{dx}\ln\left(\chi^{(n+1)}\right) = \frac{\chi^{(n+2)}}{\chi^{(n+1)}} = \frac{\lambda_n\left(\chi' + (s_n/\lambda_n)\chi\right)}{\lambda_{n-1}\left(\chi' + (s_{n-1}/\lambda_{n-1})\chi\right)}. \tag{2.11}$$

From our asymptotic limit, this reduces to

$$\frac{d}{dx}\ln\left(\chi^{(n+1)}\right) = \frac{\lambda_n}{\lambda_{n-1}}, \tag{2.12}$$

which yields

$$\chi^{(n+1)}(x) = C_1 \exp\left(\int^x \frac{\lambda_n(x')}{\lambda_{n-1}(x')}dx'\right) = C_1\lambda_{n-1}\exp\left(\int^x (\beta + \lambda_0)dx'\right), \tag{2.13}$$

where C_1 is the integration constant and the right-hand side of (2.8) and the definition of $\beta(x)$ have been used. Substituting this into (2.6), we obtain the first-order differential equation:

$$\chi' + \beta\chi = C_1 \exp\left(\int^x (\beta + \lambda_0)dx'\right), \tag{2.14}$$

which leads to the general solution:

$$\chi(x) = \exp\left[-\int^x \beta(x')dx'\right]\left(C_2 + C_1\int^x \exp\left\{\int^{x'} [\lambda_0(x'') + 2\beta(x'')]dx''\right\}dx'\right). \tag{2.15}$$

The integration constants, C_1 and C_2, can be determined by an appropriate choice of normalisation. Note that for the generation of exact solutions $C_1 = 0$.

2.2. The Improved Method

Ciftci et al. [11, 12] were among the first to note that an unappealing feature of the recursion relations in (2.8) is that at each iteration one must take the derivative of the s and λ terms of the previous iteration. This can slow the numerical implementation of the AIM down considerably and also lead to problems with numerical precision.

To circumvent these issues we developed an improved version of the AIM which bypasses the need to take derivatives at each step [13]. This greatly improves both the accuracy and speed of the method. We expand the λ_n and s_n in a Taylor series around the point at which the AIM is performed, ξ:

$$\lambda_n(\xi) = \sum_{i=0}^{\infty} c_n^i (x-\xi)^i,$$
$$s_n(\xi) = \sum_{i=0}^{\infty} d_n^i (x-\xi)^i, \tag{2.16}$$

where the c_n^i and d_n^i are the ith Taylor coefficients of $\lambda_n(\xi)$ and $s_n(\xi)$, respectively. Substituting these expressions into (2.8) leads to a set of recursion relations for the coefficients:

$$c_n^i = (i+1)c_{n-1}^{i+1} + d_{n-1}^i + \sum_{k=0}^{i} c_0^k c_{n-1}^{i-k},$$
$$d_n^i = (i+1)d_{n-1}^{i+1} + \sum_{k=0}^{i} d_0^k c_{n-1}^{i-k}. \tag{2.17}$$

In terms of these coefficients the "quantization condition" equation (2.10) can be reexpressed as

$$d_n^0 c_{n-1}^0 - d_{n-1}^0 c_n^0 = 0, \tag{2.18}$$

and thus we have reduced the AIM into a set of recursion relations which no longer require derivative operators.

Observing that the right-hand side of (2.17) involves terms of order at most $n-1$, one can recurse these equations until only c_0^i and d_0^i terms remain (i.e., the coefficients of λ_0 and s_0 only). However, for large numbers of iterations, due to the large number of terms, such expressions become impractical to compute. We avert this combinatorial problem by beginning at the $n = 0$ stage and calculating the $n+1$ coefficients sequentially until the desired number of recursions is reached. Since the quantisation condition only requires the $i = 0$ term, at each iteration n we only need to determine coefficients with $i < N - n$, where N is the maximum number of iterations to be performed. The QNMs that we calculate in this paper will be determined using this improved AIM.

2.3. Two Simple Examples

2.3.1. The Harmonic Oscillator

In order to understand the effectiveness of the AIM, it is appropriate to apply this method to a simple concrete problem: the harmonic oscillator potential in one dimension:

$$\left(-\frac{d^2}{dx^2}+x^2\right)\phi = E\phi. \tag{2.19}$$

When $|x|$ approaches infinity, the wave function ϕ must approach zero. Asymptotically the function ϕ decays like a Gaussian distribution, in which case we can write

$$\phi(x) = e^{-x^2/2}f(x), \tag{2.20}$$

where $f(x)$ is the new wave function. Substituting (2.20) into (2.19) then rearranging the equation and dividing by a common factor, one can obtain

$$\frac{d^2 f}{dx^2} = 2x\frac{df}{dx} + (1-E)f. \tag{2.21}$$

We recognise this as Hermite's equation. For convenience we let $1-E=-2j$, such that in our case $\lambda_0 = 2x$ and $s_0 = -2j$. We define

$$\delta_n = \lambda_n s_{n-1} - \lambda_{n-1}s_n, \quad \text{for } n = 1,2,3,\ldots. \tag{2.22}$$

Thus using (2.8) one can find that

$$\delta_n = 2^{n+1}\prod_{i=0}^{n}(j-i), \tag{2.23}$$

and the termination condition (2.10) can be written as $\delta_n = 0$. Hence j must be a nonnegative integer, which means that

$$E_k = 2k+1, \quad \text{for } k = 0,1,2,\ldots. \tag{2.24}$$

and this is the exact spectrum for such a potential. Moreover, the wave function $\phi(x)$ can also be derived in this method.

We will point out that in this case the termination condition, $\delta_n = 0$, is dependent only on the eigenvalue j for a given iteration number n, and this is the reason why we can obtain an exact eigenvalue. However, for the black hole cases in subsequent sections, the termination condition depends also on x, and therefore one can only obtain approximate eigenvalues by terminating the procedure after n iterations.

2.3.2. *The Poschl-Teller Potential*

To conclude this section we will also demonstrate that the AIM can be applied to the case of QNMs, which have unbounded (scattering) like potentials, by recalling that we can find QNMs for Scarf II (upside-down Poschl-Teller-like) potentials [15]. This is based on observations made by one of the current authors [16] relating QNMs from quasiexactly solvable models. Indeed bound state Poschl-Teller potentials have been used for QNM approximations previously by inverting black hole potentials [4]. However, the AIM does not require any inversion of the black hole potential as we will show.

Starting with the potential term

$$V(x) = \frac{1}{2}\operatorname{sech}^2 x, \tag{2.25}$$

and the Schrödinger equation, we obtain

$$\frac{d^2\psi}{dx^2} + \left(\omega^2 - \frac{1}{2}\operatorname{sech}^2 x\right)\psi = 0. \tag{2.26}$$

As we will also see in the following sections that it is more convenient to transform our coordinates to a finite domain. Hence, we will use the transformation $y = \tanh x$, which leads to

$$\begin{gathered}\left(1-y^2\right)\frac{d}{dy}\left[\left(1-y^2\right)\frac{d\psi}{dy}\right] + \left[\omega^2 - \frac{1}{2}\left(1-y^2\right)\right]\psi = 0,\\ \Longrightarrow \frac{d^2\psi}{dy^2} - \left(\frac{2y}{1-y^2}\right)\frac{d\psi}{dy} + \left[\frac{\omega^2}{(1-y^2)^2} - \frac{1}{2(1-y^2)}\right]\psi = 0,\end{gathered} \tag{2.27}$$

where $-1 < y < 1$. The QNM boundary conditions in (1.1) can then be implemented as follows. As $y \to 1$ we will have $\psi \sim e^{\mp i\omega x} \sim (1-y)^{\pm i\omega/2}$. Hence our boundary condition is $\psi \sim e^{i\omega x} \Rightarrow \psi \sim (1-y)^{-i\omega/2}$. Likewise, as $y \to -1$ we have $\psi \sim e^{\pm i\omega x} \sim (1+y)^{\pm i\omega/2}$ and the boundary condition $\psi \sim e^{-i\omega x} \Rightarrow \psi \sim (1+y)^{-i\omega/2}$. As such we can take the boundary conditions into account by writing

$$\psi = (1-y)^{-i\omega/2}(1+y)^{-i\omega/2}\phi, \tag{2.28}$$

and therefore we have

$$\frac{d^2\phi}{dy^2} = \frac{2y(1-i\omega)}{1-y^2}\frac{d\phi}{dy} + \frac{1-2i\omega-2\omega^2}{2(1-y^2)}\phi, \tag{2.29}$$

where

$$\begin{aligned} \lambda_0 &= \frac{2y(1 - i\omega)}{1 - y^2}, \\ s_0 &= \frac{1 - 2i\omega - 2\omega^2}{2(1 - y^2)}. \end{aligned} \tag{2.30}$$

Following the AIM procedure, that is, taking $\delta_n = 0$ successively for $n = 1, 2, \ldots$, one can obtain exact eigenvalues:

$$\omega_n = \pm\frac{1}{2} - i\left(n + \frac{1}{2}\right). \tag{2.31}$$

This exact QNM spectrum is the same as the one in [16] obtained through algebraic means.

The reader might wonder about approximate results for cases where Poschl-Teller approximations can be used, such as Schwarzschild and SdS backgrounds; for example, see [1, 4]. In fact when the black hole potential can be modeled by a Scarf like potential, the AIM can be used to find the eigenvalues exactly [15] and hence the QNMs numerically. We demonstrate this in the next section.

3. Schwarzschild (A)DS Black Holes

We will now begin the core focus of this paper, the study of black hole QNMs using the AIM. Recall that the perturbations of the Schwarzschild black holes are described by the Regge and Wheeler [17] and Zerilli [18] equations, and the perturbations of Kerr black holes are described by the Teukolsky equations [19]. The perturbation equations for Reissner-Nordström black holes were also derived by Zerilli [20] and by Moncrief [21–23]. Their radial perturbation equations all have a one-dimensional Schrödinger-like form with an effective potential.

Therefore, we will commence in the coming subsections by describing the radial perturbation equations of Schwarzschild black holes first, where our perturbed metric will be $g_{\mu\nu} = g^0_{\mu\nu} + h_{\mu\nu}$, and where $g^0_{\mu\nu}$ is spherically symmetric. As such it is natural to introduce a mode decomposition to $h_{\mu\nu}$. Typically we write

$$\Psi_{lm}(t, r, \theta, \phi) = \frac{e^{-i\omega t} u_l(r)}{r} Y_{lm}(\theta, \phi), \tag{3.1}$$

where $Y_{lm}(\theta, \phi)$ are the standard spherical harmonics. The function $u_l(r, t)$ then solves the wave equation:

$$\left(\frac{d^2}{dx^2} - \omega^2 - V_l(r)\right) u_l(r) = 0, \tag{3.2}$$

where x, defined by $dx = dr/f(r)$, are the so-called tortoise coordinates and $V(x)$ is a master potential of the following form [24]:

$$V(r) = f(r)\left[\frac{\ell(\ell+1)}{r^2} + \left(1 - s^2\right)\left(\frac{2M}{r^3} - \frac{(4-s^2)\Lambda}{6}\right)\right]. \tag{3.3}$$

In this section

$$f(r) = 1 - \frac{2M}{r} - \frac{\Lambda}{3}r^2, \tag{3.4}$$

with cosmological constant Λ. Here $s = 0, 1, 2$ denotes the spin of the perturbation: scalar, electromagnetic, and gravitational (for half-integer spin see [25–27] and Section 5.2).

3.1. The Schwarschild Asymptotically Flat Case

To explain the AIM we will start with the simplest case of the radial component of a perturbation of the Schwarzschild metric outside the event horizon [18]. For an asymptotically flat Schwarzschild solution ($\Lambda = 0$),

$$f(r) = 1 - \frac{2M}{r}, \tag{3.5}$$

where from $dx = dr/f(r)$ we have

$$x(r) = r + 2M \ln\left(\frac{r}{2M} - 1\right), \tag{3.6}$$

for the tortoise coordinate x.

Note that for the Schwarzschild background the maximum of this potential, in terms of r, is given by [28]

$$r_0 = \frac{3M}{2}\frac{1}{\ell(\ell+1)}\left[\ell(\ell+1) - \left(1-s^2\right) + \left(\ell^2(\ell+1)^2 + \frac{14}{9}\ell(\ell+1)\left(1-s^2\right) + \left(1-s^2\right)^2\right)^{1/2}\right]. \tag{3.7}$$

The choice of coordinates is somewhat arbitrary and in the next section (for SdS) we will see how an alternative choice leads to a simpler solution. Firstly, consider the change of variable:

$$\xi = 1 - \frac{2M}{r}, \tag{3.8}$$

with $0 \le \xi < 1$. In terms of ξ, our radial equation then becomes

$$\frac{d^2\psi}{d\xi^2} + \frac{1-3\xi}{\xi(1-\xi)}\frac{d\psi}{d\xi} + \left[\frac{4M^2\omega^2}{\xi^2(1-\xi)^4} - \frac{\ell(\ell+1)}{\xi(1-\xi)^2} - \frac{1-s^2}{\xi(1-\xi)}\right]\psi = 0. \tag{3.9}$$

To accommodate the out-going wave boundary condition $\psi \rightarrow e^{i\omega x} = e^{i\omega(r+2M\ln(r/2M-1))}$ as $(x, r) \rightarrow \infty$ in terms of ξ (which is the limit $\xi \rightarrow 1$) and the regular singularity at the event horizon ($\xi \rightarrow 0$), we define

$$\psi(\xi) = \xi^{-2iM\omega}(1-\xi)^{-2iM\omega}e^{2iM\omega/(1-\xi)}\chi(\xi), \tag{3.10}$$

where the Coulomb power law is included in the asymptotic behaviour (cf. [6] (5)). The radial equation then takes the following form:

$$\chi'' = \lambda_0(\xi)\chi' + s_0(\xi)\chi, \tag{3.11}$$

where

$$\begin{aligned} \lambda_0(\xi) &= \frac{4Mi\omega(2\xi^2 - 4\xi + 1) - (1-3\xi)(1-\xi)}{\xi(1-\xi)^2}, \\ s_0(\xi) &= \frac{16M^2\omega^2(\xi-2) - 8Mi\omega(1-\xi) + \ell(\ell+1) + (1-s^2)(1-\xi)}{\xi(1-\xi)^2}. \end{aligned} \tag{3.12}$$

Note that primes of χ denote derivatives with respect to ξ.

Using these expressions we have tabulated several QNM frequencies and compared them to the WKB method of [28] and the CFM of [6] in Table 1. For completeness Table 1 also includes results from an approximate semianalytic third-order WKB method [28]. More accurate semianalytic results with better agreement to Leaver's method can be obtained by extending the WKB method to 6th order [10] and indeed in Section 5 we use this to compare with the AIM for results where the CFM has not been tabulated.

It might also be worth mentioning that a different semianalytic perturbative approach has recently been discussed by Dolan and Ottewill [29], which has the added benefit of easily being extended to any order in a perturbative scheme.

3.2. The De-Sitter Case

We have presented the QNMs for Schwarzchild gravitational perturbations in Table 1 however, to further justify the use of this method, it is instructive to consider some more general cases. As such, we will now consider the Schwarzschild de Sitter (SdS) case, where we have the same WKB-like wave equation and potential as in the radial equation earlier, though now

$$f(r) = 1 - \frac{2M}{r} - \Lambda\frac{r^2}{3}, \tag{3.13}$$

where $\Lambda > 0$ is the cosmological constant. Interestingly the choice of coordinates we use here leads to a simpler AIM solution, because there is no Coulomb power law tail; however, in the limit $\Lambda = 0$ we recover the Schwarzschild results. Note that although it is possible to find an expression for the maximum of the potential in the radial equation, for the SdS case, it is the

Table 1: QNMs to 4 decimal places for gravitational perturbations ($s = 2$) where the fifth column is taken from [28]. Note that the imaginary part of the $n = 0$ and $\ell = 2$ result in [28] has been corrected to agree with [6]. [*] Note also that if the number of iterations in the AIM is increased, to say 50, then we find agreement with [6] accurate to 6 significant figures.

ℓ	n	ω_{Leaver}	ω_{AIM} (after 15 iterations)	ω_{WKB}
2	0	0.3737–0.0896i [*]	0.3737–0.0896i (<0.01%)(<0.01%)	0.3732–0.0892i (−0.13%)(0.44%)[*]
	1	0.3467–0.2739i	0.3467–0.2739i (<0.01%)(<0.01%)	0.3460–0.2749i (−0.20%)(−0.36%)
	2	0.3011–0.4783i	0.3012–0.4785i (0.03%)(−0.04%)	0.3029–0.4711i (0.60%)(1.5%)
	3	0.2515–0.7051i	0.2523–0.7023i (0.32%)(0.40%)	0.2475–0.6703i (−1.6%)(4.6%)
3	0	0.5994–0.0927i	0.5994–0.0927i (<0.01%)(<0.01%)	0.5993–0.0927i (−0.02%)(0.0%)
	1	0.5826–0.2813i	0.5826–0.2813i (<0.01%)(<0.01%)	0.5824–0.2814i (−0.03%)(−0.04%)
	2	0.5517–0.4791i	0.5517–0.4791i (<0.01%)(<0.01%)	0.5532–0.4767i (0.27%)(0.50%)
	3	0.5120–0.6903i	0.5120–0.6905i (<0.01%)(−0.03%)	0.5157–0.6774i (0.72%)(1.9%)
	4	0.4702–0.9156i	0.4715–0.9156i (0.28%)(<0.01%)	0.4711–0.8815i (0.19%)(3.7%)
	5	0.4314–1.152i	0.4360–1.147i (1.07%)(0.43%)	0.4189–1.088i (−2.9%) (5.6%)
4	0	0.8092–0.0942i	0.8092–0.0942i (<0.01%)(<0.01%)	0.8091–0.0942i (−0.01%)(0.0%)
	1	0.7966–0.2843i	0.7966–0.2843i (<0.01%)(<0.01%)	0.7965–0.2844i (−0.01%)(−0.04%)
	2	0.7727–0.4799i	0.7727–0.4799i (<0.01%)(<0.01%)	0.7736–0.4790i (0.12%)(0.19%)
	3	0.7398–0.6839i	0.7398–0.6839i (<0.01%)(<0.01%)	0.7433–0.6783i (0.47%) (0.82%)
	4	0.7015–0.8982i	0.7014–0.8985i (−0.01%)(−0.03%)	0.7072–0.8813i (0.81%)(1.9%)

solution of a cubic equation, which for brevity we refrain from presenting here. In our AIM code we use a numerical routine to find the root to make the code more general.

In the SdS case it is more convenient to change coordinates to $\xi = 1/r$ [1], which leads to the following master equation (cf. (3.3)):

$$\frac{d^2\psi}{d\xi^2} + \frac{p'}{p}\frac{d\psi}{d\xi} + \left[\frac{\omega^2}{p^2} - \frac{\ell(\ell+1) + (1-s^2)\left(2M\xi - (4-s^2)\left(\Lambda/6\xi^2\right)\right)}{p}\right]\psi = 0, \tag{3.14}$$

where we have defined

$$p = \xi^2 - 2M\xi^3 - \frac{\Lambda}{3} \Longrightarrow p' = 2\xi(1 - 3M\xi). \tag{3.15}$$

It may be worth mentioning that for SdS we can express [1]

$$e^{i\omega x} = (\xi - \xi_1)^{i\omega/2\kappa_1}(\xi - \xi_2)^{i\omega/2\kappa_2}(\xi - \xi_3)^{i\omega/2\kappa_3} \tag{3.16}$$

in terms of the roots of $f(r)$, where ξ_1 is the event horizon and ξ_2 is the cosmological horizon (and κ_n is the surface gravity at each ξ_n). This is useful for choosing the appropriate scaling behaviour for QNM boundary conditions.

Based on the pervious equation an appropriate choice for QNMs is to scale out the divergent behaviour at the cosmological horizon (note that this is opposite to the case presented in [1], where they define the QNMs as solution with boundary conditions $\psi(x) \propto e^{\mp i\omega x}$ as $x \to \pm\infty$, for $e^{i\omega x}$ time dependence):

$$\psi(\xi) = e^{i\omega x}u(\xi), \tag{3.17}$$

which implies

$$pu'' + (p' - 2i\omega)u' - \left[\ell(\ell+1) + \left(1 - s^2\right)\left(2M\xi - \left(4 - s^2\right)\frac{\Lambda}{6\xi^2}\right)\right]u = 0 \tag{3.18}$$

in term of ξ. Furthermore, based on the scaling in (3.17), the correct QNM condition at the horizon ξ_1 implies

$$u(x) = (\xi - \xi_1)^{-i\omega/\kappa_1}\chi(x), \tag{3.19}$$

where

$$\kappa_1 = \frac{1}{2}\frac{df}{dr}\bigg|_{r\to r_1} = M\xi_1^2 - \frac{1}{3}\frac{\Lambda}{\xi_1}, \tag{3.20}$$

with $\xi_1 = 1/r_1$, and r_1 is the smallest real solution of $f(r) = 0$, implying $p = 0$. The differential equation then takes the standard AIM form:

$$\chi'' = \lambda_0(\xi)\chi' + s_0(\xi)\chi, \tag{3.21}$$

where

$$\lambda_0(\xi) = -\frac{1}{p}\left[p' - \frac{2i\omega}{\kappa_1(\xi-\xi_1)} - 2i\omega\right],$$

$$s_0(\xi) = \frac{1}{p}\left[\ell(\ell+1) + \left(1-s^2\right)\left(2M\xi - \left(4-s^2\right)\frac{\Lambda}{6\xi^2}\right) + \frac{i\omega}{\kappa_1(\xi-\xi_1)^2}\left(\frac{i\omega}{\kappa_1}+1\right) + (p'-2i\omega)\frac{i\omega}{\kappa_1(\xi-\xi_1)}\right]. \quad (3.22)$$

Using these equations, we present in Table 2 results for SdS with $s = 2$.

Identical results were generated by the AIM and CFM, both after 50 iterations. Though results are presented for $n = 1, 2, 3$, and $\ell = 2, 3$ modes only, the AIM is robust enough to be applied to any other case where like the $\Lambda = 0$ case, agreement with other methods in more extreme parameter choices would only require further iterations.

As far as we are aware only [30] (who used a semianalytic WKB approach) has presented tables for general spin fields for the SdS case. We have also compared our results to those in [30] for the $s = 0, 1$ cases and find identical results (to a given accuracy in the WKB method).

It may be worth mentioning that a set of three-term recurrence relations was derived in [13] for the CFM, valid for electromagnetic and gravitational perturbations ($s = 1, 2$), while for $s = 0$ this reduces to a five-term recurrence relation. However, for the AIM we can treat the $s = 0, 1, 2$ perturbations on an equal footing; see [13] for more details. Typically n Gaussian Elimination steps are required to reduce an $n+3$ recurrence to a 3-term continued fraction; for example, for Reissner-Nördtrom see [31] and for higher-dimensional Schwarzschild backgrounds see [32] (for an application of the CFM to higher-dimensional asymptotic QNMs see [33]). However, all that is necessary in the AIM is to factor out the correct asymptotic behaviour at the horizon(s) and infinity (we showed this for higher-dimensional scalar spheroids in [34]).

3.3. *The Spin-Zero Anti-de-Sitter Case*

There are various approaches to finding QNMs for the SAdS case (an eloquent discussion is given in the appendix of [35], see also [36, 37]). One approach is that of Horowitz and Hubeny [38], which uses a series solution chosen to satisfy the SAdS QNM boundary conditions. This method can easily be applied to all perturbations ($s = 0, 1, 2$). The other approach is to use the Frobenius method of Leaver [6], but instead of developing a continued fraction the series must satisfy a boundary condition at infinity, such as a Dirichlet boundary condition [1].

The AIM does not seem easy to apply to metrics where there is an asymptotically anti-de-Sitter background, because for general spin, s, the potential at infinity is a constant and hence would include a combination of ingoing and outgoing waves, leading to a sinusoidal dependence [39]. However, for the scalar spin zero ($s = 0$) case, the potential actually blows up at infinity and is effectively a bound state problem. In this case the AIM can easily be applied as we show here in after.

Table 2: QNMs to 6 signifcant figures for Schwarzschild de Sitter gravitational perturbations ($s = 2$) for $\ell = 2$ and $\ell = 3$ modes. We only present results for the AIM method, because the results are identical to those of the CFM after a given number of iterations (in this case 50 iterations for both methods). The $n = 1, 2$ modes can be compared with the results in [30] for $s = 2$.

Λ ($\ell = 2$)	$n = 1$	$n = 2$	$n = 3$
0	0.373672–0.0889623i	0.346711–0.273915i	0.301050–0.478281i
0.02	0.338391–0.0817564i	0.318759–0.249197i	0.282732–0.429484i
0.04	0.298895–0.0732967i	0.285841–0.221724i	0.259992–0.377092i
0.06	0.253289–0.0630425i	0.245742–0.189791i	0.230076–0.319157i
0.08	0.197482–0.0498773i	0.194115–0.149787i	0.187120–0.250257i
0.09	0.162610–0.0413665i	0.160789–0.124152i	0.157042–0.207117i
0.10	0.117916–0.0302105i	0.117243–0.0906409i	0.115876–0.151102i
0.11	0.0372699–0.00961565i	0.0372493–0.0288470i	0.0372081–0.0480784i
Λ ($\ell = 3$)	$n = 1$	$n = 2$	$n = 3$
0	0.599443–0.0927030i	0.582644–0.281298i	0.551685–0.479093i
0.02	0.543115–0.0844957i	0.530744–0.255363i	0.507015–0.432059i
0.04	0.480058–0.0751464i	0.471658–0.226395i	0.455011–0.380773i
0.06	0.407175–0.0641396i	0.402171–0.192807i	0.392053–0.322769i
0.08	0.317805–0.0503821i	0.315495–0.151249i	0.310803–0.252450i
0.09	0.261843–0.0416439i	0.260572–0.124969i	0.257998–0.208412i
0.10	0.189994–0.0303145i	0.189517–0.0909507i	0.188555–0.151609i
0.11	0.0600915–0.00961888i	0.0600766–0.0288567i	0.0600469–0.0480945i

Let us consider the scalar wave equation in SAdS spacetime, where $\Lambda = -3/R^2$, and R is the AdS radius. The master equation takes the same form as for the graviational case, except that the potential becomes

$$V = \left(1 - \frac{2}{r} + r^2\right)\left(\frac{2}{r^3} + 2\right) = \frac{2(r-1)(r^2+r+2)(r^3+1)}{r^4}. \tag{3.23}$$

Here for simplicity we have taken the AdS radius $R = 1$, the mass of the black hole $M = 1$, and the angular momentum number $l = 0$. Hence the horizon radius r_+ equals 1. Thus, with this choice we can compare with the data in Table 3.2 on page 37 of [40] (see Table 3).

To implement the AIM we first look at the asymptotic behavior of ψ. As $r \to r_+ = 1$, the potential V goes to zero. In addition,

$$\psi \sim e^{\pm i[(\omega/4)\ln(r-1)]} \sim (r-1)^{\pm i\omega/4} \sim \left(1 - \frac{1}{r}\right)^{\pm i\omega/4}. \tag{3.24}$$

For QNMs we choose the *out-going* (into the black hole) boundary condition. That is,

$$\psi \sim e^{-i\omega x} \sim \left(1 - \frac{1}{r}\right)^{-i\omega/4}. \tag{3.25}$$

Table 3: Comparison of the first few QNMs to 6 significant figures for Schwarzschild anti-de-Sitter scalar perturbations ($s = 0$) for $\ell = 0$ modes with $r_+ = 1$. The second column corresponds to data [40] using the Horowitz and Hubeny (HH) method [38], while the third column is for the AIM using 70 iterations. [*] Note the mismatch for the real part of the $n = 3$ mode in [40]; we have confirmed this using the Mathematica notebook provided in [24].

n	HH method	AIM
0	2.7982–2.6712i	2.79823–2.67121i
1	4.75849–5.03757i	4.75850–5.03757i
2	6.71927–7.39449i	6.71931–7.39450i
3	8.68223[*]–9.74852i	8.68233–9.74854i
4	10.6467–12.1012i	10.6469–12.1013i
5	12.6121–14.4533i	12.6125–14.4533i
6	14.5782–16.8049i	14.5788–16.8050i
7	16.5449–19.1562i	16.5457–19.1563i

On the other extreme of our space, $r \to \infty$, the potential goes to infinity. This is a crucial difference from the case of gravitational perturbations. In that case, the potential goes to a constant. However, in the scalar case, as $r \to \infty$, $\psi \sim (1/r)^{\pm\sqrt{2}+1/2}$, and to implement the Dirichlet boundary condition, we take

$$\psi \sim \left(\frac{1}{r}\right)^{1/2+\sqrt{2}}. \tag{3.26}$$

For the AIM one possible choice of variables is

$$\xi = 1 - \frac{1}{r}, \tag{3.27}$$

and we see that to accommodate the asymptotic behaviour of the wavefunction we should take

$$\psi = \xi^{-i\omega/4}(1-\xi)^{\sqrt{2}+1/2}\chi. \tag{3.28}$$

Finally, after some work we find that the scalar perturbation equation is

$$\chi'' = \lambda_0(\xi)\chi' + s_0(\xi)\chi, \tag{3.29}$$

where

$$\lambda_0 = -\frac{-i\omega q + 2\left[-4 + 2\left(9 + 4\sqrt{2}\right)\xi - \left(21 + 10\sqrt{2}\right)\xi^2 + 4\left(2 + \sqrt{2}\right)\xi^3\right]}{2\xi q},$$

$$\begin{aligned} s_0 = \frac{1}{16\xi q^2} \Big\{ & 4i\omega\left[9 + 8\sqrt{2} - 2\left(7 + 5\sqrt{2}\right)\xi + \left(6 + 4\sqrt{2}\right)\xi^2\right]q + \omega^2(-1+\xi)^2 \\ & \times \left(-40 + 41\xi - 20\xi^2 + 4\xi^3\right) - 4\left[4 - 5\xi + 2\xi^2\right] \\ & \times \left[-8\left(3 + 2\sqrt{2}\right) + 8\left(10 + 7\sqrt{2}\right)\xi - \left(91 + 64\sqrt{2}\right)\xi^2 + \left(34 + 24\sqrt{2}\right)\xi^3\right]\Big\}, \end{aligned} \tag{3.30}$$

and $q = (-4 + 9\xi - 7\xi^2 + 2\xi^3)$. Using the AIM we find the results presented in Table 3.

4. Reissner-Nordström Black Holes

The procedure for obtaining the quasinormal frequencies of Reissner-Nordström black holes in four-dimensional spacetime is similar to that of our earlier cases. Starting with the Reissner-Nordström metric,

$$ds^2 = -f(r)dt^2 + \frac{dr^2}{f(r)} + r^2 d\theta^2 + r^2 \sin^2\theta d\phi^2, \tag{4.1}$$

where $f(r) = (1 - 1/r + Q^2/r^2)$ and $|Q| \leq 1/2$ is the charge of the black hole. If we consider perturbations exterior to the event horizon, the perturbation equations of the Reissner-Nordström (charged and nonrotating) geometry can be separated into two pairs of Schrödinger-like equations, which describe the even- and odd-parity oscillations, respectively [20–23]. They are given by

$$\left(\frac{d^2}{dx^2} - \rho^2 - V_i^{(\pm)}\right) Z_i^{(\pm)} = 0, \tag{4.2}$$

where $(+)$ corresponds to even- and $(-)$ to odd-parity modes:

$$\begin{aligned} V_i^{(-)}(r) &= \frac{\Delta}{r^5}\left(Ar - q_j + \frac{4Q^2}{r}\right), \\ V_i^{(+)}(r) &= V_i^{(-)}(r) + 2q_j \frac{d}{dx}\left(\frac{\Delta}{r^2[(l-1)(l+2)r + q_j]}\right), \end{aligned} \tag{4.3}$$

for $i = j = 1, 2$ $(i \neq j)$, where

$$\frac{dr}{dx} = \frac{\Delta}{r}, \tag{4.4}$$

$$\Delta = r^2 - r + Q^2 \equiv (r - r_+)(r - r_-),$$

$$A = l(l+1),$$

$$q_1 = \frac{1}{2}\left[3 + \sqrt{9 + 16Q^2(l-1)(l+2)}\right], \tag{4.5}$$

$$q_2 = \frac{1}{2}\left[3 - \sqrt{9 + 16Q^2(l-1)(l+2)}\right],$$

and $\rho = -i\omega$. Here ω is the frequency, l the angular momentum parameter, and r_- and r_+ the radii of the inner and outer (event) horizons of the black hole, respectively. Note that $r_+ = 1$ and $r_- = 0$ at the Schwarzschild limit ($Q = 0$); $r_+ = r_- = 1/2$ at the extremal limit ($Q = 1/2$). Here the tortoise coordinate is given by

$$x = \int \frac{r^2}{\Delta} dr = r + \frac{r_+^2}{r_+ - r_-}\ln(r - r_+) - \frac{r^2}{r_+ - r_-}\ln(r - r_-), \tag{4.6}$$

which ranges from $-\infty$ at the event horizon to $+\infty$ at spatial infinity.

The QNMs of the Reissner-Nordström black holes are ordinarily accompanied by the emission of both electromagnetic and gravitational radiation, except at the Schwarzschild limit [31, 41]. Equation (4.2) corresponds to purely gravitational perturbations for the radial wave functions $Z_2^{(\pm)}$ and purely electromagnetic perturbations for $Z_1^{(\pm)}$ at the Schwarzschild limit. Chandrasekhar [42] has shown that the solutions $Z_i^{(+)}$ for the even-parity oscillations and $Z_i^{(-)}$ for the odd-parity oscillations have the following relationship:

$$[A(A-2) - 2\rho q_j] Z_i^{(+)} = \left\{ A(A-2) + \frac{2q_j^2\Delta}{r^3[(A-2)r + q_j]} \right\} Z_i^{(-)} + 2q_j \frac{dZ_i^{(-)}}{dx}, \tag{4.7}$$

so one can just consider solutions for a specific parity, as in the Schwarzschild case, to understand the property of the black hole. Since the formalism of the effective potential $V_i^{(-)}$ in the odd-parity equation is much simpler than $V_i^{(+)}$ in the even-parity equation, it is customary to compute the QNMs for the odd-parity modes.

Note that the mass M of the Reissner-Nordström black hole has been scaled to $2M = 1$, so its quasinormal frequencies are uniquely determined by the charge Q, the angular momentum l, and the overtone number n of the mode.

The following procedure is similar to that in Section 3. At first we change r to the variable x in (4.2) for the odd-parity mode. From (4.4) we have

$$\frac{d}{dx} = \frac{\Delta}{r^2}\frac{d}{dr}, \tag{4.8}$$

$$\frac{d^2}{dx^2} = \left(\frac{\Delta}{r^2}\right)\left(\frac{r - 2Q^2}{r^3}\right)\frac{d}{dr} + \left(\frac{\Delta}{r^2}\right)^2 \frac{d^2}{dr^2}. \tag{4.9}$$

Substituting (4.9) into (4.2) for the odd-parity mode, we get

$$\left(\frac{\Delta}{r^2}\right)^2 \frac{d^2}{dr^2} Z_i^{(-)} + \left(\frac{\Delta}{r^2}\right)\left(\frac{r-2Q^2}{r^3}\right)\frac{d}{dr} Z_i^{(-)} - \left[\rho^2 + V_i^{(-)}\right] Z_i^{(-)} = 0. \qquad (4.10)$$

Considering the QNM boundary conditions in the Reissner-Nordström case

$$Z_i^{(-)} \longrightarrow \begin{cases} e^{-i\omega x}, & x \longrightarrow -\infty, \\ e^{i\omega x}, & x \longrightarrow +\infty, \end{cases} \qquad (4.11)$$

and incorporating this into the radial wave function $Z_i^{(-)}$, we have a form involving the asymptotic behaviour [31]:

$$Z_i^{(-)} = e^{-\rho r} r^{-1} (r - r_-)^{1-\rho-\rho r_+^2/(r_+-r_-)} (r - r_+)^{\rho r_+^2/(r_+-r_-)} \chi_{Z_i}(r). \qquad (4.12)$$

Differentiating (4.12) one and two times with respect to r, we have

$$Z_{i,r}^{(-)} \equiv \frac{d}{dr} Z_i^{(-)} = e^{-\rho r} r^{-1} (r - r_-)^{1-\rho-\rho r_+^2/(r_+-r_-)} (r - r_+)^{\rho r_+^2/(r_+-r_-)} \left(\chi_{Z_i,r} + \Gamma_Z \chi_{Z_i}\right),$$

$$Z_{i,rr}^{(-)} = e^{-\rho r} r^{-1} (r - r_-)^{1-\rho-\rho r_+^2/(r_+-r_-)} (r - r_+)^{\rho r_+^2/(r_+-r_-)} \left[\chi_{Z_i,rr} + 2\Gamma_Z \chi_{Z_i,r} + \left(\Gamma_Z^2 + \Gamma_{Z,r}\right)\chi_{Z_i}\right], \qquad (4.13)$$

where Γ_Z is defined by

$$\Gamma_Z = -\rho - \frac{1}{r} + \frac{(1-\rho)(r_+ - r_-) - \rho r_+^2}{(r_+ - r_-)(r - r_-)} + \frac{\rho r_+^2}{(r_+ - r_-)(r - r_+)}. \qquad (4.14)$$

Substituting (4.13) into (4.10), we obtain

$$\begin{aligned} &\left(\frac{\Delta}{r^2}\right)^2 \chi_{Z_i,rr} + \left[2\Gamma_Z \left(\frac{\Delta}{r^2}\right)^2 + \left(\frac{\Delta}{r^2}\right)\left(\frac{r-2Q^2}{r^3}\right)\right]\chi_{Z_i,r} \\ &\quad + \left\{\left(\frac{\Delta}{r^2}\right)^2 \left(\Gamma_Z^2 + \Gamma_{Z,r}\right) + \left(\frac{\Delta}{r^2}\right)\left(\frac{r-2Q^2}{r^3}\right)\Gamma_Z - \left[\rho^2 + V_i^{(-)}\right]\right\}\chi_{Z_i} = 0. \end{aligned} \qquad (4.15)$$

For the same reason as in Section 3, here we change the variable r to ξ by the definition $\xi = 1 - r_+/r$, which ranges from 0 at the event horizon to 1 at spatial infinity. Thus we have

$$\begin{aligned} \frac{d}{dr} &= \frac{(1-\xi)^2}{r_+}\frac{d}{d\xi}, \\ \frac{d^2}{dr^2} &= \frac{(1-\xi)^4}{r_+^2}\frac{d^2}{d\xi^2} - 2\frac{(1-\xi)^3}{r_+^2}\frac{d}{d\xi}. \end{aligned} \qquad (4.16)$$

Substituting (4.16) into (4.15), and rewriting the equation in the AIM form, we obtain

$$\chi_{Z_i,\xi\xi} = \lambda_{Z_i}(\xi)\chi_{Z_i,\xi} + s_{Z_i}(\xi)\chi_{Z_i}, \tag{4.17}$$

where

$$\begin{aligned}
\lambda_{Z_i}(\xi) &= \frac{2}{1-\xi} - \frac{2r_+\Gamma_Z}{(1-\xi)^2} - \frac{r_+ - 2Q^2(1-\xi)}{\Delta(1-\xi)^2},\\
s_{Z_i}(\xi) &= \frac{r_+^6\left[\rho^2 + V_i^{(-)}\right]}{\Delta^2(1-\xi)^8} - \frac{r_+\Gamma_Z}{\Delta(1-\xi)^4}\left[r_+ - 2Q^2(1-\xi)\right] - \frac{r_+^2\Gamma_Z^2}{\Delta(1-\xi)^4} - \frac{r_+\Gamma_{Z,\xi}}{(1-\xi)^2},\\
\Gamma_Z &= -\rho - \frac{1-\xi}{r_+} + \frac{\left[(1-\rho)(r_+ - r_-) - \rho r_+^2\right](1-\xi)}{(r_+ - r_-)\left[r_+ - r_-(1-\xi)\right]} + \frac{\rho r_+(1-\xi)}{(r_+ - r_-)\xi},\\
V_i^{(-)} &= \Delta\frac{(1-\xi)^5}{r_+^5}\left[\frac{Ar_+}{1-\xi} - q_j + \frac{4Q^2(1-\xi)}{r_+}\right],\\
\Delta &= \frac{r_+\xi\left[r_+ - r_-(1-\xi)\right]}{(1-\xi)^2}.
\end{aligned} \tag{4.18}$$

The numerical results to four decimal places are presented in Tables 4, 5, and 6. They are compared with ρ_{Leaver} and ρ_{WKB} from [31] and [41], respectively. The quasinormal frequencies appear as complex conjugate pairs in ρ; we list only the ones with $\text{Im}(\rho) > 0$. Note that we arrange ρ as $(\text{Im}(\rho), \text{Re}(\rho))$. In Table 6 the quasinormal frequencies obtained by the WKB method are not available. It is apparent that the quasinormal frequencies obtained by the AIM are very accurate except for $n = 2$ in the extremal case $Q = 1/2$ in Tables 4 and 6.

The QNMs of $l = 2$ and $i = 2$ in Table 4 reduce to the purely gravitational QNMs in the Schwarzschild case at $Q = 0$, while the QNMs of $l = 2$ and $i = 1$ in Table 5 reduce to the purely electromagnetic QNMs at $Q = 0$.

Some comments on the higher ($n = 2$) overtones for the Reissner-Nordström black hole for the extremal limit ($Q = 1/2$) are perhaps necessary. In general, much like the CFM the AIM begins to break down for larger overtones, requiring more iterations. However, near the extremal limit ($Q = 1/2$) the horizons become degenerate and the singularity structure of the corresponding differential (radial) equation changes [43] (the number of singular points is different in the nonextremal and the extremal cases) and causes the current implementation of the AIM (cf. (4.12)) to break down. Thus, we see in Tables 4–6 that for $Q = 0.495$ some of the values have large errors when compared to the CFM.

5. Kerr Black Holes

A rotating black hole carrying angular momentum is described by the Kerr metric (in Boyer-Lindquist coordinates) as

$$ds^2 = -\left(1 - \frac{r}{\Sigma}\right)dt^2 - \frac{2ar\sin^2\theta}{\Sigma}dt\,d\phi + \frac{\Sigma}{\Delta}dr^2\Sigma d\theta^2 + \left(r^2 + a^2 + \frac{a^2 r\sin^2\theta}{\Sigma}\right)\sin^2\theta d\phi, \tag{5.1}$$

Table 4: Reissner-Nordström quasinormal frequency parameter values ($\rho = -i\omega$) for the fundamental ($n = 0$) and two lowest overtones for $l = 2$ and $i = 2$.

n	Q	ρ_{Leaver}	ρ_{AIM}	ρ_{WKB}
0	0	(0.7473, −0.1779)	(0.7473, −0.1779) (<0.01%)(<0.01%)	(0.7463, −0.1784) (−0.13%)(−0.28%)
0	0.2	(0.7569, −0.1788)	(0.7569, −0.1788) (<0.01%)(<0.01%)	(0.7558, −0.1793) (−0.15%)(−0.28%)
0	0.4	(0.8024, −0.1793)	(0.8024, −0.1793) (<0.01%)(<0.01%)	(0.8011, −0.1797) (−0.16%)(−0.22%)
0	0.495	(0.8586, −0.1685)	(0.8586, −0.1685) (<0.01%)(<0.01%)	(0.8566, −0.1706) (−0.23%)(−1.25%)
1	0	(0.6934, −0.5478)	(0.6934, −0.5478) (<0.01%)(<0.01%)	(0.6920, −0.5478) (−0.20%)(−0.37%)
1	0.2	(0.7035, −0.5503)	(0.7035, −0.5502) (<0.01%)(0.02%)	(0.7020, −0.5522) (−0.21%)(−0.36%)
1	0.4	(0.7538, −0.5499)	(0.7538, −0.5499) (<0.01%)(<0.01%)	(0.7510, −0.5525) (−0.37%)(−0.47%)
1	0.495	(0.8070, −0.5140)	(0.8067, −0.5164) (−0.04%)(0.47%)	(0.8068, −0.5287) (0.01%)(−2.21%)
2	0	(0.6021, −0.9566)	(0.6021, −0.9566) (<0.01%)(<0.01%)	(0.6059, −0.9421) (0.63%)(1.52%)
2	0.2	(0.6129, −0.9599)	(0.6128, −0.9599) (0.02%)(<0.01%)	(0.6164, −0.9458) (0.57%)(1.47%)
2	0.4	(0.6703, −0.9531)	(0.6703, −0.9531) (<0.01%)(<0.01%)	(0.6717, −0.9455) (0.21%)(0.80%)
2	0.495	(0.7078, −0.8872)	(0.8350, −0.8347) (17.97%)(5.92%)	(0.7344, −0.9135) (2.66%)(−2.96%)

with

$$\Delta = r^2 + a^2 - 2Mr \equiv (r - r_-)(r - r_+), \qquad \Sigma = r^2 + a^2\cos^2\theta, \tag{5.2}$$

where a is the Kerr rotation parameter with $0 \leq a \leq M$, M being included as a general black hole mass. The horizons r_- and r_+ are again the inner and the outer (event) horizons, respectively. Teukolsky [19] showed that the perturbation equations in the Kerr geometry are separable, where the separated equations for the angular wave function ${}_sS_{lm}(\theta)$ and the radial wave function $R(r)$ are given by

$$\left[\left(1-u^2\right)S_{,u}\right]_{,u} + \left[a^2\omega^2u^2 - 2a\omega su + s + {}_sA_{lm} - \frac{(m+su)^2}{1-u^2}\right]{}_sS_{lm} = 0, \tag{5.3}$$

$$\Delta R_{,rr} + (s+1)(2r-1)R_{,r} + K(r)R = 0, \tag{5.4}$$

Table 5: Reissner-Nordström quasinormal frequency parameter values ($\rho = -i\omega$) for the fundamental ($n = 1$) and two lowest overtones for $l = 2$ and $i = 1$.

n	Q	ρ_{Leaver}	ρ_{AIM}	ρ_{WKB}
0	0	(0.9152, −0.1900)	(0.9152, −0.1900) (<0.01%)(<0.01%)	(0.9143, −0.1901) (−0.10%)(−0.05%)
0	0.2	(0.9599, −0.1929)	(0.9599, −0.1929) (<0.01%)(<0.01%)	(0.9590, −0.1930) (−0.09%)(−0.05%)
0	0.4	(1.1403, −0.1984)	(1.1403, −0.1981) (<0.01%)(0.15%)	(1.1395, −0.1980) (−0.07%)(−0.20%)
0	0.495	(1.3855, −0.1773)	(1.3855, −0.1773) (<0.01%)(<0.01%)	(1.3850, −0.1783) (−0.04%) (−0.56%)
1	0	(0.8731, −0.5814)	(0.8731, −0.5814) (<0.01%)(<0.01%)	(0.8717, −0.5819) (−0.16%)(−0.09%)
1	0.2	(0.9200, −0.5894)	(0.9200, −0.5894) (<0.01%)(<0.01%)	(0.9186, −0.5897) (−0.15%)(−0.05%)
1	0.4	(1.1100, −0.6021)	(1.1100, −0.6021) (<0.01%)(<0.01%)	(1.1081, −0.6014) (−0.17%)(0.12%)
1	0.495	(1.3573, −0.5350)	(1.3573, −0.5350) (<0.01%)(<0.01%)	(1.3579, −0.5423) (0.04%)(−1.36%)
2	0	(0.8024, −1.0032)	(0.8024, −1.0032) (<0.01%)(<0.01%)	(0.8046, −0.9917) (0.27%)(1.15%)
2	0.2	(0.8530, −1.0143)	(0.8530, −1.0143) (<0.01%)(<0.01%)	(0.8548, −1.0037) (0.21%)(1.05%)
2	0.4	(1.0582, −1.0263)	(1.0582, −1.0263) (<0.01%)(<0.01%)	(1.0568, −1.0181) (−0.13%)(0.80%)
2	0.495	(1.3019, −0.9024)	(1.3019, −0.9024) (<0.01%)(<0.01%)	(1.3141, −0.9222) (0.94%)(−2.19%)

where we have the following function:

$$K(r) = \frac{1}{\Delta}\left\{\left(r^2 + a^2\right)^2\omega^2 - 2am\omega r + a^2m^2 + is\left[am(2r-1) - \omega\left(r^2 - a^2\right)\right]\right\}$$
$$+2is\omega r - a^2\omega^2 - {}_sA_{lm}. \tag{5.5}$$

In $u = \cos\theta$, s is the spin weight, ${}_sA_{lm}$ is the spin-weighted separation constant for the angular equation, and m is another angular momentum parameter. For completeness the evaluation of the separation constant ${}_sA_{lm}$ using the AIM is discussed in Appendix A.

In order to use the AIM we need to solve for the angular solution in the radial equation. However, for nonzero s the effective potential of the radial equation is in general complex. A straight forward application of the AIM does not give the correct answer. In fact a similar problem occurs in both numerical [44] and WKB [7–9] methods. For this reason we will look at each of the spin cases $(0, 1/2, 2)$ separately in the following subsections.

Table 6: Reissner-Nordström quasinormal frequency parameter values ($\rho = -i\omega$) for the fundamental ($n = 1$) and two lowest overtones for $l = 1$ and $i = 1$.

n	Q	ρ_{Leaver}	ρ_{AIM}
0	0	(0.4965, −0.1850)	(0.4965, −0.1850) (<0.01%)(<0.01%)
0	0.2	(0.5238, −0.1883)	(0.5238, −0.1883) (<0.01%)(<0.01%)
0	0.4	(0.6470, −0.1965)	(0.6470, −0.1965) (<0.01%)(<0.01%)
0	0.495	(0.8428, −0.1742)	(0.8428, −0.1742) (<0.01%)(<0.01%)
1	0	(0.4290, −0.5873)	(0.4292, −0.5873) (<0.01%) (0.05%)
1	0.2	(0.4598, −0.5953)	(0.4598, −0.5953) (<0.01%)(<0.01%)
1	0.4	(0.5980, −0.6107)	(0.5980, −0.6107) (<0.01%)(<0.01%)
1	0.495	(0.7979, −0.5293)	(0.7978, −0.5280) (0.01%)(0.25%)
2	0	(0.3496, −1.0504)	(0.3495, −1.0504) (0.03%)(<0.01%)
2	0.2	(0.3832, −1.0596)	(0.3832, −1.0596) (<0.01%)(<0.01%)
2	0.4	(0.5340, −1.0660)	(0.5340, −1.0659) (<0.01%)(<0.01%)
2	0.495	(0.7104, −0.9055)	(0.6248, −1.0574) (−12.05%)(−16.78%)

5.1. The Spin-Zero Case

Because the AIM works better on a compact domain, we define a new variable $y = 1 - r_+/r$, which ranges from 0 at the event horizon ($r = r_+$) to 1 at spatial infinity. It is then necessary to incorporate the boundary conditions, expressed in the new compact domain, where

$$R(r) = \left(r^2 + a^2\right)^{-1/2} \psi(r) \tag{5.6}$$

is

$$\psi(y) = \left(1 - \frac{r_-}{r_+}(1-y)\right)^{-i\sigma_-} y^{-\sigma_+} (1-y)^{-r_+\omega} e^{i\omega(r_+/(1-y))} \chi(y). \tag{5.7}$$

By making the change of coordinates and change of function, (5.7) takes the following form:

$$\chi(y) = \lambda_0(y) + s_0(y), \tag{5.8}$$

where

$$\lambda_0 = -2\frac{1}{g}\frac{dg}{dy} - \frac{1}{f}\frac{df}{dy},$$
$$s_0 = -\frac{1}{g}\frac{d^2g}{dy^2} - \frac{1}{f}\frac{df}{dy} \times \frac{1}{g}\frac{dg}{dy} - \frac{1}{f^2}\left(\omega^2 - V\big|_{r=r_+(1-y)^{-1}}\right). \tag{5.9}$$

As aforementioned, we have defined

$$f = \left(\frac{\Delta}{r^2 + a^2}\frac{dy}{dr}\right)\Bigg|_{r=r_+(1-y)^{-1}},$$
$$g = (1-y)^{-2i\omega}\left(1 - \frac{r_-}{r_+}(1-y)\right)^{i\sigma_-} y^{-i\sigma_+} e^{i\omega r_+(1-y)^{-1}}, \tag{5.10}$$

where $\Delta = r^2 + a^2 - 2Mr$,

$$\sigma_\pm = \frac{1}{r_+ - r_-}\left[\left(r_\pm^2 + a^2\right)\omega + am\right], \tag{5.11}$$

and a is again our rotation parameter. Equation (5.8) is now in the correct form to use the AIM for QNM frequency calculations. Note that the potential V is

$$V = -\frac{1}{\Delta}\left(\left(K - 2\left(r^2 + a^2\right)\right)^2 - \Delta\lambda\right), \tag{5.12}$$

where the angular separation constant is defined via $\lambda = A_{l,m} + a^2\omega^2 - 2am\omega$. (Even though the radial and angular equations are coupled via the separation constant, ${}_0A_{l,m}$, we are able to find excellent agreement with the CFM by starting from the Schwarzschild ($a = 0$) result for our initial guess of the Kerr ($a \neq 0$) QNM solution using FindRoot in Mathematica in our AIM code (at least for $a \leq 1$)).

Presented in Tables 7 and 8 are the QNM frequencies for the scalar perturbations of the Kerr black hole with the two *extreme* (minimum and maximum) values of the angular momentum per unit mass, that is, $a = 0.00$ and $a = 0.80$. m was set to 0, while l was given values of 0, 1, and 2 and n varied accordingly.

Included in Table 7 are the numerically determined QNM frequencies published by Leaver in 1985 [6]. The percentages bracketed under each QNM frequency via the AIM are the percentage differences between the calculated value and the numerical value published by Leaver. With the exceptions of the QNM frequencies for $l = 0$, $n = 0$, and $l = 2$, $n = 2$, the AIM values correspond to the CFM up to four decimal places and even those *anomalies* differ by less than 0.30%, proving, at least in this case that the AIM is a precise semianalytical technique.

In Table 8, all three values were calculated in this work, even though published values are available for the third-order WKB(J), at least graphically, where numerical values using the CFM were taken from [45, 46]. Since the WKB(J) is a generally accepted semianalytical

Table 7: The QNM frequencies for the Scalar Perturbations of the Kerr Black Hole, with $a = 0.00$, that is, the Schwarzschild limit ($M = 1, m = 0$). Numerical data via the CFM were taken from [45, 46], where the AIM was set to run at 15 iterations.

l	n	Numerical	Third-order WKB(J)	Sixth-order WKB(J)	AIM
0	0	0.1105–0.1049i	0.1046–0.1001i	0.1105–0.1008i	0.1103–0.1046i
			(−5.34%, 9.82%)	(<0.01%, −2.91%)	(−0.18%, −0.29%)
1	0	0.2929–0.0977i	0.2911–0.0989i	0.2929–0.0978i	0.2929–0.0977i
			(−0.61%, 1.23%)	(<0.01%, 0.10%)	(<0.01%, <0.01%)
	1	0.2645–0.3063i	0.2622–0.3074i	0.2645–0.3065i	0.2645–0.3063i
			(−0.87%, 0.36%)	(<0.01%, 0.07%)	(<0.01%, <0.01%)
2	0	0.4836–0.0968i	0.4832–0.0968i	0.4836–0.0968i	0.4836–0.0968i
			(−0.08%, <0.01%)	(<0.01%, <0.01%)	(<0.01%, <0.01%)
	1	0.4639–0.2956i	0.4632–0.2958i	0.4638–0.2956i	0.4639–0.2956i
			(−0.15%, 0.07%)	(−0.02%, <0.01%)	(<0.01%, <0.01%)
	2	0.4305–0.5086i	0.4317–0.5034i	0.4304–0.5087i	0.4306–0.5086i
			(0.28%, −1.02%)	(−0.02%, 0.02%)	(0.02%, <0.01%)

Table 8: The QNM frequencies for the Scalar Perturbations of the Kerr Black Hole, with $a = 0.80$ ($M = 1$, $m = 0$). Numerical data via the CFM taken from [45, 46], where the AIM was set to run at 15 iterations.

l	n	Numerical	Third-order WKB(J)	Sixth-order WKB(J)	AIM
0	0	0.1145–0.0957i	0.1005–0.1007i	0.1211–0.0897i	0.1141–0.0939i
			(−12.2%, 5.22%)	(5.76%, −6.23%)	(−0.35%, −1.88%)
1	0	0.3067–0.0901i	0.3029–0.0891i	0.3053–0.0893i	0.3052–0.0892i
			(−1.24%, −1.11%)	(−0.46%, −0.89%)	(−0.49%, −1.00%)
	1	0.2820–0.2783i	0.2758–0.2779i	0.2821–0.2755i	0.2817–0.2756i
			(−2.20%, −0.14%)	(0.04%, −1.01%)	(−0.11%, −0.97%)
2	0	0.5071–0.0897i	0.5035–0.0885i	0.5041–0.0886i	0.5041–0.0886i
			(−0.71%, −1.34%)	(−0.59%, −1.23%)	(−0.59%, −1.23%)
	1	0.4906–0.2722i	0.4866–0.2693i	0.4885–0.2690i	0.4885–0.2689i
			(−0.82%, −1.07%)	(−0.43%, 1.18%)	(0.43%, −1.21%)
	2	0.4609–0.4634i	0.4585–0.4570i	0.4607–0.4581i	0.4606–0.4579i
			(−0.52%, −1.38%)	(−0.04%, −1.14%)	(−0.06%, −1.19%)

technique for QNM frequency calculations, the percentages below the AIM values are the differences to the sixth-order WKB(J) values. Only in the case of $l = 0$ and $n = 0$ does the AIM QNM frequency significantly differ from the sixth-order WKB(J) value. Note that in an upcoming work, further values will be presented for values of $a = 0.20$, $a = 0.40$, and $a = 0.60$, with the same variations of l and n with $M = 1$ [47].

5.2. *The Spin-Half Case*

For the spin-1/2 case we would like to know how the AIM can be used to derive an appropriate form of the Dirac equation in this spacetime background using the basis set-up by four null vectors which are the basis of the Newman-Penrose formalism; for further details see [47]. That is, in the Kerr background we adopt the following vectors as the null tetrad:

$$
\begin{aligned}
l_j &= \frac{1}{\Delta}\left(\Delta, -\rho^2, 0, -a\Delta\sin^2\theta\right),\\
n_j &= \frac{1}{2\rho^2}\left(\Delta, \rho^2, 0, -a\Delta\sin^2\theta\right),\\
m_j &= \frac{1}{\sqrt{2}\overline{\rho}}\left(ia\sin\theta, 0, -\rho^2, -i\left(r^2+a^2\right)\sin\theta\right),\\
l^j &= \frac{1}{\Delta}\left(r^2+a^2, \Delta, 0, a\right),\\
n^j &= \frac{1}{\sqrt{2}\overline{\rho}}\left(r^2+a^2, -\Delta, 0, a\right),\\
m^j &= \frac{1}{\sqrt{2}\overline{\rho}}\left(ia\sin\theta, 0, 1, -i\frac{1}{\sin\theta}\right),
\end{aligned}
\tag{5.13}
$$

where $\overline{m}_j$ and $\overline{m}^j$ are nothing but complex conjugates of m_j and m^j, respectively.

It is clear that the basis vectors basically become derivative operators when these are applied as tangent vectors to the function $e^{i(\omega t+m\phi)}$. Therefore we can write

$$
\vec{l} = D = \mathfrak{D}_0, \qquad \vec{n} = D^* = -\frac{\Delta}{2\rho^2}\mathfrak{D}_0^\dagger, \qquad \vec{m} = \delta = \frac{1}{\sqrt{2}\overline{\rho}}\mathcal{L}_0^\dagger, \qquad \vec{\overline{m}} = \delta^* = \frac{1}{\sqrt{2}\overline{\rho}^*}\mathcal{L}_0, \tag{5.14}
$$

where

$$
\begin{aligned}
\mathfrak{D}_n &= \partial_r + i\frac{K}{\Delta} + 2n\frac{r-M}{\Delta},\\
\mathfrak{D}_n^\dagger &= \partial_r - i\frac{K}{\Delta} + 2n\frac{r-M}{\Delta},\\
\mathcal{L}_n &= \partial_\theta + Q + n\cot\theta,\\
\mathcal{L}_n^\dagger &= \partial_\theta - Q + n\cot\theta,
\end{aligned}
\tag{5.15}
$$

and $K = (r^2+a^2)\omega + am$ with $Q = a\omega\sin\theta + m\csc\theta$.

The spin coefficients can be written as a combination of basis vectors in the Newman-Penrose formalism which are now expressed in terms of the elements of different components of the Kerr metric. So by combining these different components of basis vectors in a suitable manner we get the spin coefficients as

$$\begin{gathered}\kappa=\sigma=\lambda=\nu=\varepsilon=0,\\ \tilde{\rho}=-\frac{1}{\bar{\rho}^{*}},\qquad \beta=\frac{\cot\theta}{2\sqrt{2}\bar{\rho}^{*}},\qquad \pi=\frac{ia\sin\theta}{\sqrt{2}(\bar{\rho}^{*})^{2}},\\ \tau=-\frac{ia\sin\theta}{\sqrt{2}\rho^{2}},\qquad \mu=-\frac{\Delta}{2\bar{\rho}^{*}\rho^{2}},\qquad \gamma=\mu+\frac{r-M}{2\rho^{2}},\\ \alpha=\pi-\beta^{*}.\end{gathered} \tag{5.16}$$

Using the aformentioned definitions, and by choosing $f_1 = \bar{\rho}^{*}F_1$, $g_2 = \bar{\rho}G_2$, $f_2 = F_2$, and $g_1 = G_1$ (where $F_{1,2}$ and $G_{1,2}$ are a pair of spinors), the Dirac equation reduces to

$$\begin{gathered}\mathfrak{D}_0 f_1+\frac{1}{\sqrt{2}}\mathcal{L}_{1/2}f_2=0,\\ \Delta\mathfrak{D}^{\dagger}_{1/2}f_2-\sqrt{2}\mathcal{L}^{\dagger}_{1/2}f_1=0,\\ \mathfrak{D}_0 g_2-\frac{1}{\sqrt{2}}\mathcal{L}^{\dagger}_{1/2}f_1=0,\\ \Delta\mathfrak{D}^{\dagger}_{1/2}g_1+\sqrt{2}\mathcal{L}_{1/2}g_2=0.\end{gathered} \tag{5.17}$$

We separate the Dirac equation into radial and angular parts by choosing

$$\begin{aligned}f_1(r,\theta)&=R_{-1/2}(r)S_{-1/2}(\theta), & f_2(r,\theta)&=R_{1/2}(r)S_{1/2}(\theta),\\ g_1(r,\theta)&=R_{1/2}(r)S_{-1/2}(\theta), & g_2(r,\theta)&=R_{-1/2}(r)S_{1/2}(\theta).\end{aligned} \tag{5.18}$$

Replacing these f_j and g_j $(j = 1,2)$ and using λ as the separation constant, we get

$$\begin{gathered}\mathcal{L}_{1/2}S_{1/2}=-\lambda S_{-1/2},\\ \mathcal{L}^{\dagger}_{1/2}S_{-1/2}=\lambda S_{1/2},\end{gathered} \tag{5.19}$$

$$\begin{gathered}\Delta^{1/2}\mathfrak{D}_0 R_{-1/2}=\lambda\Delta^{1/2}R_{1/2},\\ \Delta^{1/2}\mathfrak{D}^{\dagger}_0\Delta^{1/2}R_{1/2}=\lambda R_{-1/2},\end{gathered} \tag{5.20}$$

where $2^{1/2}R_{-1/2}$ is redefined as $R_{-1/2}$.

Equations (5.19) and (5.20) are the angular and radial Dirac equation respectively, in a coupled form with the separation constant λ [42]. Decoupling (5.19) gives the eigenvalue/angular equation for spin half particles as

$$\left[\mathcal{L}_{1/2}\mathcal{L}^{\dagger}_{1/2}+\lambda^{2}\right]S_{-1/2}=0, \tag{5.21}$$

and $S_{1/2}$ satisfies the *adjoint* equation (obtained by replacing θ by $\pi - \theta$). (Note that this angular equation is that given by (5.3) for $s = 1/2$ and hence using the method in Appendix A, we could solve this numerically). Decoupling (5.20) then gives the radial equation for spin-half particles as

$$\left[\Delta \mathfrak{D}_{1/2}^{\dagger} \mathfrak{D}_0 - \lambda^2\right] R_{-1/2} = 0, \tag{5.22}$$

and $\Delta^{1/2} R_{1/2}$ satisfies the complex-conjugate equation. Furthermore, unlike the case of a scalar particle, a spin-half particle is not capable of extracting energy from a rotating black hole; that is, there is no Penrose Process (superradiance) equivalent scenario [47].

Returning now to the AIM, recall that it will work better on a compact domain, where we define a new variable $y^2 = 1 - r_+/r$, which ranges from 0 at the event horizon ($r = r_+$) to 1 at spatial infinity. It is then necessary to incorporate the boundary conditions, expressed in the new compact domain, where

$$\psi(y) = \left(1 - \frac{r_-}{r_+}\left(1 - y^2\right)\right)^{-1/2 - i\sigma_-} \left(y^2\right)^{1/2 - \sigma_+} \left(1 - y^2\right)^{-r_+\omega} e^{i\omega(r_+/(1-y^2))} \chi(y), \tag{5.23}$$

where we have defined

$$\psi = \sqrt{\Delta} R_{1/2} + R_{-1/2}, \tag{5.24}$$

ψ satisfies the WKB(J)-like equation:

$$\frac{d^2\psi}{dy^2} + \left(\omega^2 - V\right)\psi = 0 \tag{5.25}$$

with potential

$$V = \lambda^2 \frac{\Delta}{\overline{K}^2} + \lambda \frac{d}{dx}\left(\frac{\sqrt{\Delta}}{\overline{K}}\right), \tag{5.26}$$

and $\overline{K} = K/\omega = (r^2 + a^2) + am/\omega$ (for more details see [47]).

By making the change of coordinates and change of functions, our equation takes the following form:

$$\chi(y) = \lambda_0(y) + s_0(y), \tag{5.27}$$

where as in Section 5.1 we have

$$\begin{aligned} \lambda_0 &= -2\frac{1}{g}\frac{dg}{dy} - \frac{1}{f}\frac{df}{dy}, \\ s_0 &= -\frac{1}{g}\frac{d^2 g}{dy^2} - \frac{1}{f}\frac{df}{dy} \times \frac{1}{g}\frac{dg}{dy} - \frac{1}{f^2}\left(\omega^2 - V\Big|_{r = r_+(1-y^2)^{-1}}\right). \end{aligned} \tag{5.28}$$

Presented in Tables 9 and 10 are the QNM frequencies for the spin-half perturbations of the Kerr black hole with the two *extreme* values of the angular momentum per unit mass, that is, $a = 0.00$ and $a = 0.80$, m was set to 0, while l was given values of 0, 1, and 2 and n varied accordingly.

Included in Table 9 are the numerically determined QNM frequencies published by Jing and Pan [48]. Even though the WKB method has been used to calculate the Schwarzschild limit QNM frequencies before [27], the sixth-order WKB values and AIM values are novel to this work and will be explored more fully in [47]. The percentages bracketed under each QNM frequency is the percentage difference between the calculated value and the numerical value published by Jing and Pan [48].

As expected, since there are additional correction terms, the sixth-order WKB QNM frequencies are closer to the numerical values than the third-order WKB values. While the AIM does not prove as accurate in its calculation of the spin-half QNM frequencies as it did with the scalar values (both for 15 iterations), none of the differences between the AIM values and the numerical values exceed 0.30%, except for when $l = 2$ and $n = 2$ (better accuracy can be achieved by increasing the number of iterations).

Similarly in Table 10 are the numerically determined QNM frequencies published by Jing and Pan [48]. Both the third- and sixth-order WKB values along with the AIM values are novel to this work and will also be explored more fully in [47]. The percentages bracketed under each QNM frequency are the percentage differences between the calculated value and the numerical value published by Jing and Pan, at least for $l = 0$ and $l = 1$. For $l = 2$, the AIM values are compared to the sixth-order WKB values. As already noted, since there are additional correction terms, the sixth-order WKB QNM frequencies are closer to the numerical values than the third-order WKB values. Again the AIM does not appear to be as precise in calculating the QNM frequencies for spin-half perturbations of the Kerr black hole as it was for the scalar perturbations (at least for 15 iterations). As we mentioned, additional tables and plots of these Kerr processes will constitute a future work [47].

5.3. The Spin-Two Case

As we have mentioned earlier, the radial equation for nonzero spin s is in general complex. In fact, it does not even reduce to the Regge-Wheeler and Zerilli equations when the rotation parameter is $a \to 0$. Detweiler [44] has found a way to overcome this problem, where he defined a new function

$$X = \Delta^{s/2}\left(r^2 + a^2\right)^{1/2}\left[\alpha(r)R + \beta(r)\Delta^{s+1}\frac{dR}{dr}\right]. \tag{5.29}$$

If the functions $\alpha(r)$ and $\beta(r)$ are required to satisfy

$$\alpha^2 - \alpha'\beta\Delta^{s+1} + \alpha\beta'\Delta^{s+1} - \beta^2\Delta^{2s+1}K = \text{constant}, \tag{5.30}$$

then it can be shown that the radial equation in (5.4) becomes

$$\frac{d^2X}{dx^2} - VX = 0, \tag{5.31}$$

Table 9: The QNM frequencies for the spin half perturbations of the Kerr black hole, with $a = 0.00$, that is, the Schwarzschild limit ($M = 1, m = 0$). Numerical data via the CFM were taken from [48], where the AIM was set to run at 15 iterations.

l	n	Numerical	Third-order WKB	Sixth-order WKB	AIM
0	0	0.1830–0.0970i	0.1765–0.1001i (−3.55%, 3.20%)	0.1827–0.0949i (−0.16%, −2.16%)	0.1830–0.0969i (<0.01%, −0.10%)
1	0	0.3800–0.0964i	0.3786–0.0965i (−0.37%, 0.10%)	0.3801–0.0964i (0.03%, <0.01%)	0.3800–0.0964i (<0.01%, <0.01%)
	1	0.3558–0.2975i	0.3536–0.2987i (−0.62%, 0.40%)	0.3559–0.2973i (0.03%, −0.07%)	0.3568–0.2976i (0.28%, 0.03%)
2	0	0.5741–0.0963i	0.5737–0.0963i (−0.07%,<0.01%)	0.5741–0.0963i (<0.01%, <0.01%)	0.5741–0.0963i (<0.01%, <0.01%)
	1	0.5570–0.2927i	0.5562–0.2930i (−0.14%, 0.10%)	0.5570–0.2927i (<0.01%, <0.01%)	0.5573–0.2928i (0.05%, 0.03%)
	2	0.5266–0.4997i	0.5273–0.4972i (0.13%, −0.50%)	0.5265–0.4997i (−0.02%, <0.01%)	0.5189–0.5213i (−1.46%, 4.32%)

Table 10: The QNM frequencies for the spin half perturbations of the Kerr black hole, with $a = 0.80$ ($M = 1, m = 0$). Numerical data via the CFM were taken from [48], where the AIM was set to run at 15 iterations.

l	n	Numerical	Third-order WKB(J)	Sixth-order WKB(J)	AIM
0	0	0.1932–0.0891i	0.1883–0.0896i (−2.54%, 0.56%)	0.1914 –0.0865i (−0.93%, −2.92%)	0.1920–0.0872i (−0.62%, −2.13%)
1	0	0.3993–0.0893i	0.3956–0.0881i (−0.93%, −1.34%)	0.3967–0.0880i (−0.65%, −1.46%)	0.3965–0.0880i (−0.70%, −1.46%)
	1	0.3789–0.2728i	0.3751–0.2701i (−1.00%, −0.99%)	0.3777–0.2687i (−0.32%, −1.50%)	0.3764 − 0.2517i (−0.66%, −7.73%)
2	0		0.5984–0.0881i	0.5987–0.0881i	0.5987–0.0882i (<0.01%, 0.11%)
	1		0.5844–0.2669i	0.5855–0.2667i	0.5846–0.2644i (−0.15%, −0.86%)
	2		0.5600–0.4512i	0.5609–0.4517i	0.6023–0.4260i (7.38%, −5.70%)

where

$$V = \frac{\Delta U}{(r^2 + a^2)^2} + G^2 + \frac{dG}{dx},$$
$$G = \frac{s(2r-1)}{2(r^2+a^2)} + \frac{r\Delta}{(r^2+a^2)^2},$$
$$U = K + \frac{2\alpha' + (\beta'\Delta^{s+1})'}{\beta\Delta^s}, \tag{5.32}$$
$$x = r + \frac{r_+}{r_+ - r_-}\ln(r - r_+) - \frac{r_-}{r_+ - r_-}\ln(r - r_-).$$

As Detweiler has indicated, it is possible to choose the functions $\alpha(r)$ and $\beta(r)$ so that the resulting effective potential $V(r)$ is real and has the following form:

$$V = \frac{\rho^2\Delta}{(r^2+a^2)^2}\left\{\frac{f(f+2)}{g+b\Delta} - \frac{b\Delta}{\rho^4} + \frac{[\kappa\rho\Delta - (g'\Delta - g\Delta')][\kappa\rho g - b(g'\Delta - g\Delta')]}{\rho^2(g+b\Delta)(g-b\Delta)^2}\right\}$$
$$+ \left[\frac{ram\Delta}{\omega\rho(r^2+a^2)^2}\right]^2 - \frac{\Delta}{r^2+a^2}\frac{d}{dr}\left[\frac{ram\Delta}{\omega\rho(r^2+a^2)^2}\right] - \frac{Y^2}{(r^2+a^2)^2}, \tag{5.33}$$

where

$$g = f\rho^2 + 3\rho\left(r^2+a^2\right) - 3r^2\Delta,$$
$$\rho = r^2 + a^2 - \frac{am}{\omega}, \tag{5.34}$$

$$\kappa = \pm\left\{9 - 2f\left[\left(a^2 - \frac{am}{\omega}\right)(5f+6) - 12a^2\right] + 2bf(f+2)\right\}^{1/2},$$
$$b = \pm 3\left(a^2 - \frac{am}{\omega}\right), \tag{5.35}$$

$$Y = am - \left(r^2+a^2\right)\omega,$$
$$f = A + a^2\omega^2 - 2am\omega. \tag{5.36}$$

When the Kerr rotation parameter a approaches zero, the potential V in (5.33) coincides with the Regge-Wheeler potential for negative κ and coincides with the Zerilli potential for positive κ. Here we choose κ to be negative, where the choice of the sign in (5.35) is determined by the sign of m [7–9, 44].

The QNM boundary conditions for X are

$$X \longrightarrow \begin{cases} e^{-ikx} & x \longrightarrow -\infty, \\ e^{i\omega x} & x \longrightarrow \infty, \end{cases} \tag{5.37}$$

where

$$k = \omega - \frac{am}{r_+}. \tag{5.38}$$

Hence, we write

$$X = e^{i\omega r} r^{i\omega/(r_+-r_-)} \left[\frac{(r-r_+)^{r_+}}{(r-r_-)^{r_-}}\right]^{-ik/(r_+-r_-)} \chi_G. \tag{5.39}$$

Substituting this into (5.31), we have

$$\chi_{G,rr} + \left[2\Gamma_G + \frac{r^2 - a^2}{\Delta(r^2 + a^2)}\right]\chi_{G,r} + \left[\Gamma_G^2 + \Gamma_{G,r} + \frac{\Gamma_G(r^2 - a^2)}{\Delta(r^2 + a^2)} - \left(\frac{r^2 + a^2}{\Delta}\right)^2 V\right]\chi_G = 0, \tag{5.40}$$

where

$$\Gamma_G = i\omega + \frac{i\omega}{r(r_+ - r_-)} - \frac{ikr}{\Delta}. \tag{5.41}$$

As we did earlier, we define the variable $\xi = 1 - r_+/r$ which has a compact domain $0 < \xi < 1$. Equation (5.40) can then be written in the AIM form:

$$\chi_{G,\xi\xi} = \lambda_G(\xi)\chi_{G,\xi} + s_G(\xi)\chi_G, \tag{5.42}$$

where

$$\begin{aligned}\lambda_G &= \frac{2}{1-\xi} - \frac{r_+}{(1-\xi)^2}\left[2\Gamma_G + \frac{1}{\Delta}\frac{r_+^2 - a^2(1-\xi)^2}{r_+^2 + a^2(1-\xi)^2}\right],\\ s_G &= \frac{r_+^2}{(1-\xi)^4}\left\{\left[\frac{r_+^2 + a^2(1-\xi)^2}{\Delta(1-\xi)^2}\right]^2 V - \frac{\Gamma_G}{\Delta}\frac{r_+^2 - a^2(1-\xi)^2}{r_+^2 + a^2(1-\xi)^2} - \Gamma_G^2\right\} - \frac{r_+}{(1-\xi)^2}\Gamma_{G,\xi}.\end{aligned} \tag{5.43}$$

The results for the gravitational (spin-two) case are presented in Tables 11 and 12. In general the error in the separation constant is smaller than that of the quasinormal frequencies. As for the quasinormal frequencies the error in the Kerr case is larger than that of either the Schwarzschild or the Reissner-Nordström cases, where this is due to our consideration of the angular and the radial equations simultaneously. The number of iterations that can be performed in the code is relatively small: much like the number of continued fractions in the CFM is typically smaller due to the coupling between radial and angular equations.

6. Doubly Rotating Kerr (A)DS Black Holes

Rotating black holes in higher dimensions were first discussed in the seminal paper by Myers and Perry [49]. One of the unexpected results to come from this work was that some families of solutions were shown to have event horizons for arbitrarily large values of their rotation parameters. The stability of such black holes is certainly in question [50, 51], with numerical evidence recently provided by Shibata and Yoshino [52, 53].

Another new feature of the Myers-Perry (MP) solutions is that they in general have $\lfloor (D-1)/2 \rfloor$ spin parameters, making them more complex than the four-dimensional Kerr solution. The first asymptotically nonflat five-dimensional MP metric was given in [54].

Table 11: Spin-2 angular separation constants and Kerr gravitational quasinormal frequencies for the fundamental mode corresponding to $l = 2$ and $m = 0$ compared with the CFM [6] ($M = 1/2$).

a	A_{Leaver}	A_{AIM}	ω_{Leaver}	ω_{AIM}
0	(4.0000, 0.0000)	(4.0000, 0.0000) (<0.01%)(<0.01%)	(0.7473, −0.1779)	(0.7413, −0.1780) (−0.80%)(−0.06%)
0.1	(3.9972, 0.0014)	(3.9973, 0.0014) (<0.01%)(<0.01%)	(0.7502, −0.1774)	(0.7444, −0.1775) (−0.77%, −0.06%)
0.2	(3.9886, 0.0056)	(3.9887, 0.0056) (<0.01%)(<0.01%)	(0.7594, −0.1757)	(0.7540, −0.1763) (−0.71%)(−0.03%)
0.3	(3.9730, 0.0126)	(3.9733, 0.0126) (<0.01%)(<0.01%)	(0.7761, −0.1720)	(0.7715, −0.1722) (−0.59%)(−0.12%)
0.4	(3.9480, 0.0223)	(3.9482, 0.0222) (<0.01%)(−0.45%)	(0.8038, −0.1643)	(0.8025, −0.1639) (−0.16%)(0.24%)
0.45	(3.9304, 0.0276)	(3.9303, 0.0280) (<0.01%)(1.45%)	(0.8240, −0.1570)	(0.8250, −0.1591) (0.12%)(−1.34%)

Subsequent generalizations to arbitrary dimensions were done in [55], and finally the most general Kerr-(A)dS-NUT metric was found by Chen et al. [56].

In this section we review how the AIM can be used to solve the $D \geq 6$ two-rotation scalar perturbation equations (for more details on the metric and resulting separation see [57]). The scalar field master equations are found to be [57]

$$0 = \frac{1}{r^{D-6}}\frac{d}{dr}\left(r^{D-6}\Delta_r\frac{dR_r}{dr}\right) + \left(\frac{(r^2+a_1^2)^2(r^2+a_2^2)^2}{\Delta_r}\tilde{\omega}_r^2 - \frac{a_1^2a_2^2 j(j+D-7)}{r^2} - b_1r^2 - b_2\right)R_r, \tag{6.1}$$

$$0 = \left(\frac{a_i}{y_i}\right)^{D-6}\frac{d}{dy_i}\left[\left(\frac{y_i}{a_i}\right)^{D-6}\Delta_{y_i}\frac{dR_{\theta_i}}{dy_i}\right] - \left\{\frac{(a_1^2-y_i^2)^2(a_2^2-y_i^2)^2}{\Delta_{y_i}}\tilde{\omega}_{y_i}^2 + \frac{a_1^2a_2^2 j(j+D-7)}{y_i^2} + b_1y_i^2 - b_2\right\}R_{\theta_i}, \tag{6.2}$$

where

$$\Delta_r = \left(1+g^2r^2\right)\left(r^2+a_1^2\right)\left(r^2+a_2^2\right) - 2Mr^{7-D},$$
$$\Delta_{y_i} = \left(1-g^2y_i^2\right)\left(a_1^2-y_i^2\right)\left(a_2^2-y_i^2\right), \tag{6.3}$$

Table 12: Spin-2 angular separation constants and Kerr gravitational quasinormal frequencies for the fundamental mode corresponding to $l = 2$ and $m = 1$ compared with the CFM [6] ($M = 1/2$).

a	A_{Leaver}	A_{AIM}	ω_{Leaver}	ω_{AIM}
0	(4.0000, 0.0000)	(4.0000, 0.0000) (<0.01%)(<0.01%)	(0.7473, −0.1779)	(0.7413, −0.1780) (−0.80%)(−0.06%)
0.1	(3.8932, 0.0252)	(3.8937, 0.0250) (−0.01%)(−0.89%)	(0.7765, −0.1770)	(0.7726, −0.1755) (−0.51%, 0.86%)
0.2	(3.7676, 0.0532)	(3.7681, 0.0526) (0.01%)(−1.12%)	(0.8160, −0.1745)	(0.8143, −0.1726) (−0.02%)(1.09%)
0.3	(3.6125, 0.0835)	(3.6123, 0.0826) (<0.01%)(−0.99%)	(0.8719, −0.1693)	(0.8722, −0.1674) (0.03%)(1.00%)
0.4	(3.4023, 0.1122)	(3.4011, 0.1110) (−0.03%)(−1.00%)	(0.9605, −0.1559)	(0.9620, −0.1543) (0.16%)(1.05%)
0.45	(3.2535, 0.1195)	(3.2491, 0.1173) (−0.03%)(−1.82%)	(1.0326, −0.1396)	(1.0376, −0.1369) (0.48%)(1.97%)

the radial and angular frequencies are defined by

$$\tilde{\omega}_r = \omega - \left(1 + g^2 r^2\right)\left(\frac{m_1 a_1}{r^2 + a_1^2} + \frac{m_2 a_2}{r^2 + a_2^2}\right),$$
$$\tilde{\omega}_{y_i} = \omega - \left(1 - g^2 y_i^2\right)\left(\frac{m_1 a_1}{a_1^2 - y_i^2} + \frac{m_2 a_2}{a_2^2 - y_i^2}\right), \tag{6.4}$$

and $i = 1, 2$. In the aformentioned g is the curvature of the spacetime satisfying $R_{\mu\nu} = -3g^2 g_{\mu\nu}$ (e.g., see [56]), a_1 and a_2 are the two rotation parameters, and for later reference we define $\epsilon = a_2/a_1$.

Doubly rotating black holes are more complicated than simply rotating black holes (cf. [58]), because two rotation planes lead to two coupled spheroids which are also needed for the solution of the radial equation.

6.1. Radial Quasinormal Modes

For simplicity we will consider the flat case, setting $g = 0$, which leads to easier QNM boundary conditions (cf. Schwarzschild to Schwarzschild-dS). These satisfy the boundary condition that there are only waves ingoing at the black hole horizon and outgoing waves at asymptotic infinity.

As we have shown with the previous examples, it is easier to work on a compact domain and define the variable $x = 1/r$, so that infinity is mapped to zero and the outer horizon stays at $x_h = 1/r_h = 1$. The domain of x will therefore be $[0,1]$. Thus the QNM boundary condition is translated into the statement that the waves move leftward at $x = 0$ and rightward at $x = 1$. We again choose the AIM point in the middle of the domain, that is, at $x = 1/2$.

In terms of x the radial equation in (6.1) becomes

$$0 = -x^{D-4}\frac{d}{dx}\left(-x^{8-D}\Delta_x\frac{dR}{dx}\right) + \left(\frac{(x^{-2}+a_1^2)^2(x^{-2}+a_2^2)^2}{\Delta_x}\tilde{\omega}_x^2 - a_1^2a_2^2j(j+D-7)x^2 - \frac{b_1}{x^2} - b_2\right)R, \tag{6.5}$$

where $\Delta_x(x) \equiv \Delta_r(r = 1/x)$ and $\omega_x(x) \equiv \omega_r(r = 1/x)$.

After performing some asymptotic analysis, we find that for the solutions to satisfy the QNM boundary conditions we must have

$$R \sim (1-x)^{i\tilde{\omega}_h\alpha_h}x^{(D-2)/2}e^{i\omega_x/x}y(x), \tag{6.6}$$

where

$$\tilde{\omega}_h \equiv \omega_x\ (x=1), \qquad \alpha_h \equiv \frac{(1+\alpha_1^2)(1+\alpha_2^2)}{\Delta_x'(x=1)}. \tag{6.7}$$

We then substitute this anstaz into (6.5) and rewrite into the AIM form:

$$y'' = \lambda_0 y' + s_0 y. \tag{6.8}$$

This final step can be performed in Mathematica, where the resulting expressions for λ_0 and s_0 are fed into the AIM routine. The method we use to find thr QNMs proceeds in a fashion similar to that used in [59, 60] (see also Section 5.3) except we use the AIM instead of the CFM.

First we set the number of AIM iterations in both the eigenvalue and QNM calculations to sixteen. We start with the schwarzschild values (b_1, b_2, ω), that is, at the point $(a_1, a_2) \sim 0$ and then increment a_1 and a_2 by some small value. We take the initial eigenvalues (b_1, b_2), insert them into the radial equation (6.6), and then use the AIM to find the new QNM that is closest to ω using the Mathematica routine `FindRoot`.

Taking this new value of omega, ω', we insert it into the two angular equations (at the same value of a_1 and a_2) and then solve using the AIM, searching closest to the pervious b_1 and b_2 values, thereby obtaining the new eigenvalues b_1' and b_2'. We then repeat this process with the new (ω', b_1', b_2') as the starting point until the results converge and we have achieved four decimal places of accuracy. When this occurs, we increment a_1 and a_2 again and repeat the proces. In this way we are able to find the QNMs and along lines passing approximately through the origin (i.e, starting from the near Scwharzschild values) in the (a_1, a_2) parameter space.

As an example, we have plotted various values of $\epsilon = a_2/a_1 (= 0, 0.2, 0.4, 0.6, 0.8, 1)$ against a_1 and used an interpolating function to interpolate batween these values as shown in Figure 1 (for further details see [57]).

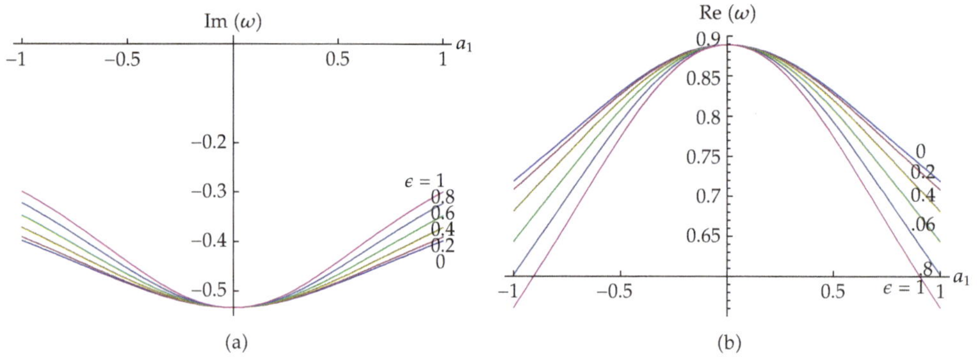

Figure 1: An example of the $D = 6$ fundamental $(j, m_1, m_2, n_1, n_2) = (0, 0, 0, 0, 0)$ QNM. (a) is a plot of the imaginary part and (b) is a plot of the real part.

7. Summary and Outlook

In this paper we have shown that the AIM can be used to calculate the radial QNMs of a variety of black hole spacetimes. In particular, we have used it to calculate perturbations of Schwarzschild (in asymptotically flat, de Sitter and anti-de-Sitter), RN, and Kerr (for spin 0, 1, and 2 perturbations) black holes in four dimensions. We argued that the method will be of use in studies of extra dimensional black holes and gave an explicit example of this in the case of the doubly rotating Myers Perry black hole.

We have hopefully demonstrated how the AIM can also be applied to radial QNMs and not just to spheroidal eigenvalue problems [14, 34]. Given the fact that the AIM can be used in both the radial and angular wave equations [34], we expect no problems in obtaining QNMs for Kerr-dS black holes in four and higher dimensions. Note that this was only recently accomplished via the CFM in [61] using Heun's equation [62] to reduce the problem to a 3-term recurrence relation. In higher dimensions a similar method was used for simply rotating Kerr-AdS black holes [58].

It remains to be seen if the AIM can be tailored to handle asymptotic QNMs (see [63] for an adapted version of the CFM). However, given the close relationship between the AIM and the exact WKB approach [64], it might be possible to adapt the AIM to find asymptotic QNMs [65–68] numerically or even semianalytically.

The AIM might be of some topical use, for example, in the angular spheroids/QNMs needed in the phenomenology of Hawking radiation from spinning higher-dimensional black holes; for a recent review, see [69]. We recently used a combination of all the techniques discussed in this work to evaluate the angular eigenvalues, ${}_0A_{kjm}$, for real $c = a\omega$, which are needed for the tensor graviton emission rates on a *simply* rotating Kerr-de Sitter black hole background in $(n+4)$-dimensions [70] (also see [71]) and it might also be interesting to find QNMs of doubly rotating Kerr-(A)dS black holes (for asymptotically flat Kerr see [57]). Finally, attempting to solve the QNMs for all spins on the Schwarzschild-AdS background via the AIM also seems an interesting problem.

As such we hope to have provided the reader with enough technical details and to have addressed some of the possible questions to allow them to pursue the study of QNMs with the AIM (source code and other information for some of the cases presented here can be found on the AIM link at http://www-het.phys.sci.osaka-u.ac.jp/~naylor/).

Appendices

A. Angular Eigenvalues for Spin-Weighted Spheroidal Harmonics

As mentioned in Section 5.3, aside from radial QNMs, the AIM can also be applied to various kinds of spin-weighted spheroidal harmonics, ${}_sS_{lm}(\theta)$; for example, see [60]. Therefore, in this appendix, we briefly compare the AIM with the CFM for the four-dimensional spin-weighted spheroids.

With the regular boundary conditions, the angular wave function can be written as [6]

$$S = e^{a\omega u}(1+u)^{(1/2)|m-s|}(1-u)^{(1/2)|m+s|}\chi_A(u). \tag{A.1}$$

Putting this back into (5.3) and rewriting the equation in AIM form, we have

$$\chi_{A,uu} = \lambda_A(u)\chi_{A,u} + s_A(u)\chi_A, \tag{A.2}$$

where

$$\lambda_A(u) = \frac{2u}{1-u^2} - 2N,$$
$$s_A(u) = \frac{1}{1-u^2}\left[\frac{(m+su)^2}{1-u^2} + 2a\omega su + 2uN - \left(a^2\omega^2u^2 + s + {}_sA_{lm}\right)\right] - N^2 - N_{,u}, \tag{A.3}$$
$$N = a\omega + \frac{|m-s|}{2(1+u)} - \frac{|m+s|}{2(1-u)}.$$

These are the relevant equations for calculating the eigenvalues of the spheroidal harmonics in the four-dimensional case. It was noticed in [14] that the AIM converges fastest at the maximum of the potential, when the AIM is written in WKB form. In four dimensions this occurs at $x = 0$ and is true for general spin-s as we have verified. Note that for higher-dimensional generalizations it is not easy to explicitly find a maximum [34].

It may be worth mentioning that for the case where $c = 0$, an exact analytic solution of the pervious equations leads to [60]

$${}_sA_{lm} = l(l+1) - s(s+1). \tag{A.4}$$

In the exact limit $c = 0$, the AIM recovers the result for spherical harmonics, (A.4) above, while for the CFM taking $c = 0$ leads to singularities [60]; however for $c \ll 1$ we find agreement with the CFM and (A.4).

For the purposes of consistency we have calculated (see Table 13) the ${}_2A_{l1}$ eigenvalues for the lowest $l = 2, \ldots, 7$ modes to 10 significant figures and have also compared this with the results of the CFM. In both the AIM and CFM larger l modes require more iterations/recursions to achieve convergence in a given l eigenvalue to the required precision. Care should be taken when comparing the number of iterations in the AIM with that of the number of recursions in the CFM, because one iteration of the AIM is not equivalent to one iteration of the CFM. In fact although we typically need to iterate the improved AIM on

Table 13: Selected spin two eigenvalues, ${}_2A_{lm}$, obtained from the AIM for a Kerr black hole with varying values of $c = a\omega$ and $m = 1$: the same number of iterations in the AIM, n_A, and the number of recursions in the CFM, n_C, for $c = 0.1, 0.8$ at a working precision of 15 digit precision, where results are presented to 10 s.f.

l	$c = 0.1$ ($n_A = n_C = 15$,)	$c = 0.8$ ($n_A = 35$, $n_C = 70$)	$c = -10i$ ($n_A = 80$, $n_C = 130$ CFM)	$c = 10$ ($n_A = 75$, $n_C = 145$)
2	−0.1391483511	−1.462479552	(12.44128209, 0.8956143162)	−101.8949078
3	5.929826236	5.247141863	(32.31138608, 1.302608040)	−63.74900642
4	13.95640426	13.45636668	(51.27922784, 1.946041848)	−30.35607486
5	23.96944247	23.54163307	(69.25750923, 3.012877154)	−4.557015739
6	35.97681567	35.58524928	(85.86796852, 4.990079008)	−2.555206382
7	9.98139515	49.61077286	(99.20081385, 6.801617108)	12.32203552

average a lesser number of times, the CFM is typically faster for smaller values of c. However, for larger values of c both methods can be faster or slower.

The results of the first few l eigenvalues for different values of $c = a\omega$, with $m = 1$, are presented in Table 13. As far as we are aware this is the first time tables of spin-2 spheroids (for general complex c) have been presented using the AIM. Further results are presented in Section 5.3 along with the radial QNMs for the spin two perturbations of the Kerr black hole in Table 12.

B. Higher-Dimensional Scalar Spheroidal Harmonics with Two Rotation Parameters

The two Equations (6.2), for $i = 1, 2, \ldots$, are in fact the two-rotation generalization of the higher-dimensional spheroidal harmonics (HSHs) studied in [59]. In this case, the existence of two rotation parameters leads to a system of two coupled second-order ODEs. (For the moment we are considering ω to be an independent parameter). In general, one would expect that the generalizations of the HSHs to $\lfloor (D-1)/2 \rfloor$ rotation parameters would lead to even larger systems of equations. While these systems would also be useful generally in studies of MP black holes, here we will only focus on the two rotation case.

The angular equations can be written in the Sturm-Liouville form (assuming momentarily that ω and b_2 are real):

$$\lambda w(\xi_i) R_{\theta_i}(\xi_i) = -\frac{d}{d\xi_i}\left(p(\xi_i)\frac{d}{d\xi_i}R_{\theta_i}(\xi_i)\right) + q(\xi_i) R_{\theta_i}(\xi_i) \tag{B.1}$$

with the weight function $w_1(\xi_i) = (1/4)\xi_i^{(D-5)/2}$, the eigenvalue $\lambda = -b_1$, and

$$p(\xi_i) = \xi_i^{(D-5)/2}\Delta_{\xi_i},$$
$$q(\xi_i) = \frac{1}{4}\xi_i^{(D-7)/2}\left(\frac{\left(a_1^2-\xi_i\right)^2\left(a_2^2-\xi_i\right)^2}{\Delta_{\xi_i}}\widetilde{\omega}_{\xi_i}^2 + \frac{a_1^2a_2^2 j(j+D-7)}{\xi_i} - b_2\right), \tag{B.2}$$

where Δ_{ξ_i} and $\widetilde{\omega}_{\xi_i}$ are defined in the obvious way under the change of coordinates. Since $w(\xi) > 0$, we can define the two norms:

$$N_1(R_{\theta_1}) \propto \int_{a_2^2}^{a_1^2} \xi_1^{(D-5)/2}|R_{\theta_1}|^2 d\xi_1,$$
$$N_2(R_{\theta_2}) \propto \int_{0}^{a_2^2} \xi_2^{(D-5)/2}|R_{\theta_2}|^2 d\xi_2. \tag{B.3}$$

For further details see [57].

The regular solutions are found to be

$$R_1 \sim \left(\xi_1 - a_2^2\right)^{|m_2|/2}\left(a_1^2-\xi_1\right)^{|m_1|/2}\Psi_1; \quad \xi_1 \in \left(a_2^2, a_1^2\right),$$
$$R_2 \sim \xi_2^{j/2}\left(a_2^2-\xi_2\right)^{|m_2|/2}\Psi_2; \quad \xi_2 \in \left(0, a_2^2\right). \tag{B.4}$$

Now for a given value of ω we can determine b_1 and b_2 simply by performing the improved AIM [13] on both of the angular equations separately. This will result in two equations in the two unknowns b_1 and b_2 which we can then be solved using a numerical routine such as the built-in Mathematica functions `NSolve` or `FindRoot`. More specifically we rewrite (6.2) using (B.4) and transform them into the AIM form:

$$\frac{d^2\Psi_1}{d\xi_1^2} - \lambda_{01}\frac{d\Psi_1}{d\xi_1} + s_{01}\Psi_1,$$
$$\frac{d^2\Psi_2}{d\xi_2^2} = \lambda_{02}\frac{d\Psi_2}{d\xi_2} + s_{02}\Psi_2. \tag{B.5}$$

The AIM requires that a special point be taken about which the λ_{0i} and s_{0i} coefficients are expanded. As was shown in [34] different choices of this point can worsen or improve the speed of the convergence. In the absence of a clear selection criterion we simply choose this point conveniently in the middle of the domains:

$$\xi_{01} = \frac{a_1^2 + a_2^2}{2}, \qquad \xi_{02} = \frac{a_2^2}{2}. \tag{B.6}$$

Some results are plotted in Figures 2 and 3. This method can now be used in the radial QNM equation in Section 6.1.

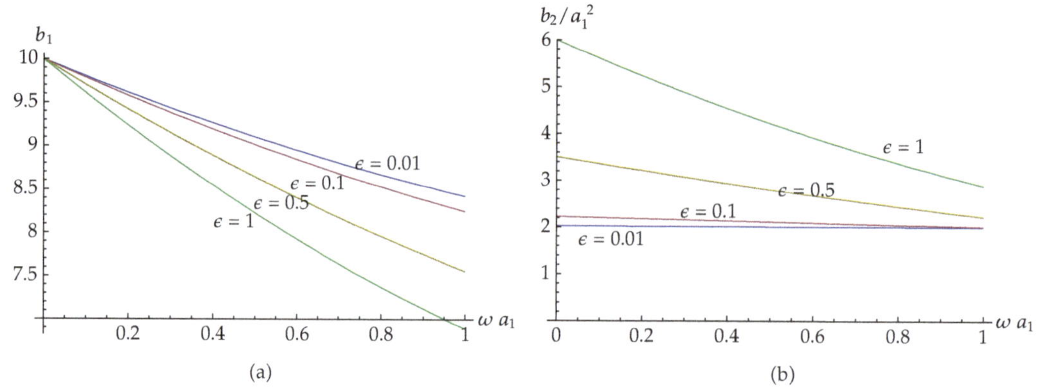

Figure 2: (Color online) $D = 6$, $g = 0$, $(j, m_1, m_2, n_1, n_2) = (0, 1, 1, 0, 0)$: A plot of the eigenvalues for various choices of $\epsilon \equiv a_2/a_1$. Note that the dependence on a_1 has been scaled into the other quantities.

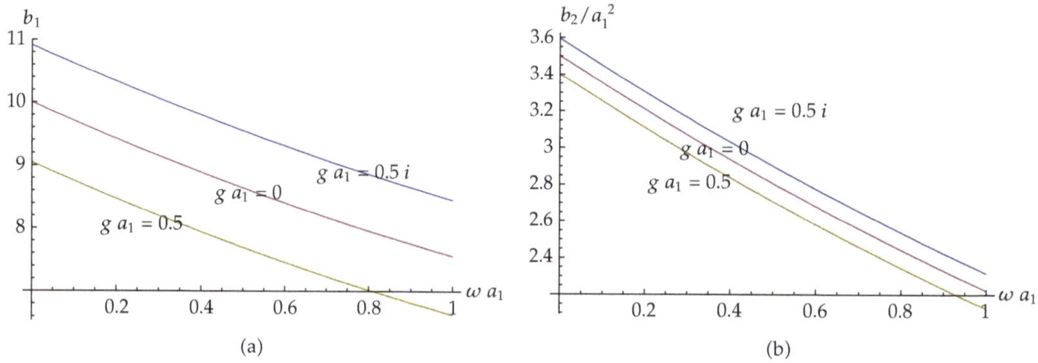

Figure 3: (Color Online) $D = 6$, $\epsilon \equiv a_2/a_1 = 1/2$, $(j, m_1, m_2, n_1, n_2) = (0, 1, 1, 0, 0)$: A plot of the eigenvalues for $ga_1 = 0.5i, 0, 0.5$, corresponding to de-Sitter, flat, and anti-deSitter spacetimes respectively. Note that the dependence on a_1 has been scaled into the other quantities.

Acknowledgments

H. T. Cho was supported in part by the National Science Council of the Republic of China under the Grant NSC 99-2112-M-032-003-MY3 and the National Science Centre for Theoretical Sciences. The work of J. Doukas was supported by the Japan Society for the Promotion of Science (JSPS), under fellowship no. P09749. W. Naylor would like to thank the Particle Physics Theory Group, Osaka University, for computing resources.

References

[1] I. G. Moss and J. P. Norman, "Gravitational quasinormal modes for anti-de Sitter black holes," *Classical and Quantum Gravity*, vol. 19, no. 8, pp. 2323–2332, 2002.

[2] C. V. Vishveshwara, "Scattering of gravitational radiation by a schwarzschild black-hole," *Nature*, vol. 227, no. 5261, pp. 936–938, 1970.

[3] S. Chandrasekhar and S. L. Detweiler, "The quasi-normal modes of the Schwarzschild black hole," *Proceedings of the Royal Society A*, vol. 344, pp. 441–452, 1975.

[4] V. Ferrari and B. Mashhoon, "New approach to the quasinormal modes of a black hole," *Physical Review D*, vol. 30, no. 2, pp. 295–304, 1984.

[5] D. L. Gunter, "A study of the coupled gravitational and electromagnetic perturbations to the Reissner-Nordström black hole: the scattering matrix, energy conversion, and quasi-normal modes," *Philosophical Transactions of the Royal Society of London A*, vol. 296, no. 1422, pp. 497–526, 1980.

[6] E. W. Leaver, "An analytic representation for the quasinormal modes of Kerr black holes," *Proceedings of the Royal Society. London A*, vol. 402, no. 1823, pp. 285–298, 1985.

[7] B. Mashhoon, *Proceedings of the Third Marcel Grossmann Meeting on Recent Developments of General Relativity, Edited*, vol. by H. Ning, North-Holland, Amsterdam, The Netherlands.

[8] B. F. Schutz and C. M. Will, "Black Hole Normal Modes: A Semianalytic Approach," *The Astrophysical Journal*, vol. L291, p. 33, 1985.

[9] E. Seidel and S. Iyer, "Black-hole normal modes: a WKB approach. IV. Kerr black holes," *Physical Review D*, vol. 41, no. 2, pp. 374–382, 1990.

[10] R. A. Konoplya, "Quasinormal behavior of the D-dimensional Schwarzschild black hole and the higher order WKB approach," *Physical Review D*, vol. 68, no. 2, Article ID 024018, 8 pages, 2003.

[11] H. Ciftci, R. L. Hall, and N. Saad, "Asymptotic iteration method for eigenvalue problems," *Journal of Physics A*, vol. 36, no. 47, pp. 11807–11816, 2003.

[12] H. Ciftci, R. L. Hall, and N. Saad, "Perturbation theory in a framework of iteration methods," *Physics Letters A*, vol. 340, no. 5-6, pp. 388–396, 2005.

[13] H. T. Cho, A. S. Cornell, J. Doukas, and W. Naylor, "Black hole quasinormal modes using the asymptotic iteration method," *Classical and Quantum Gravity*, vol. 27, no. 15, Article ID 155004, 2010.

[14] T. Barakat, "The asymptotic iteration method for Dirac and Klein-Gordon equations with a linear scalar potential," *International Journal of Modern Physics A*, vol. 21, no. 19-20, pp. 4127–4135, 2006.

[15] O. Ozer and P. Roy, "The asymptotic iteration method applied to certain quasinormal modes and non Hermitian systems," *Central European Journal of Physics*, vol. 7, no. 4, pp. 747–752, 2009.

[16] H.-T. Cho and C.-L. Ho, "Quasi-exactly solvable quasinormal modes," *Journal of Physics A*, vol. 40, no. 6, pp. 1325–1331, 2007.

[17] T. Regge and J. A. Wheeler, "Stability of a Schwarzschild singularity," vol. 108, pp. 1063–1069, 1957.

[18] F. J. Zerilli, "Gravitational field of a particle falling in a Schwarzschild geometry analyzed in tensor harmonics," *Physical Review D*, vol. 2, pp. 2141–2160, 1970.

[19] S. A. Teukolsky, "Rotating black holes: Separable wave equations for gravitational and electromagnetic perturbations," *Physical Review Letters*, vol. 29, no. 16, pp. 1114–1118, 1972.

[20] F. J. Zerilli, "Perturbation analysis for gravitational and electromagnetic radiation in a Reissner-Nordström geometry," *Physical Review D*, vol. 9, no. 4, pp. 860–868, 1974.

[21] V. Moncrief, "Odd-parity stability of a Reissner-Nordström black hole," *Physical Review D*, vol. 9, no. 10, pp. 2707–2709, 1974.

[22] V. Moncrief, "Stability of Reissner-Nordström black holes," *Physical Review D*, vol. 10, no. 4, pp. 1057–1059, 1974.

[23] V. Moncrief, "Gauge-invariant perturbations of Reissner-Nordström black holes," *Physical Review D*, vol. 12, no. 6, pp. 1526–1537, 1975.

[24] E. Berti, V. Cardoso, and A. O. Starinets, "Quasinormal modes of black holes and black branes," *Classical and Quantum Gravity*, vol. 26, no. 16, Article ID 163001, 2009.

[25] D. R. Brill and J. A. Wheeler, "Interaction of neutrinos and gravitational fields," *Reviews of Modern Physics*, vol. 29, pp. 465–479, 1957.

[26] A. J. M. Medved, D. Martin, and M. Visser, "Dirty black holes: quasinormal modes," *Classical and Quantum Gravity*, vol. 21, no. 6, pp. 1393–1405, 2004.

[27] H. T. Cho, "Dirac quasinormal modes in Schwarzschild black hole spacetimes," *Physical Review D*, vol. 68, no. 2, Article ID 024003, 2003.

[28] S. Iyer, "Black-hole normal modes: a WKB approach. II. Schwarzschild black holes," *Physical Review D*, vol. 35, no. 12, pp. 3632–3636, 1987.

[29] S. R. Dolan and A. C. Ottewill, "On an expansion method for black hole quasinormal modes and Regge poles," *Classical and Quantum Gravity*, vol. 26, no. 22, Article ID 225003, 2009.

[30] A. Zhidenko, "Quasi-normal modes of Schwarzschild de Sitter black holes," *Classical and Quantum Gravity*, vol. 21, no. 1, pp. 273–280, 2004.

[31] E. W. Leaver, "Quasinormal modes of Reissner-Nordström black holes," *Physical Review D*, vol. 41, no. 10, pp. 2986–2997, 1990.

[32] A. Zhidenko, "Massive scalar field quasinormal modes of higher dimensional black holes," *Physical Review D*, vol. 74, no. 6, Article ID 064017, 2006.
[33] V. Cardoso, J. P. S. Lemos, and S. Yoshida, "Quasinormal modes of Schwarzschild black holes in four and higher dimensions," *Physical Review D*, vol. 69, no. 4, Article ID 044004, 2004.
[34] H. T. Cho, A. S. Cornell, J. Doukas, and W. Naylor, "Asymptotic iteration method for spheroidal harmonics of higher-dimensional Kerr-(A)dS black holes," *Physical Review D*, vol. 80, no. 6, Article ID 064022, 2009.
[35] E. Berti and K. D. Kokkotas, "Quasinormal modes of Reissner-Nordström-anti-de Sitter black holes: scalar, electromagnetic, and gravitational perturbations," *Physical Review D*, vol. 67, no. 6, Article ID 064020, 2003.
[36] J. S. F. Chan and R. B. Mann, "Scalar wave falloff in asymptotically anti-de Sitter backgrounds," *Physical Review D*, vol. 55, no. 12, pp. 7546–7562, 1997.
[37] J. S. F. Chan and R. B. Mann, "Scalar wave falloff in topological black hole backgrounds," *Physical Review D*, vol. 59, no. 6, pp. 1–23, 1999.
[38] G. T. Horowitz and V. E. Hubeny, "Quasinormal modes of AdS black holes and the approach to thermal equilibrium," *Physical Review D*, vol. 62, no. 2, Article ID 024027, 2000.
[39] V. Cardoso and J.P.S. Lemos, "Quasinormal modes of Schwarzschild-anti-de Sitter black holes: Electromagnetic and gravitational perturbations," *Physical Review D*, vol. 64, no. 8, Article ID 084017, 2001.
[40] V. Cardoso, *Quasinormal modes and gravitational radiation in black hole spacetimes*, Ph.D. thesis, http://arxiv.org/abs/gr-qc/0404093.
[41] K. D. Kokkotas and B. F. Schutz, "Black-hole normal modes: a WKB approach. III. the reissner-nordström black hole," *Physical Review D*, vol. 37, no. 12, pp. 3378–3387, 1988.
[42] S. Chandrasekhar, *The Mathematical Theory of Black Holes*, vol. 69 of *International Series of Monographs on Physics*, The Clarendon Press, New York, NY, USA, 1992.
[43] N. Andersson and H. Onozawa, "Quasinormal modes of nearly extreme Reissner-Nordström black holes," *Physical Review D*, vol. 54, no. 12, pp. 7470–7475, 1996.
[44] S. Detweiler, "Resonant oscillations of a rapidly rotating black hole," *Proceedings of the Royal Society A*, vol. 352, p. 381, 1977.
[45] E. Berti and K. D. Kokkotas, "Quasinormal modes of Kerr-Newton black holes: coupling of electromagnetic and gravitational perturbations," *Physical Review D*, vol. 71, no. 12, Article ID 124008, 2005.
[46] http://www.phy.olemiss.edu/~berti/qnms.html.
[47] W. A. Carlson, A. S. Cornell, and B. Jordan, "Fermion Quasi-normal modes of the Kerr Black-Hole," In press, http://arxiv.org/abs/1201.3267.
[48] J.-l. Jing and Q.-y. Pan, "Dirac quasinormal frequencies of the Kerr-Newman black hole," *Nuclear Physics B*, vol. 728, no. 1–3, pp. 109–120, 2005.
[49] R. C. Myers and M. J. Perry, "Black holes in higher-dimensional space-times," *Annals of Physics*, vol. 172, no. 2, pp. 304–347, 1986.
[50] R. Emparan and R. C. Myers, "Instability of ultra-spinning black holes," *Journal of High Energy Physics A*, vol. 2003, no. 9, article 025, 2003.
[51] R. A. Konoplya and A. Zhidenko, "Quasinormal modes of black holes: from astrophysics to string theory," *Reviews of Modern Physics*, vol. 83, no. 3, pp. 793–836, 2011.
[52] M. Shibata and H. Yoshino, "Nonaxisymmetric instability of rapidly rotating black hole in five dimensions," *Physical Review D*, vol. 81, no. 2, Article ID 021501, 2010.
[53] M. Shibata and H. Yoshino, "Bar-mode instability of a rapidly spinning black hole in higher dimensions: numerical simulation in general relativity," *Physical Review D*, vol. 81, no. 10, Article ID 104035, 2010.
[54] S. W. Hawking, C. J. Hunter, and M. M. Taylor-Robinson, "Rotation and the AdS-CFT correspondence," *Physical Review D*, vol. 59, no. 6, Article ID 064005, 1999.
[55] G. W. Gibbons, H. Lu, D. N. Page, and C. N. Pope, *Physical Review Letters*, vol. 93, no. 17, Article ID 171102, 2004.
[56] W. Chen, H. Lü, and C. N. Pope, "General Kerr-NUT-AdS metrics in all dimensions," *Classical and Quantum Gravity*, vol. 23, no. 17, pp. 5323–5340, 2006.
[57] H. T. Cho, J. Doukas, W. Naylor, and A. S. Cornell, "Quasinormal modes for doubly rotating black holes," *Physical Review D*, vol. 83, no. 12, Article ID 124034, 2011.

[58] H. Kodama, R. A. Konoplya, and A. Zhidenko, "Gravitational instability of simply rotating AdS black holes in higher dimensions," *Physical Review D*, vol. 79, no. 4, Article ID 044003, 2009.

[59] E. Berti, V. Cardoso, and M. Casals, "Eigenvalues and eigenfunctions of spin-weighted spheroidal harmonics in four and higher dimensions," *Physical Review D*, vol. 73, no. 2, Article ID 024013, 2006.

[60] E. Berti, V. Cardoso, and C. M. Will, "Gravitational-wave spectroscopy of massive black holes with the space interferometer LISA," *Physical Review D*, vol. 73, no. 6, Article ID 064030, 2006.

[61] S. Yoshida, N. Uchikata, and T. Futamase, "Quasinormal modes of Kerr-de Sitter black holes," *Physical Review D*, vol. 81, no. 4, Article ID 044005, 2010.

[62] H. Suzuki, E. Takasugi, and H. Umetsu, "Perturbations of Kerr-de Sitter black holes and Heun's equations," *Progress of Theoretical Physics*, vol. 100, no. 3, pp. 491–505, 1998.

[63] H.-P. Nollert, "Quasinormal modes of Schwarzschild black holes: The determination of quasinormal frequencies with very large imaginary parts," *Physical Review D*, vol. 47, no. 12, pp. 5253–5258, 1993.

[64] A. R. Matamala, F. A. Gutierrez, and J. Díaz-Valdés, "A simple algebraic approach to a nonlinear quantum oscillator," *Physics Letters A*, vol. 361, no. 1-2, pp. 16–17, 2007.

[65] L. Motl and A. Neitzke, "Asymptotic black hole quasinormal frequencies," *Advances in Theoretical and Mathematical Physics*, vol. 7, no. 2, pp. 307–330, 2003.

[66] N. Andersson and C. J. Howls, "The asymptotic quasinormal mode spectrum of non-rotating black holes," *Classical and Quantum Gravity*, vol. 21, no. 6, pp. 1623–1642, 2004.

[67] S. Das and S. Shankaranarayanan, "High-frequency quasi-normal modes for black holes with generic singularities," *Classical and Quantum Gravity*, vol. 22, no. 3, pp. L7–L21, 2005.

[68] A. Ghosh, S. Shankaranarayanan, and S. Das, "High frequency quasi-normal modes for black holes with generic singularities. II. Asymptotically non-flat spacetimes," *Classical and Quantum Gravity*, vol. 23, no. 6, pp. 1851–1874, 2006.

[69] J. A. Frost, J. R. Gaunt, M. O. P. Sampaio et al., "Phenomenology of production and decay of spinning extra-dimensional black holes at hadron colliders," *Journal of High Energy Physics*, vol. 2009, no. 10, article 014, 2009.

[70] J. Doukas, H. T. Cho, A. S. Cornell, and W. Naylor, "Graviton emission from simply rotating Kerr-de Sitter black holes: transverse traceless tensor graviton modes," *Physical Review D*, vol. 80, no. 4, Article ID 045021, 2009.

[71] P. Kanti, H. Kodama, R. A. Konoplya, N. Pappas, and A. Zhidenko, "Graviton emission in the bulk by a simply rotating black hole," *Physical Review D*, vol. 80, Article ID 084016, 2009.

Nonlinear Bi-Integrable Couplings of Multicomponent Guo Hierarchy with Self-Consistent Sources

Hongwei Yang,[1] Huanhe Dong,[1] Baoshu Yin,[2,3] and Zhenyu Liu[4]

[1] *Information School, Shandong University of Science and Technology, Qingdao 266590, China*
[2] *Institute of Oceanology, China Academy of Sciences, Qingdao 266071, China*
[3] *Key Laboratory of Ocean Circulation and Wave, Chinese Academy of Sciences, Qingdao 266071, China*
[4] *Department of Mathematics, Zaozhuang College, Zaozhuang 277160, China*

Correspondence should be addressed to Baoshu Yin, baoshuyin@126.com

Academic Editor: D. E. Pelinovsky

Based on a well-known Lie algebra, the multicomponent Guo hierarchy with self-consistent sources is proposed. With the help of a set of non-semisimple Lie algebra, the nonlinear bi-integrable couplings of the multicomponent Guo hierarchy with self-consistent sources are obtained. It enriches the content of the integrable couplings of hierarchies with self-consistent sources. Finally, the Hamiltonian structures are worked out by employing the variational identity.

1. Introduction

Since the notion of integrable couplings was proposed in view of Virasoro symmetric algebras [1, 2] and the soliton theory, considerable research has been reported on the integrable couplings and associated properties of some known interesting integrable hierarchies, such as the AKNS hierarchy, the KN hierarchy, and the Burger hierarchy [3–6]. In order to produce multicomponent integrable systems, a type of multicomponent loop algebra $\widetilde{G}_M$ was structured, and it followed that some multicomponent integrable systems were given and Hamiltonian structures of multicomponent systems were constructed through the component-trace identity in [7–13] and a general structure between Lie algebras and integrable coupling was given in [14, 15]. Here it is necessary to point out that the above-mentioned integrable couplings are linear for the supplementary variable, so they are called linear integrable couplings. Recently, Ma proposed the notion of nonlinear integrable couplings and gave the general scheme to construct the nonlinear integrable couplings of hierarchies [16]. In [17, 18], based on the general scheme of constructing nonlinear integrable couplings, Zhang introduced some new explicit Lie algebras and obtained the nonlinear integrable couplings of the GJ hierarchy, the Yang hierarchy, and the CBB hierarchy. In [19], Ma employed a class of non-semisimple matrix loop algebras to generate bi-integrable couplings of soliton equation from the zero curvature equations.

On the other hand, with the development of soliton theory, people began to focus on the soliton equation with self-consistent sources [20, 21]. The soliton equations with self-consistent sources have important physical meaning and are often used to express interactions between different solitary waves and are relevant to some problems of hydrodynamics, solid state physics, plasma physics, and so on [22, 23]. In [24–26], Yu and Xia discussed the integrable couplings of hierarchies with self-consistent sources and obtained a series of integrable hierarchies with self-consistent sources and their couplings systems as well as Hamiltonian structures. In these papers, the integrable couplings are linear for supplementary variables.

As mentioned above, we will find that there are very few researches on the nonlinear integrable couplings of hierarchies with self-consistent sources available up till now, especially on the nonlinear bi-integrable couplings of multicomponent hierarchy. In our work, we will derive the nonlinear bi-integrable couplings of multicomponent Guo hierarchy with self-consistent sources and their Hamiltonian structures. It will enrich the content of the integrable couplings of hierarchies with self-consistent sources. The plan of the paper is as follows. In Section 2, the multicomponent Guo hierarchy with self-consistent sources is given. By employing the variational identity, the Hamiltonian structure of the above hierarchy is obtained. The notion of bi-integrable couplings is introduced and the nonlinear bi-integrable couplings of the multicomponent Guo hierarchy with self-consistent sources and their Hamiltonian structures are derived in Section 3. Finally, some conclusions are given in Section 4.

2. Multicomponent Guo Hierarchy with Self-Consistent Sources

In this section, we recall briefly some basic notions from the theory of the multicomponent integrable system. If $F_M = \{a = (a_{ij})_{M\times 3} = (a_1, a_2, a_3)\}$ denotes a class of matrices where $a_i\,(i = 1, 2, 3)$ is the ith column of the matrix a, M stands for an arbitrary positive integer, then F_M is a linear space.

Definition 2.1 (see [7]). Let $\alpha = (\alpha_1, \alpha_2, \ldots, \alpha_M)^T$ and $\beta = (\beta_1, \beta_2, \ldots, \beta_M)^T$, their product is defined as

$$\alpha * \beta = \beta * \alpha = (\alpha_1\beta_1, \alpha_2\beta_2, \ldots, \alpha_M\beta_M)^T, \qquad \frac{\alpha}{\beta} = \left(\frac{\alpha_1}{\beta_1}, \frac{\alpha_2}{\beta_2}, \ldots, \frac{\alpha_M}{\beta_M}\right)^T. \tag{2.1}$$

Definition 2.2 (see [7]). Set $a = (a_1, a_2, a_3)$ and $b = (b_1, b_2, b_3) \in F_M$, a commutation operation is defined as

$$[a, b] = (a_2 * b_3 - a_3 * b_2,\ 2(a_1 * b_2 - a_2 * b_1),\ 2(a_3 * b_1 - a_1 * b_3))^T. \tag{2.2}$$

It is easy to verify that F_M is a Lie algebra. Its corresponding loop algebra $\widetilde{F}_M$ is

$$\widetilde{F}_M = \{a\lambda^n,\ a \in F_M,\ n = 0, \pm 1, \pm 2, \ldots\}, \tag{2.3}$$

equipped with a commutation operation as follows:

$$[a\lambda^m, b\lambda^n] = [a,b]\lambda^{m+n}, \quad \forall a, b \in F_M. \tag{2.4}$$

Consider the following isospectral problem:

$$\varphi_x = [U, \varphi], \qquad \lambda_t = 0, \qquad U = \left(\frac{\lambda I_M}{2}, \frac{(u_1 + u_2)}{2}, \frac{\lambda(u_1 - u_2)}{2}\right)^T, \tag{2.5}$$

where u_1, u_2 are potentials, λ is the spectral parameter, $u_i = (u_{i1}, u_{i2}, \ldots, u_{iM})^T$, $i = 1, 2$, and $I_M = \underbrace{(1, 1, \ldots, 1)}_{M}^T$. By means of constructing a proper time evolution

$$\varphi_t = V\varphi, \qquad V = (\lambda v_1, v_2 + v_3, \lambda(v_2 - v_3))^T, \tag{2.6}$$

where $v_k = \sum_{m=0}^{\infty}(v_{k_1,m}, v_{k_2,m}, \ldots, v_{k_M,m})^T \lambda^{-m}$, $k = 1, 2, 3$. Then solving stationary zero curvature equation $V_x = [U, V]$, one arrives at the recursion relation as follows:

$$\begin{aligned}
&v_{1,mx} = u_2 * v_{2,m} - u_1 * v_{3,m}, \\
&v_{2,mx} = v_{3,m+1} - u_2 * v_{1,m+1}, \\
&v_{3,mx} = v_{2,m+1} - u_1 * v_{1,m+1}, \\
&v_{1,0} = 0I_M, \qquad v_{2,0} = u_1, \qquad v_{3,0} = u_2, \qquad v_{1,1} = \frac{(u_2 * u_2 - u_1 * u_1)}{2}, \\
&v_{2,1} = u_{2,x} + u_1 * \frac{(u_2 * u_2 - u_1 * u_1)}{2}, \qquad v_{3,1} = u_{1,x} + u_2 * \frac{(u_2 * u_2 - u_1 * u_1)}{2}.
\end{aligned} \tag{2.7}$$

For an arbitrary natural number n, denoting

$$V_+^{(n)} = \sum_{m=0}^{n}(\lambda v_{1,m}, v_{2,m} + v_{3,m}, \lambda(v_{2,m} - v_{3,m}))^T \lambda^{n-m} = \lambda^n V - V_-^{(n)}, \tag{2.8}$$

then the stationary zero curvature equation can be written as

$$-V_{+x}^{(n)} + \left[U, V_+^{(n)}\right] = V_{-x}^{(n)} - \left[U, V_-^{(n)}\right], \tag{2.9}$$

a direct calculation reads

$$-V_{+x}^{(n)} + \left[U, V_{+}^{(n)}\right] = -\left(0, \frac{(v_{2,nx} + v_{3,nx})}{2}, \frac{\lambda(v_{2,nx} - v_{3,nx})}{2}\right). \tag{2.10}$$

Then the zero curvature equation

$$U_t - V_x^{(n)} + \left[U, V^{(n)}\right] = 0 \tag{2.11}$$

gives the integrable system

$$U_t = \begin{pmatrix} u_1 \\ u_2 \end{pmatrix}_t = \begin{pmatrix} v_{2,nx} \\ v_{3,nx} \end{pmatrix} = \begin{pmatrix} \partial & 0 \\ 0 & -\partial \end{pmatrix} \begin{pmatrix} v_{2,n} \\ -v_{3,n} \end{pmatrix} = J_1 L_1^n \begin{pmatrix} u_1 \\ u_2 \end{pmatrix}, \tag{2.12}$$

where J_1 is a Hamiltonian operator, and the recurrence operator L_1 is given from (2.7) by

$$L_1 = \begin{pmatrix} -u_1 * \partial^{-1} u_1 * \partial & -\partial - u_1 * \partial^{-1} u_2 * \partial \\ -\partial + u_2 * \partial^{-1} u_1 * \partial & u_2 * \partial^{-1} u_2 * \partial \end{pmatrix}. \tag{2.13}$$

Accordingly, when $M = 1$, the system (2.12) reduces to the Guo hierarchy in [5]; when $M > 1$, the system (2.12) is a multicomponent Guo hierarchy. We observe that $J_1 L_1 = L_1^* J_1$, the Hamiltonian functions H_n $(n \geq 1)$ are involutive with each other, and each H_n is a common conserved density of the expression (2.12). Therefore, the multicomponent Guo hierarchy (2.12) is Liouville integrable.

In order to deduce the Hamiltonian structure of (2.12), we rewrite (2.2) as follows:

$$[a, b] = a^T * R_1(b), \tag{2.14}$$

where

$$R_1(b) = \begin{pmatrix} 0 & 2b_2 & -2b_3 \\ b_3 & -2b_1 & 0 \\ -b_2 & 0 & 2b_1 \end{pmatrix}. \tag{2.15}$$

Solving the matrix equation for the constant matrix F: $R(b) * F = -(R(b) * F)^T$, $F^T = F$, presents

$$F = \begin{pmatrix} 2I_M & 0 & 0 \\ 0 & 0 & I_M \\ 0 & I_M & 0 \end{pmatrix}. \tag{2.16}$$

Then in terms of F, define a linear functional:

$$\{a, b\} = a^T * F * b = 2a_1 * b_1 + a_3 * b_2 + a_2 * b_3. \tag{2.17}$$

It is easy to find that $\{a, b\}$ satisfies the variational identity

$$\frac{\delta}{\delta u}\{V, U_\lambda\} = \lambda^{-\gamma}\frac{\partial}{\partial\lambda}\lambda^{\gamma}\{V, U_u\}. \tag{2.18}$$

By using the above linear functional (2.17), according the variational identity, we have

$$\frac{\delta}{\delta u}\left(v_1 + (u_1 - u_2) * \frac{(v_2 + v_3)}{2}\right) = \lambda^{-\gamma}\frac{\partial}{\partial\lambda}\lambda^{\gamma}(v_2, -v_3)^T. \tag{2.19}$$

Comparing the coefficients of λ^{-n-1} yields

$$\frac{\delta}{\delta u}\left(v_{1,n+1} + (u_1 - u_2) * \frac{(v_{2,n} + v_{3,n})}{2}\right) = (\gamma - n)(v_{2,n}, -v_{3,n})^T. \tag{2.20}$$

Taking $n = 0$ in the above equation gives $\gamma = 0$.

Hence, the Guo hierarchy (2.12) can be written as a Hamiltonian form

$$U_t = \begin{pmatrix} u_1 \\ u_2 \end{pmatrix}_t = J_1 L_1^n \begin{pmatrix} u_1 \\ u_2 \end{pmatrix} = J_1 \begin{pmatrix} \dfrac{\delta}{\delta u_1} \\ \dfrac{\delta}{\delta u_2} \end{pmatrix} H_n = J_1 \frac{\delta H_n}{\delta u}, \tag{2.21}$$

where $H_n = -(1/n)[v_{1,n+1} + (u_1 - u_2) * (v_{2,n} + v_{3,n})/2]$.

Next, we will derive the Guo hierarchy with self-consistent sources. Consider the linear problem

$$\begin{pmatrix} \varphi_{1,j} \\ \varphi_{2,j} \end{pmatrix}_x = U(u, \lambda_j)\begin{pmatrix} \varphi_{1,j} \\ \varphi_{2,j} \end{pmatrix}, \qquad \begin{pmatrix} \varphi_{1,j} \\ \varphi_{2,j} \end{pmatrix}_t = V(u, \lambda_j)\begin{pmatrix} \varphi_{1,j} \\ \varphi_{2,j} \end{pmatrix}. \tag{2.22}$$

The following equation was given in [21]

$$\frac{\delta H_k}{\delta u} + \sum_{j=1}^{N}\alpha_j \frac{\delta \lambda_j}{\delta u} = 0, \tag{2.23}$$

where α_j are constants, and $\delta H_k/\delta u$ determines a finite-dimensional invariant set for the flow (2.21). Here we get [27]

$$\frac{\delta \lambda_j}{\delta u} = \frac{1}{2}\left\{\Psi_j, \frac{\partial U(u, \lambda_j)}{\partial u}\right\}, \quad \Psi_j = (\phi_{1,j} * \phi_{2,j}, -\phi_{1,j} * \phi_{1,j}, \phi_{2,j} * \phi_{2,j}), \tag{2.24}$$

$$\sum_{j=1}^{N}\frac{\delta \lambda_j}{\delta u} = \frac{1}{4}\begin{pmatrix} -\wedge\langle\Phi_1, \Phi_1\rangle + \langle\Phi_2, \Phi_2\rangle \\ \wedge\langle\Phi_1, \Phi_1\rangle + \langle\Phi_2, \Phi_2\rangle \end{pmatrix}, \tag{2.25}$$

where $\phi_{i,j} = (\phi_{i_1,j}, \phi_{i_2,j}, \ldots, \phi_{i_M,j})^T$, $\wedge = \mathrm{diag}(\lambda_1, \ldots, \lambda_N)$, $\Phi_i = (\phi_{i,1}, \ldots, \phi_{i,N})^T$, $i = 1, 2; j = 1, \ldots, N$; $\langle\cdot,\cdot\rangle$ stands for the inner product in R^N.

Then, according to (2.23) and taking $\alpha_i = -4$, the Guo hierarchy with self-consistent sources can be rewritten as follows:

$$\begin{pmatrix} u_1 \\ u_2 \end{pmatrix}_t = J_1 \frac{\delta H_n}{\delta u} + J_1 \begin{pmatrix} -\wedge \langle \Phi_1, \Phi_1 \rangle + \langle \Phi_2, \Phi_2 \rangle \\ \wedge \langle \Phi_1, \Phi_1 \rangle + \langle \Phi_2, \Phi_2 \rangle \end{pmatrix}. \tag{2.26}$$

Taking $n = 1$, we obtain the following multicomponent equations with self-consistent sources

$$\begin{aligned} u_{1,t} - u_{2,x} - u_1 * \frac{(u_2 * u_2 - u_1 * u_1)}{2} &= -\partial \sum_{j=1}^{N} \lambda_j \phi_{1j}^2 + \partial \sum_{j=1}^{N} \phi_{2j}^2, \\ u_{2,t} - u_{1,x} - u_2 * \frac{(u_2 * u_2 - u_1 * u_1)}{2} &= -\partial \sum_{j=1}^{N} \lambda_j \phi_{1j}^2 - \partial \sum_{j=1}^{N} \phi_{2j}^2. \end{aligned} \tag{2.27}$$

3. Nonlinear Bi-Integrable Couplings of Multicomponent Guo Hierarchy with Self-Consistent Sources

The multicomponent Guo hierarchy has been given in the above. In the section, we will construct the nonlinear bi-integrable couplings of multicomponent Guo hierarchy with self-consistent sources by following the way in [19], which is introduced as follows.

Given an integrable equation

$$u_t = K(u), \tag{3.1}$$

which possesses a zero curvature representation

$$U_t - V_x + [U, V] = 0, \tag{3.2}$$

where $U = U(u, \lambda)$, $V = V(u, \lambda)$, and taking $\overline{U} = \overline{U}(\overline{u}, \lambda) = (U(u, \lambda), U_1(u_1, \lambda), U_2(u_2, \lambda))$, $\overline{V} = \overline{V}(\overline{u}, \lambda) = (V(u, \lambda), V_1(u_1, \lambda), V_2(u_2, \lambda))$, then the enlarged zero-curvature equation

$$\overline{U}_t - \overline{V}_x + \left[\overline{U}, \overline{V}\right] = 0 \tag{3.3}$$

equivalently yields

$$\begin{aligned} U_t - V_x + [U, V] &= 0, \\ U_{1,t} - V_{1,x} + [U, V_1] + [U_1, V] + [U_1, V_1] &= 0, \\ U_{2,t} - V_{2,x} + [U, V_2] + [U_2, V] + [U_1, V_1] + [U_1, V_2] + [U_2, V_1] &= 0. \end{aligned} \tag{3.4}$$

We call (3.4) bi-integrable couplings of (3.1). Note that it is nonlinear with respect to the two supplementary variables u_1, u_2, thereby providing candidates for nonlinear bi-integrable couplings.

Next, we will construct an expanding Lie algebra Q_M of the Lie algebra F_M and its corresponding loop algebra. Consider a vector space $Q_M = \text{span}\{a = (a_{ij})_{M\times 9} = (a_1, \dots, a_9)\}$, where a_i $(i = 1, \dots, 9)$ is the ith column of the matrix a.

Definition 3.1. Set a, b as two elements of Q_M, their commutation operation is defined as

$$\begin{aligned}[a,b] = (&a_2 * b_3 - a_3 * b_2, 2(a_1 * b_2 - a_2 * b_1), 2(a_3 * b_1 - a_1 * b_3), a_2 * b_6 - a_6 * b_2 + a_5 * b_3 \\ &- a_3 * b_5 + a_5 * b_6 - a_6 * b_5, 2(a_1 * b_5 - a_5 * b_1 + a_4 * b_2 - a_2 * b_4 + a_4 * b_5 - a_5 * b_4), \\ &2(a_6 * b_1 - a_1 * b_6 + a_3 * b_4 - a_4 * b_3 + a_6 * b_4 - a_4 * b_6), a_2 * b_9 - a_9 * b_2 + a_8 * b_3 \\ &- a_3 * b_8 + a_5 * b_6 - a_6 * b_5 + a_5 * b_9 - a_9 * b_5 + a_8 * b_6 - a_6 * b_8, \\ &2(a_1 * b_8 - a_8 * b_1 + a_7 * b_2 - a_2 * b_7 + a_4 * b_5 - a_5 * b_4 + a_4 * b_8 - a_8 * b_4 \\ &+ a_7 * b_5 - a_5 * b_7), \\ &2(a_9 * b_1 - a_1 * b_9 + a_3 * b_7 - a_7 * b_3 + a_6 * b_4 - a_4 * b_6 + a_9 * b_4 - a_4 * b_9 \\ &+ a_6 * b_7 - a_7 * b_6))^T.\end{aligned} \tag{3.5}$$

It is easy to verify that Q_M consists of a Lie algebra along with (3.5). Its corresponding loop algebra $\widetilde{Q}_M$ is given by

$$\widetilde{Q}_M = \{a\lambda^n,\ a \in Q_M,\ n = 0, \pm 1, \pm 2, \dots\}, \tag{3.6}$$

equipped with a commutation operation as follows:

$$[a\lambda^m, b\lambda^n] = [a,b]\lambda^{m+n}, \quad \forall a, b \in Q_M. \tag{3.7}$$

Denoting $\widetilde{Q}_M(1) = \{(a_1, a_2, a_3, 0, 0, 0, 0, 0, 0)\lambda^n\}$ and $\widetilde{Q}_M(2) = \{(0, 0, 0, a_4, a_5, a_6, a_7, a_8, a_9)\lambda^n\}$, then we have

$$\widetilde{Q}_M = \widetilde{Q}_M(1) \oplus \widetilde{Q}_M(2), \qquad \widetilde{Q}_M(1) \cong \widetilde{F}_M, \qquad \left[\widetilde{Q}_M(1), \widetilde{Q}_M(2)\right] \subset \widetilde{Q}_M(2). \tag{3.8}$$

Here we note that the subalgebras $\widetilde{Q}_M(1)$ and $\widetilde{Q}_M(2)$ are both non-semisimple, which is very important for deriving the nonlinear integrable couplings of hierarchy.

In terms of the loop algebra $\widetilde{Q}_M$, consider an isospectral problem as follows:

$$\varphi_x = U\varphi, \qquad U = \frac{1}{2}(\lambda I_M, u_1 + u_2, \lambda(u_1 - u_2), 0, u_3 + u_4, \lambda(u_3 - u_4), 0, u_5 + u_6, \lambda(u_5 - u_6))^T. \tag{3.9}$$

Set

$$V = (\lambda v_1, v_2 + v_3, \lambda(v_2 - v_3), \lambda v_4, v_5 + v_6, \lambda(v_5 - v_6), \lambda v_7, v_8 + v_9, \lambda(v_8 - v_9))^T, \tag{3.10}$$

where $v_k = \sum_{m=0}^{\infty}(v_{k_1,m}, v_{k_2,m}, \ldots, v_{k_M,m})^T \lambda^{-m}$, $k = 1, \ldots, 9$. A solution to $V_x = [U, V]$ gives

$$\begin{aligned}
v_{1,mx} &= u_2 * v_{2,m} - u_1 * v_{3,m},\\
v_{2,mx} &= v_{3,m+1} - u_2 * v_{1,m+1},\\
v_{3,mx} &= v_{2,m+1} - u_1 * v_{1,m+1},\\
v_{4,mx} &= u_2 * v_{5,m} - u_1 * v_{6,m} + u_4 * v_{2,m} - u_3 * v_{3,m} + u_4 * v_{5,m} - u_3 * v_{6,m},\\
v_{5,mx} &= 2v_{6,m+1} - u_2 * v_{4,m+1} + v_{3,m+1} - u_4 * v_{1,m+1} - u_4 * v_{4,m+1},\\
v_{6,mx} &= 2v_{5,m+1} - u_1 * v_{4,m+1} + v_{2,m+1} - u_3 * v_{1,m+1} - u_3 * v_{4,m+1},\\
v_{7,mx} &= u_2 * v_{8,m} - u_1 * v_{9,m} + u_6 * v_{2,m} - u_5 * v_{3,m} + u_4 * v_{5,m} - u_3 * v_{6,m}\\
&\quad + u_4 * v_{8,m} - u_3 * v_{9,m} + u_6 * v_{5,m} - u_5 * v_{6,m},\\
v_{8,mx} &= 2v_{9,m+1} - u_2 * v_{7,m+1} + v_{3,m+1} - u_6 * v_{1,m+1} + 2v_{6,m+1} - u_4 * v_{4,m+1}\\
&\quad - u_4 * v_{7,m+1} - u_6 * v_{4,m+1},\\
v_{9,mx} &= 2v_{8,m+1} - u_1 * v_{7,m+1} + v_{2,m+1} - u_5 * v_{1,m+1} + 2v_{5,m+1} - u_3 * v_{4,m+1}\\
&\quad - u_3 * v_{7,m+1} - u_5 * v_{4,m+1},\\
v_{1,0} &= 0I_M, \qquad v_{2,0} = u_1, \qquad v_{3,0} = u_2, \qquad v_{4,0} = 0I_M,\\
v_{5,0} &= u_1 + 2u_3, \qquad v_{6,0} = u_2 + 2u_4, \qquad v_{7,0} = 0I_M,\\
v_{8,0} &= u_1 + 2u_3 + 2u_5, \qquad v_{9,0} = u_2 + 2u_4 + 2u_6, \qquad v_{1,1} = \frac{(u_2 * u_2 - u_1 * u_1)}{2},\\
v_{2,1} &= u_{2,x} + u_1 * \frac{(u_2 * u_2 - u_1 * u_1)}{2}, \qquad v_{3,1} = u_{1,x} + u_2 * \frac{(u_2 * u_2 - u_1 * u_1)}{2},\\
v_{4,1} &= \frac{[(u_2 + u_4) * (u_2 + u_4) - (u_1 + u_3) * (u_1 + u_3)]}{2} - \frac{(u_2 * u_2 - u_1 * u_1)}{2},\\
v_{5,1} &= u_{4,x} - \frac{u_1 * (u_2 * u_2 - u_1 * u_1)}{2} + \frac{(u_1 + u_3) * [(u_2 + u_4) * (u_2 + u_4) - (u_1 + u_3) * (u_1 + u_3)]}{4},\\
v_{6,1} &= u_{3,x} - \frac{u_2 * (u_2 * u_2 - u_1 * u_1)}{2} + \frac{(u_2 + u_4) * [(u_2 + u_4) * (u_2 + u_4) - (u_1 + u_3) * (u_1 + u_3)]}{4},\\
v_{7,1} &= \frac{(u_4 * u_4 - u_3 * u_3)}{2} + [(u_2 + u_4) * u_6 - (u_1 + u_3) * u_5],\\
v_{8,1} &= u_{6,x} + \frac{(u_5 - u_1) * [(u_2 + u_4) * (u_2 + u_4) - (u_1 + u_3) * (u_1 + u_3)]}{4}\\
&\quad + \frac{(u_1 - u_3) * (u_2 * u_2 - u_1 * u_1)}{4} + \frac{(u_1 + u_3) * (u_4 * u_4 - u_3 * u_3)}{4}\\
&\quad + \frac{(u_1 + u_3) * [(u_2 + u_4) * u_6 - (u_1 + u_3) * u_5]}{2},\\
v_{9,1} &= u_{5,x} + \frac{(u_6 - u_2) * [(u_2 + u_4) * (u_2 + u_4) - (u_1 + u_3) * (u_1 + u_3)]}{4}\\
&\quad + \frac{(u_2 - u_4) * (u_2 * u_2 - u_1 * u_1)}{4} + \frac{(u_2 + u_4) * (u_4 * u_4 - u_3 * u_3)}{4}\\
&\quad + \frac{(u_2 + u_4) * [(u_2 + u_4) * u_6 - (u_1 + u_3) * u_5]}{2}.
\end{aligned} \tag{3.11}$$

Similar to (2.9), we obtain

$$-V_{+x}^{(n)} + \left[U, V_{+}^{(n)}\right] = \left(0, \frac{(v_{2,nx} + v_{3,nx})}{2}, \frac{\lambda(v_{2,nx} - v_{3,nx})}{2}, 0, v_{5,nx} + v_{6,nx} + \frac{(v_{2,nx} + v_{3,nx})}{2}, \right.$$
$$\lambda(v_{5,nx} - v_{6,nx}) + \frac{\lambda(v_{2,nx} - v_{3,nx})}{2}, 0, (v_{8,nx} + v_{9,nx} + v_{5,nx} + v_{6,nx})$$
$$\left. + \frac{(v_{2,nx} + v_{3,nx})}{2}, (v_{8,nx} - v_{9,nx} + v_{5,nx} - v_{6,nx}) + \frac{(v_{2,nx} - v_{3,nx})}{2}\right). \tag{3.12}$$

Therefore, the zero-curvature equation

$$U_t - V_x^{(n)} + \left[U, V^{(n)}\right] = 0 \tag{3.13}$$

admits the following bigger integrable system:

$$U_t = \begin{pmatrix} u_1 \\ u_2 \\ u_3 \\ u_4 \\ u_5 \\ u_6 \end{pmatrix}_t = \begin{pmatrix} \partial & 0 & -\partial & 0 & \partial & 0 \\ 0 & -\partial & 0 & \partial & 0 & -\partial \\ -\partial & 0 & \partial & 0 & \partial & 0 \\ 0 & \partial & 0 & -\partial & 0 & -\partial \\ \partial & 0 & \partial & 0 & -\partial & 0 \\ 0 & -\partial & 0 & -\partial & 0 & \partial \end{pmatrix} \begin{pmatrix} v_{2,n} + v_{5,n} + v_{8,n} \\ -v_{3,n} - v_{6,n} - v_{9,n} \\ v_{2,n} + 2v_{5,n} + v_{8,n} \\ -v_{3,n} - 2v_{6,n} - v_{9,n} \\ v_{2,n} + v_{5,n} \\ -v_{3,n} - v_{6,n} \end{pmatrix}$$
$$= J_2 P_n = J_2 L_2^n \begin{pmatrix} v_{2,0} + v_{5,0} + v_{8,0} \\ -v_{3,0} - v_{6,0} - v_{9,0} \\ v_{2,0} + 2v_{5,0} + v_{8,0} \\ -v_{3,0} - 2v_{6,0} - v_{9,0} \\ v_{2,0} + v_{5,0} \\ -v_{3,0} - v_{6,0} \end{pmatrix}, \tag{3.14}$$

where J_2 is the Hamiltonian operator. Similar to (2.13), by employing (3.11), after tedious calculation, the recurrence operator L_2 can be obtained and here we omit it.

Taking $u_3 = u_4 = 0$ in (3.14) reduces to an integrable coupling of the Guo hierarchy; taking $u_5 = u_6 = 0$ in (3.14) reduces to another integrable coupling of the Guo hierarchy. Here we note that except the first variables u_1, u_2, all integrable couplings presented above are nonlinear, since the supplementary systems (3.14) with $n \geq 1$ are nonlinear with respect to the four dependent variables u_3, u_4, u_5, u_6. So, in term of the definition of integrable couplings, we conclude that the system (3.14) is nonlinear bi-integrable couplings of the multicomponent Guo hierarchy.

In order to deduce to the Hamiltonian structure of (3.14), we rewrite (3.5) as follows:

$$[a, b] = a^T R(b), \tag{3.15}$$

where

$$R(b) = \begin{pmatrix} 0 & 2b_2 & -2b_3 & 0 & 2b_5 & -2b_6 & 0 & 2b_8 & -2b_9 \\ b_3 & -2b_1 & 0 & b_6 & -2b_4 & 0 & b_9 & -2b_7 & 0 \\ -b_2 & 0 & 2b_1 & -b_5 & 0 & 2b_4 & -b_8 & 0 & 2b_7 \\ 0 & 0 & 0 & 0 & 2(b_2+b_5) & -2(b_3+2b_6) & 0 & 2(b_5+b_8) & -2(b_6+b_9) \\ 0 & 0 & 0 & b_3+b_6 & -2(b_1+b_4) & 0 & b_6+b_9 & -2(b_4+b_7) & 0 \\ 0 & 0 & 0 & -b_2-b_5 & 0 & 2(b_1+b_4) & -b_5-b_8 & 0 & 2(b_4+b_7) \\ 0 & 0 & 0 & 0 & 0 & 0 & 0 & 2(b_2+b_5) & -2(b_3+b_6) \\ 0 & 0 & 0 & 0 & 0 & 0 & b_3+b_6 & -2(b_1+b_4) & 0 \\ 0 & 0 & 0 & 0 & 0 & 0 & -(b_2+b_5) & 0 & 2(b_1+b_4) \end{pmatrix}. \quad (3.16)$$

Solving the matrix equation for the constant matrix F: $R(b) * F = -(R(b) * F)^T$, $F^T = F$, presents

$$F = \begin{pmatrix} 2I_M & 0 & 0 & 2I_M & 0 & 0 & 2I_M & 0 & 0 \\ 0 & 0 & I_M & 0 & 0 & I_M & 0 & 0 & I_M \\ 0 & I_M & 0 & 0 & I_M & 0 & 0 & I_M & 0 \\ 2I_M & 0 & 0 & 4I_M & 0 & 0 & 2I_M & 0 & 0 \\ 0 & 0 & I_M & 0 & 0 & 2I_M & 0 & 0 & I_M \\ 0 & I_M & 0 & 0 & 2I_M & 0 & 0 & I_M & 0 \\ 2I_M & 0 & 0 & 2I_M & 0 & 0 & 0 & 0 & 0 \\ 0 & 0 & I_M & 0 & 0 & I_M & 0 & 0 & 0 \\ 0 & I_M & 0 & 0 & I_M & 0 & 0 & 0 & 0 \end{pmatrix}. \quad (3.17)$$

Then in terms of F, define a linear functional

$$\begin{aligned} \{a, b\} &= a^T * F * b \\ &= 2(a_1 + a_4 + a_7) * b_1 + (a_3 + a_6 + a_9) * b_2 \\ &\quad + (a_2 + a_5 + a_8) * b_3 + 2(a_1 + 2a_4 + a_7) * b_4 + (a_3 + 2a_6 + a_9) * b_5 \\ &\quad + (a_2 + 2a_5 + a_8) * b_6 + 2(a_1 + a_4) * b_7 + (a_3 + a_6) * b_8 + (a_2 + a_5) \ * b_9. \end{aligned} \quad (3.18)$$

It is easy to find that $\{a, b\}$ satisfies the variational identity (2.18). By using the above linear functional, we have

$$\begin{aligned} \{V, U_\lambda\} &= \lambda(v_1 + v_4 + v_7) + \frac{1}{2}(v_2 + v_3 + v_5 + v_6 + v_8 + v_9) * (u_1 - u_2) \\ &\quad + \frac{1}{2}(v_2 + v_3 + 2v_5 + 2v_6 + v_8 + v_9) * (u_3 - u_4) + \frac{1}{2}(v_2 + v_3 + v_5 + v_6) * (u_5 - u_6), \\ \{V, U_{u_1}\} &= \lambda(v_2 + v_5 + v_8), \qquad \{V, U_{u_2}\} = \lambda(-v_3 - v_6 - v_9), \qquad \{V, U_{u_3}\} = \lambda(v_2 + 2v_5 + v_8), \\ \{V, U_{u_4}\} &= \lambda(-v_3 - v_6 - v_9), \qquad \{V, U_{u_5}\} = \lambda(v_2 + v_5), \qquad \{V, U_{u_6}\} = \lambda(-v_3 - v_6). \end{aligned} \quad (3.19)$$

Substituting (3.19) into the variational identity (2.18), we have

$$\begin{aligned}
&\frac{\delta}{\delta u}\Bigg(\lambda(v_1+v_4+v_7)+\frac{1}{2}(v_2+v_3+v_5+v_6+v_8+v_9)*(u_1-u_2)\\
&\quad+\frac{1}{2}(v_2+v_3+2v_5+2v_6+v_8+v_9)*(u_3-u_4)+\frac{1}{2}(v_2+v_3+v_5+v_6)*(u_5-u_6)\Bigg)\\
&=\lambda^{-\gamma}\frac{\partial}{\partial\lambda}\lambda^{\gamma}(\lambda(v_2+v_5+v_8),\lambda(-v_3-v_6-v_9),\lambda(v_2+2v_5+v_8),\lambda(-v_3-v_6-v_9),\\
&\qquad\lambda(v_2+v_5),\lambda(-v_3-v_6))^T.
\end{aligned}\tag{3.20}$$

Comparing the coefficients of λ^{-n-1} yields

$$\begin{aligned}
&\frac{\delta}{\delta u}\Bigg(v_{1,n+1}+v_{4,n+1}+v_{7,n+1}+\frac{1}{2}(v_{2,n}+v_{3,n}+v_{5,n}+v_{6,n}+v_{8,n}+v_{9,n})*(u_1-u_2)\\
&\quad+\frac{1}{2}(v_{2,n}+v_{3,n}+2v_{5,n}+2v_{6,n}+v_{8,n}+v_{9,n})(u_3-u_4)\\
&\quad+\frac{1}{2}(v_{2,n}+v_{3,n}+v_{5,n}+v_{6,n})*(u_5-u_6)\Bigg)\\
&=(\gamma-n)((v_{2,n}+v_{5,n}+v_{8,n}),(-v_{3,n}-v_{6,n}-v_{9,n}),(v_{2,n}+2v_{5,n}+v_{8,n}),\\
&\qquad(-v_{3,n}-2v_{6,n}-v_{9,n}),(v_{2,n}+v_{5,n}),(-v_{3,n}-v_{6,n}))^T.
\end{aligned}\tag{3.21}$$

Taking $n = 0$ in the above equation gives $\gamma = 0$.

Hence, the nonlinear bi-integrable couplings of the multicomponent Guo hierarchy (3.14) can be written as a Hamiltonian form

$$U_t=\begin{pmatrix}u_1\\u_2\\u_3\\u_4\\u_5\\u_6\end{pmatrix}_t=J_2\begin{pmatrix}\frac{\delta}{\delta u_1}\\\frac{\delta}{\delta u_2}\\\frac{\delta}{\delta u_3}\\\frac{\delta}{\delta u_4}\\\frac{\delta}{\delta u_5}\\\frac{\delta}{\delta u_6}\end{pmatrix}\overline{H}_n=J_2\frac{\delta\overline{H}_n}{\delta u},\tag{3.22}$$

where $\overline{H}_n=-(1/n)[v_{1,n+1}+v_{4,n+1}+v_{7,n+1}+(1/2)(v_{2,n}+v_{3,n}+v_{5,n}+v_{6,n}+v_{8,n}+v_{9,n})*(u_1-u_2)+(1/2)(v_{2,n}+v_{3,n}+2v_{5,n}+2v_{6,n}+v_{8,n}+v_{9,n})(u_3-u_4)+(1/2)(v_{2,n}+v_{3,n}+v_{5,n}+v_{6,n})*(u_5-u_6)]$.

Next, we will derive the nonlinear bi-integrable couplings of the multicomponent Guo hierarchy with self-consistent sources. Consider the linear problem

$$\begin{pmatrix} \psi_{1,j} \\ \psi_{2,j} \\ \psi_{3,j} \\ \psi_{4,j} \\ \psi_{5,j} \\ \psi_{6,j} \end{pmatrix}_x = U(u,\lambda_j)\begin{pmatrix} \psi_{1,j} \\ \psi_{2,j} \\ \psi_{3,j} \\ \psi_{4,j} \\ \psi_{2,j} \\ \psi_{6,j} \end{pmatrix}, \quad \begin{pmatrix} \psi_{1,j} \\ \psi_{2,j} \\ \psi_{3,j} \\ \psi_{4,j} \\ \psi_{5,j} \\ \psi_{6,j} \end{pmatrix}_t = V(u,\lambda_j)\begin{pmatrix} \psi_{1,j} \\ \psi_{2,j} \\ \psi_{3,j} \\ \psi_{4,j} \\ \psi_{5,j} \\ \psi_{6,j} \end{pmatrix}. \tag{3.23}$$

According to (2.24) and the definition of $\{a,b\}$, we get

$$\begin{aligned}
\Psi_j &= \big(\phi_{1,j} * \phi_{2,j}, -\phi_{1,j} * \phi_{1,j}, \phi_{2,j} * \phi_{2,j}, \phi_{3,j} * \phi_{4,j}, -\phi_{3,j} * \phi_{3,j}, \phi_{4,j} * \phi_{4,j}, \phi_{5,j} * \phi_{6,j}, \\
&\qquad -\phi_{5,j} * \phi_{5,j}, \phi_{6,j} * \phi_{6,j}\big), \\
\sum_{j=1}^{N}\frac{\delta\lambda_j}{\delta u} &= \frac{1}{4}W \\
&= \frac{1}{4}\begin{pmatrix}
\langle\Phi_2,\Phi_2\rangle + \langle\Phi_4,\Phi_4\rangle + \langle\Phi_6,\Phi_6\rangle - \Lambda(\langle\Phi_1,\Phi_1\rangle + \langle\Phi_3,\Phi_3\rangle + \langle\Phi_5,\Phi_5\rangle) \\
\langle\Phi_2,\Phi_2\rangle + \langle\Phi_4,\Phi_4\rangle + \langle\Phi_6,\Phi_6\rangle + \Lambda(\langle\Phi_1,\Phi_1\rangle + \langle\Phi_3,\Phi_3\rangle + \langle\Phi_5,\Phi_5\rangle) \\
\langle\Phi_2,\Phi_2\rangle + 2\langle\Phi_4,\Phi_4\rangle + \langle\Phi_6,\Phi_6\rangle - \Lambda(\langle\Phi_1,\Phi_1\rangle + 2\langle\Phi_3,\Phi_3\rangle + \langle\Phi_5,\Phi_5\rangle) \\
\langle\Phi_2,\Phi_2\rangle + 2\langle\Phi_4,\Phi_4\rangle + \langle\Phi_6,\Phi_6\rangle + \Lambda(\langle\Phi_1,\Phi_1\rangle + 2\langle\Phi_3,\Phi_3\rangle + \langle\Phi_5,\Phi_5\rangle) \\
\langle\Phi_2,\Phi_2\rangle + \langle\Phi_4,\Phi_4\rangle - \Lambda(\langle\Phi_1,\Phi_1\rangle + \langle\Phi_3,\Phi_3\rangle) \\
\langle\Phi_2,\Phi_2\rangle + \langle\Phi_4,\Phi_4\rangle - \Lambda(\langle\Phi_1,\Phi_1\rangle + \langle\Phi_3,\Phi_3\rangle)
\end{pmatrix},
\end{aligned} \tag{3.24}$$

where $\phi_{i,j} = (\phi_{i_1,j}, \phi_{i_2,j}, \ldots, \phi_{i_M,j})^T$, $\Lambda = \operatorname{diag}(\lambda_1, \ldots, \lambda_N)$, $\Phi_i = (\phi_{i,1}, \ldots, \phi_{i,N})^T$, $i = 1, \ldots, 6$; $j = 1, \ldots, N$; $\langle\cdot,\cdot\rangle$ stands for the inner product in R^N.

Then, according to (2.23) and taking $\alpha_i = -4$, the nonlinear bi-integrable couplings of the multicomponent Guo hierarchy with self-consistent sources can be rewritten as follows:

$$(u_1, u_2, u_3, u_4, u_5, u_6)_t^T = J_2\frac{\delta\overline{H}_n}{\delta u} + J_2 W. \tag{3.25}$$

4. Conclusions

In this paper, we derive the multicomponent Guo hierarchy with self-consistent sources and its nonlinear bi-integrable couplings by using the Lie algebra Q_M. As far as we know, the existing papers are not designed for constructing the nonlinear bi-integrable couplings of multicomponent hierarchy with self-consistent sources. This method can be used to derive the other multicomponent soliton hierarchies with self-consistent sources and their nonlinear bi-integrable couplings.

Acknowledgments

This work was supported by Innovation Project of Chinese Academy of Sciences (no. KZCX2-EW-209) and National Natural Science Foundation of China (no. 11271007).

References

[1] W. X. Ma and B. Fuchssteiner, "Integrable theory of the perturbation equations," *Chaos, Solitons and Fractals*, vol. 7, no. 8, pp. 1227–1250, 1996.

[2] W. X. Ma, "Integrable couplings of soliton equations by perturbations. I: a general theory and application to the KdV hierarchy," *Methods and Applications of Analysis*, vol. 7, no. 1, pp. 21–55, 2000.

[3] G. Z. Tu, "The trace identity, a powerful tool for constructing the Hamiltonian structure of integrable systems," *Journal of Mathematical Physics*, vol. 30, no. 2, pp. 330–338, 1989.

[4] W. X. Ma, "A new hierarchy of Liouville integrable generalized Hamiltonian equations and its reduction," *Chinese Journal of Contemporary Mathematics*, vol. 13, no. 1, pp. 79–89, 1992.

[5] F. K. Guo, "Two hierarchies of integrable Hamiltonian equations," *Mathematica Applicata*, vol. 9, no. 4, pp. 495–499, 1996.

[6] Y. F. Zhang, "Two types of new Lie algebras and corresponding hierarchies of evolution equations," *Physics Letters A*, vol. 310, no. 1, pp. 19–24, 2003.

[7] F. K. Guo and Y. F. Zhang, "A new loop algebra and a corresponding integrable hierarchy, as well as its integrable coupling," *Journal of Mathematical Physics*, vol. 44, no. 12, pp. 5793–5803, 2003.

[8] W. X. Ma, "A Hamiltonian structure associated with a matrix spectral problem of arbitrary-order," *Physics Letters A*, vol. 367, no. 6, pp. 473–477, 2007.

[9] T. C. Xia, F. C. You, and W. Y. Zhao, "Multi-component Levi hierarchy and its multi-component integrable coupling system," *Communications in Theoretical Physics*, vol. 44, no. 6, pp. 990–996, 2005.

[10] H.-H. Dong and N. Zhang, "The quadratic-form identity for constructing Hamiltonian structures of the Guo hierarchy," *Chinese Physics*, vol. 15, no. 9, pp. 1919–1926, 2006.

[11] T. Xia, F. Yu, and Y. Zhang, "The multi-component coupled Burgers hierarchy of soliton equations and its multi-component integrable couplings system," *Physica A*, vol. 343, no. 15, pp. 238–246, 2004.

[12] Z. Li and H. Dong, "Integrable couplings of the multi-component dirac hierarchy and its Hamiltonian structure," *Chaos, Solitons & Fractals*, vol. 36, no. 2, pp. 290–295, 2008.

[13] W. X. Ma and Y. Zhang, "Component-trace identities for Hamiltonian structures," *Applicable Analysis*, vol. 89, no. 4, pp. 457–472, 2010.

[14] W. X. Ma, X. X. Xu, and Y. F. Zhang, "Semi-direct sums of Lie algebras and continuous integrable couplings," *Physics Letters A*, vol. 351, no. 3, pp. 125–130, 2006.

[15] W. X. Ma, "Variational identities and Hamiltonian structures," *AIP Conference Proceedings*, vol. 1212, no. 1, pp. 1–27, 2010.

[16] W. X. Ma, "Nonlinear continuous integrable Hamiltonian couplings," *Applied Mathematics and Computation*, vol. 217, no. 17, pp. 7238–7244, 2011.

[17] Y. F. Zhang, "Lie algebras for constructing nonlinear integrable couplings," *Communications in Theoretical Physics*, vol. 56, no. 5, pp. 805–812, 2011.

[18] Y. F. Zhang and B. L. Feng, "A few Lie algebras and their applications for generating integrable hierarchies of evolution types," *Communications in Nonlinear Science and Numerical Simulation*, vol. 16, no. 8, pp. 3045–3061, 2011.

[19] W. X. Ma, "Loop algebras and bi-integrable couplings," *Chinese Annals of Mathematics B*, vol. 33, no. 2, pp. 207–224, 2012.

[20] M. Antonowicz, "Gelfand-Dikii hierarchies with the sources and Lax representation for restricted flows," *Physics Letters A*, vol. 165, no. 1, pp. 47–52, 1992.

[21] Y. B. Zeng, W. X. Ma, and R. Lin, "Integration of the soliton hierarchy with self-consistent sources," *Journal of Mathematical Physics*, vol. 41, no. 8, pp. 5453–5489, 2000.

[22] V. K. Mel'nikov, "New method for deriving nonlinear integrable systems," *Journal of Mathematical Physics*, vol. 31, no. 5, pp. 1106–1113, 1990.

[23] C. Claude, A. Latifi, and J. Leon, "Nonlinear resonant scattering and plasma instability: an integrable model," *Journal of Mathematical Physics*, vol. 32, no. 12, pp. 3310–3321, 1991.

[24] F. J. Yu, "A kind of integrable couplings of soliton equations hierarchy with self-consistent sources associated with $\widetilde{sl}(4)$," *Physics Letters A*, vol. 372, no. 44, pp. 6613–6621, 2008.

[25] T. C. Xia, "Two new integrable couplings of the soliton hierarchies with self-consistent sources," *Chinese Physics B*, vol. 19, no. 10, Article ID 100303, 2010.
[26] F. J. Yu and L. Li, "Integrable coupling system of JM equations hierarchy with self-consistent sources," *Communications in Theoretical Physics*, vol. 53, no. 1, pp. 6–12, 2010.
[27] G. Z. Tu, "An extension of a theorem on gradients of conserved densities of integrable systems," *Northeastern Mathematical Journal*, vol. 6, no. 1, pp. 26–32, 1990.

10

Nonlinear Effects of Electromagnetic TM Wave Propagation in Anisotropic Layer with 0Kerr Nonlinearity

Yu G. Smirnov and D. V. Valovik

Department of Mathematics and Supercomputer Modeling, Penza State University, Krasnaya Street 40, Penza 440026, Russia

Correspondence should be addressed to D. V. Valovik, dvalovik@mail.ru

Academic Editor: Vladimir B. Taranov

The problem of electromagnetic TM wave propagation through a layer with Kerr nonlinearity is considered. The layer is located between two half-spaces with constant permittivities. This electromagnetic problem is reduced to the nonlinear boundary eigenvalue problem for ordinary differential equations. It is necessary to find eigenvalues of the problem (propagation constants of an electromagnetic wave). The dispersion equation (DE) for the eigenvalues is derived. The DE is applied to nonlinear metamaterial as well. Comparison with a linear case is also made. In the nonlinear problem there are new eigenvalues and new eigenwaves. Numerical results are presented.

1. Introduction

Problems of electromagnetic wave propagation in nonlinear waveguide structures are intensively investigated during several decades. First known studies about nonlinear optics' problems are given in the monographs [1, 2]. Propagation of electromagnetic wave in a layer and a circle cylindrical waveguide are among such problems. Phenomena of electromagnetic wave propagation in nonlinear media have original importance and also find a lot of applications, for example, in plasma physics, microelectronics, optics, and laser technology. There are a lot of different nonlinear phenomena in media when an electromagnetic wave propagates, such as self-focusing, defocusing, and self-channeling [1–5].

Investigation of nonlinear phenomena leads us to solve nonlinear differential equations. In some cases it is necessary to solve nonlinear boundary eigenvalue problems (NBEPs), which rarely can be solved analytically. One of the important nonlinear phenomenan is the case when the permittivity of the sample depends on electric field intensity. And one of the simplest nonlinearities is a Kerr nonlinearity [4, 6, 7]. When we speak about NBEPs we mean that differential equations and boundary conditions nonlinearly depend on the spectral parameter and also the differential equations nonlinearly depend on the sought-for functions. These facts do not allow to apply well-known methods of spectral problems' investigation.

Here we consider electromagnetic TM wave propagation in a layer with Kerr nonlinearity. Perhaps, the papers [6, 7] were the first studies where some problems of electromagnetic wave propagation are considered in a strong electromagnetic statement. Propagation of polarized electromagnetic waves in a layer and in a circle cylindrical waveguide with Kerr nonlinearity is considered in this paper. When one says that the permittivity ε is described by Kerr law this means that (for an isotropic material) $\varepsilon = \varepsilon_{\text{const}} + \alpha|\mathbf{E}|^2$, $\varepsilon_{\text{const}}$ is the constant part of the permittivity ε; α is the nonlinearity coefficient; $|\mathbf{E}|^2 = E_x^2 + E_y^2 + E_z^2$, where $\mathbf{E} = (E_x, E_y, E_z)$ is an electric field. Below we consider an anisotropic case (some results were presented in [8]). The first approximation for eigenvalues of the problem is presented in [9].

Problems of electromagnetic wave propagation in a linear layer (with constant permittivity) and in a linear circle cylindrical waveguide were deeply studied many years ago, see, for example, [10]. Such problems are formulated as boundary eigenvalue problems for ordinary differential equations. Indeed, the main interest in this problem is the value of the spectral parameter (eigenvalues) which corresponds to the propagating wave. If an eigenvalue is known it is easy to solve differential equations numerically. Otherwise numerical methods cannot be successfully applied. However, in nonlinear cases it is often paid more attention to solve the differential equations (see, e.g., [11–13]). Though the first problem is to find eigenvalues therefore to find Des, from the mathematical standpoint the DE is an equation with respect to the spectral parameter. Analysis of this equation allows us to make conclusions about problem's solvability, eigenvalues' localization, and so forth. In most cases the equations of the problem cannot be integrated in an explicit form. Of course, if one has the explicit solutions of the differential equations it is easy to derive the DE. Therefore, when the equations cannot be integrated things do not go to a DE. However, in some cases the DE can be found in an explicit form and it is not necessary to have explicit solutions of differential equations.

Let us discuss in detail the case of Kerr nonlinearity. The work [4] contains a wide range of details of third-order nonlinear electromagnetic TE and TM guided waves. Problems of surface wave propagation along the interface between two semi-infinite linear or/and nonlinear media were studied completely (see the results in [4]). At the same time we should notice that problems of wave propagation in a nonlinear layer that is located between two semi-infinite linear or/and nonlinear media are much more difficult than (and cannot be reduced to) the problems where surface waves are considered only at the interface between two semi-infinite linear or/and nonlinear media. Propagation of TE waves was more studied. The work [14] is devoted to the problem of electromagnetic wave propagation in a nonlinear dielectric layer with absorption and the case of Kerr nonlinearity is considered separately. One of the most interesting works about propagation of TE waves in a layered structure with Kerr nonlinearity is the paper [15]. Also the reader can see the work [16], where a layer with Kerr nonlinearity without absorption is considered.

The case of TM wave propagation in a nonlinear medium is more complicated. This is due to the fact that two components of the electric field make the analysis much harder [17].

In the work [18] a linear dielectric layer is considered. The layer is located between two half-spaces. The half-spaces are filled by nonlinear medium with Kerr nonlinearity. This problem for TE waves is solved analytically [19, 20]. For the TM case in [18] obtained DE is an algebraic equation. It should be noticed that in [18] authors simplify the problem. Earlier in [21] the DE is obtained with other simplifying assumption (authors take into account only one component E_x of the electric field). Later in [22] it is proved that the dominating nonlinear contribution in the permittivity is proportional to the transversal component E_z. In the works [11] propagation of TM waves in a nonlinear half-space with Kerr nonlinearity is considered. Formal solutions of differential equations in quadratures are obtained. In the paper [11] DEs are presented for isotropic and anisotropic media in a half-space with nonlinear permittivity. The DEs are rational functions with respect to the value of field's components at the interface. Authors found the first integral of the system of differential equations (so called a conservation law). This is also very interesting work to study, another way to simplify the problem pointed out in [23].

In the case of TE wave you can see the papers [24–26]. Propagation of TM wave in terms of the magnetic component is studied in [12, 13]. The paper [21] is devoted to the question (from physical standpoint) why it is possible to take into account only one component of the electric field in the expression for permittivity in the case of TM waves in a nonlinear layer. The results are compared with the case of TE waves.

The most important results about TM wave propagation in a layer with Kerr nonlinearity (system of differential equations, first integral) and a circle cylindrical waveguide (system of differential equations) were obtained in [6, 7]. In some papers (e.g., [12]) polarized wave propagation in a layer with arbitrary nonlinearity is considered. However, DEs were not obtained and no results about solvability of the boundary eigenvalue problem were obtained as well. The problem of TM wave propagation in a layer with Kerr nonlinearity is solved at first for a thin layer and then for a layer of arbitrary thickness [27–29]. Theorems of existence and localization of eigenvalues are proved in [30, 31]. Some numerical results are shown in [8, 9].

In this paper the DE is an equation with additional conditions. Only for linear media (when permittivity is a constant) in a layer or in a circle cylindrical waveguide the DEs are sufficiently simple (but even for these cases the DEs are transcendental equations). For a nonlinear layer the DE is quite complicated nonlinear integral equation, where the integrand is defined by implicit algebraic function. It should be stressed that in spite of the fact that the DE is complicated it can be rather easily solved numerically.

This DE allows to study both nonlinear materials and nonlinear metamaterials. It should be noticed that in this paper materials with nonlinear permittivity and constant positive permeability are studied. But it is not difficult to take into account the sign of the permeability.

Problems of propagation of TE and TM waves in a nonlinear circle cylindrical waveguide are also close to the problem considered here. These problems are more complicated in comparison with corresponding problems in nonlinear layers. And even in the case of Kerr nonlinearity the results are not so complete as in the case in layers [30, 32, 33].

2. Statement of the Problem

Let us consider electromagnetic wave propagation through a homogeneous anisotropic nonmagnetic dielectric layer. The layer is located between two half-spaces: $x < 0$ and $x > h$ in Cartesian coordinate system $Oxyz$. The half-spaces are filled with isotropic nonmagnetic media without any sources and characterized by permittivities $\varepsilon_1 \geq \varepsilon_0$ and $\varepsilon_3 \geq \varepsilon_0$, respectively, where ε_0 is the permittivity of free space. Assume that everywhere $\mu = \mu_0$ is the permeability of free space (see Figure 1).

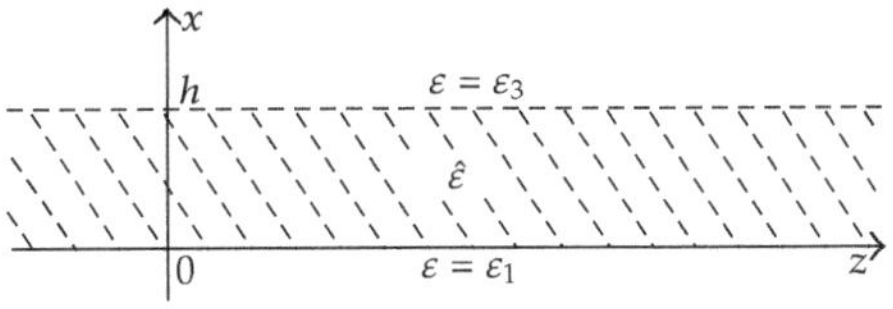

Figure 1: The geometry of the problem.

It should be noticed that conditions $\varepsilon_1 \geq \varepsilon_0$, $\varepsilon_3 \geq \varepsilon_0$ are not necessary. They are not used for derivation of DEs, but they are useful for DEs' solvability analysis.

The electromagnetic field depends on time harmonically [6]:

$$\begin{aligned}\widetilde{\mathbf{E}}(x,y,z,t) &= \mathbf{E}_+(x,y,z)\cos\omega t + \mathbf{E}_-(x,y,z)\sin\omega t,\\ \widetilde{\mathbf{H}}(x,y,z,t) &= \mathbf{H}_+(x,y,z)\cos\omega t + \mathbf{H}_-(x,y,z)\sin\omega t,\end{aligned} \tag{2.1}$$

where ω is the circular frequency; $\widetilde{\mathbf{E}}$, $\mathbf{E}_+$, $\mathbf{E}_-$, $\widetilde{\mathbf{H}}$, $\mathbf{H}_+$, $\mathbf{H}_-$ are real functions. Everywhere below the time multipliers are omitted.

Form complex amplitudes of the electromagnetic field

$$\mathbf{E} = \mathbf{E}_+ + i\mathbf{E}_-, \qquad \mathbf{H} = \mathbf{H}_+ + i\mathbf{H}_-, \tag{2.2}$$

where $\mathbf{E} = (E_x, E_y, E_z)^T$, $\mathbf{H} = (H_x, H_y, H_z)^T$, and $(\,\cdot\,)^T$ denotes the operation of transposition, and each component of the fields is a function of three spatial variables.

Electromagnetic field $(\mathbf{E}, \mathbf{H})$ satisfies the Maxwell equations

$$\operatorname{rot}\mathbf{H} = -i\omega\varepsilon\mathbf{E}, \qquad \operatorname{rot}\mathbf{E} = i\omega\mu\mathbf{H}, \tag{2.3}$$

the continuity condition for the tangential field components on the media interfaces $x = 0$, $x = h$, and the radiation condition at infinity: the electromagnetic field exponentially decays as $|x| \to \infty$ in the domains $x < 0$ and $x > h$.

The permittivity inside the layer is described by the diagonal tensor

$$\widehat{\varepsilon} = \begin{pmatrix} \varepsilon_{xx} & 0 & 0 \\ 0 & \varepsilon_{yy} & 0 \\ 0 & 0 & \varepsilon_{zz} \end{pmatrix}, \tag{2.4}$$

where

$$\varepsilon_{xx} = \varepsilon_{2x} + b|E_x|^2 + a|E_z|^2, \qquad \varepsilon_{zz} = \varepsilon_{2z} + a|E_x|^2 + b|E_z|^2 \tag{2.5}$$

and $a, b, \varepsilon_2 > \max(\varepsilon_1, \varepsilon_3)$ are positive constants (below the solutions are sought under more general conditions). It does not matter what a form ε_{yy} has. Since ε_{yy} is not contained in the equations below for the TM case, it should be noticed that $\widehat{\varepsilon}$ describes tensor Kerr nonlinearity. When $a = b$ we obtain scalar Kerr nonlinearity. Moreover, chosen nonlinearity satisfies the condition $\partial\varepsilon xx/\partial E_z^2 = \partial\varepsilon zz/\partial E_x^2$. This equation is satisfied by almost every known nonlinear Kerr mechanism, such as electronic distortion, molecular orientation, electrostriction, and Kerr nonlinearities described within the uniaxial approximation mentioned in the paper [11]. The case when $\varepsilon_{2x} = \varepsilon_{2z}$ is studied in [8]. Pay heed to the fact that the problem considered here is not studied in [31].

The solutions to the Maxwell equations are sought in the entire space.

3. TM Waves

Let us consider TM waves

$$\mathbf{E} = (E_x, 0, E_z)^T, \qquad \mathbf{H} = (0, H_y, 0)^T, \tag{3.1}$$

and E_x, E_z, H_y are functions of three spatial variables. It is easy to show that the components of the fields do not depend on y. Waves propagating along medium interface z depend on z harmonically. This means that the fields components have the form

$$E_x = E_x(x)e^{i\gamma z}, \qquad E_z = E_z(x)e^{i\gamma z}, \qquad H_y = H_y(x)e^{i\gamma z}, \tag{3.2}$$

where γ is the spectral parameter of the problem.

So we obtain from system (2.3) [6]

$$\begin{aligned} i\gamma E_x(x) - E_z'(x) &= i\omega\mu H_y(x), \\ H_y'(x) &= -i\omega\varepsilon_{zz}E_z(x), \\ i\gamma H_y(x) &= i\omega\varepsilon_{xx}E_x(x), \end{aligned} \tag{3.3}$$

where $(\;\cdot\;)' \equiv d/dx$.

The following equation can be easily derived from the previous system:

$$H_y(x) = \frac{1}{i\omega\mu}\left(i\gamma E_x(x) - E_z'(x)\right). \tag{3.4}$$

Differentiating (3.4) and using the second and the third equations of system (3.3) we obtain

$$\begin{aligned} \gamma(iE_x(x))' - E_z''(x) &= \omega^2\mu\varepsilon_{zz}E_z(x), \\ \gamma^2(iE_x(x)) - \gamma E_z'(x) &= \omega^2\mu\varepsilon_{xx}(iE_x(x)). \end{aligned} \tag{3.5}$$

Let us denote $k_0^2 := \omega^2\mu_0\varepsilon_0$ and perform the normalization according to the formulas $\tilde{x} = k_0 x$, $d/dx = k_0(d/d\tilde{x})$, $\tilde{\gamma} = \gamma/k_0$, $\tilde{\varepsilon}_j = \varepsilon_j/\varepsilon_0$ $(j = 1, 2, 3)$, $\tilde{a} = a/\varepsilon_0$, $\tilde{b} = b/\varepsilon_0$. Denoting by

$Z(\tilde{x}) := E_z$, $X(\tilde{x}) := iE_x$ and omitting the tilde symbol, from system (3.5) we obtain

$$\begin{aligned} -Z'' + \gamma X' &= \varepsilon_{zz} Z, \\ -Z' + \gamma X &= \gamma^{-1} \varepsilon_{xx} X. \end{aligned} \tag{3.6}$$

It is necessary to find eigenvalues γ of the problem that correspond to surface waves propagating along boundaries of the layer $0 < x < h$. We seek the real values of the spectral parameter γ such that real solutions $X(x)$ and $Z(x)$ to system (3.6) exist. Indeed, in this case $|\mathbf{E}|^2$ does not depend on z. Since $\mathbf{E} = (E_x(x)e^{i\gamma z}, 0, E_z(x)e^{i\gamma z}) = e^{i\gamma z}(E_x, 0, E_z)$, therefore, $|\mathbf{E}| = |e^{i\gamma z}| \cdot \sqrt{|E_x|^2 + |E_z|^2}$. It is known that $|e^{i\gamma z}| = 1$ as $\operatorname{Im} \gamma = 0$. Let $\gamma = \gamma' + i\gamma''$. Then, we obtain $|e^{i\gamma z}| = |e^{i\gamma' z}| \cdot |e^{-\gamma'' z}| = e^{-\gamma'' z}$. If $\gamma'' \neq 0$, then $e^{-\gamma'' z}$ is a function on z. In this case the components E_x and E_z depend on z, but it contradicts to the choice of $E_x(x)$ and $E_z(x)$. So we can consider only real values of γ.

We consider that

$$\varepsilon = \begin{cases} \varepsilon_1, & x < 0, \\ \hat{\varepsilon}, & 0 < x < h, \\ \varepsilon_3, & x > h. \end{cases} \tag{3.7}$$

Also we assume that $\max(\varepsilon_1, \varepsilon_3) < \gamma^2 < \min(\varepsilon_{2x}, \varepsilon_{2z})$. This two-sided inequality naturally appears for the problem in a layer with a constant permittivity tensor.

Functions X, Z are supposed to be sufficiently smooth due to physical nature of the problem

$$\begin{aligned} X(x) &\in C(-\infty,0] \cap C[0,h] \cap C[h,+\infty) \cap C^1(-\infty,0] \cap C^1[0,h] \cap C^1[h,+\infty), \\ Z(x) &\in C(-\infty,+\infty) \cap C^1(-\infty,0] \cap C^1[0,h] \cap C^1[h,+\infty) \cap C^2(-\infty,0) \cap C^2(0,h) \cap C^2(h,+\infty). \end{aligned} \tag{3.8}$$

It is clear that system (3.6) is an autonomous one. System (3.6) can be rewritten in a normal form. This system in the normal form can be considered as a dynamical system with analytical with respect to X and Z right-hand sides. Of course, in the domain where these right-hand sides are analytical with respect to X and Z, it is well known (see, e.g., [34]) that the solutions X and Z of such a system are analytical functions with respect to the independent variable as well. This is an important fact for DEs' derivation.

We consider that $\gamma^2 > \max(\varepsilon_1, \varepsilon_3)$.

4. Differential Equations of the Problem

In the domain $x < 0$ we have $\varepsilon = \varepsilon_1$. From system (3.6) we obtain $X' = \gamma Z$, $Z' = \gamma^{-1}(\gamma^2 - \varepsilon_1)X$. In accordance with the radiation condition we obtain

$$\begin{aligned} X(x) &= A e^{x\sqrt{\gamma^2-\varepsilon_1}}, \\ Z(x) &= -A\gamma^{-1}\sqrt{\gamma^2 - \varepsilon_1}\, e^{x\sqrt{\gamma^2-\varepsilon_1}}. \end{aligned} \tag{4.1}$$

We assume that $\gamma^2 - \varepsilon_1 > 0$; otherwise it will be impossible to satisfy the radiation condition.

In the domain $x > h$ we have $\varepsilon = \varepsilon_3$. From system (3.6) we obtain $X' = \gamma Z$, $Z' = \gamma^{-1}(\gamma^2 - \varepsilon_3)X$. In accordance with the radiation condition we obtain

$$\begin{aligned} X(x) &= Be^{-(x-h)\sqrt{\gamma^2-\varepsilon_3}}, \\ Z(x) &= -B\gamma^{-1}\sqrt{\gamma^2 - \varepsilon_3}e^{-(x-h)\sqrt{\gamma^2-\varepsilon_3}}. \end{aligned} \tag{4.2}$$

Here for the same reason as above we consider that $\gamma^2 - \varepsilon_3 > 0$.

Constants A and B in (4.1) and (4.2) are defined by transmission conditions and initial conditions.

Inside the layer $0 < x < h$ system (3.6) takes the form

$$\begin{aligned} -\frac{d^2Z}{dx^2} + \gamma\frac{dX}{dx} &= \left[\varepsilon_{2z} + aX^2 + bZ^2\right]Z, \\ -\frac{dZ}{dx} + \gamma X &= \gamma^{-1}\left[\varepsilon_{2x} + bX^2 + aZ^2\right]X. \end{aligned} \tag{4.3}$$

Differentiating the second equation and substituting its right-hand side instead of left-hand side into the first equation we can rewritten system (4.3) in the following form:

$$\begin{aligned} \frac{dX}{dx} &= \frac{2a}{\gamma}\frac{\varepsilon_{2x} - \gamma^2 + bX^2 + aZ^2}{\varepsilon_{2x} + 3bX^2 + aZ^2}X^2Z + \gamma\frac{\varepsilon_{2z} + aX^2 + bZ^2}{\varepsilon_{2x} + 3bX^2 + aZ^2}Z, \\ \frac{dZ}{dx} &= -\gamma^{-1}\left[\varepsilon_{2x} - \gamma^2 + bX^2 + aZ^2\right]X. \end{aligned} \tag{4.4}$$

Now system (4.4) is written in a normal form. If the right-hand sides are analytic functions with respect to X and Z, then the solutions are analytic functions with respect to its independent variable.

Dividing the first equation in system (4.4) to the second one we obtain the ordinary differential equation

$$-\left(\varepsilon_{2x} + 3bX^2 + aZ^2\right)\frac{dX}{dZ} = 2aXZ + \gamma^2\frac{\varepsilon_{2z} + aX^2 + bZ^2}{\varepsilon_{2x} - \gamma^2 + bX^2 + aZ^2}\frac{Z}{X}. \tag{4.5}$$

Equation (4.5) can be transformed into a total differential equation.

Its solution (first integral of system (4.4)) can be easily found and be written in the following form:

$$C = b^2X^6 + 2abX^4Z^2 + a^2X^2Z^4 + \frac{1}{2}\left(4\varepsilon_{2x} - 3\gamma^2\right)bX^4 + \left(2\varepsilon_{2x} - \gamma^2\right)aX^2Z^2 + \frac{1}{2}\gamma^2 bZ^4 \\ + \gamma^2\left(\varepsilon_{2x} - \gamma^2\right)X^2 + \left(\varepsilon_{2x} - \gamma^2\right)^2 X^2 + \gamma^2\varepsilon_{2z}Z^2, \tag{4.6}$$

where C is a constant of integration.

5. Transmission Conditions and the Transmission Problem

Tangential components of an electromagnetic field are known to be continuous at media interfaces. In this case the tangential components are H_y and E_z. Hence, we obtain

$$\begin{aligned} H_y(h+0) &= H_y(h-0), \qquad & H_y(0-0) &= H_y(0+0), \\ E_z(h+0) &= E_z(h-0), \qquad & E_z(0-0) &= E_z(0+0). \end{aligned} \tag{5.1}$$

From the continuity conditions for the tangential components of the fields **E** and **H** and using (3.4) we obtain

$$\begin{aligned} \gamma X(h) - Z'(h) &= H_y^{(h)}, \qquad & \gamma X(0) - Z'(0) &= H_y^{(0)}, \\ Z(h) = E_z(h+0) &= E_z^{(h)}, \qquad & Z(0) = E_z(0-0) &= E_z^{(0)}, \end{aligned} \tag{5.2}$$

where $H_y^{(h)} := i(\sqrt{\mu}/\sqrt{\varepsilon_0})H_y(h+0)$, $H_y^{(0)} := i(\sqrt{\mu}/\sqrt{\varepsilon_0})H_y(0-0)$.

The constant $E_z^{(h)} := E_z(h+0)$ is supposed to be known (initial condition). Let us denote $X_0 := X(0)$, $X_h := X(h)$, $Z_0 := Z(0)$, and $Z_h := Z(h)$. So we obtain that $A = (\gamma/\sqrt{\gamma^2 - \varepsilon_1})Z_0$, $B = (\gamma/\sqrt{\gamma^2 - \varepsilon_3})Z_h$.

Then from conditions (5.2) we obtain

$$H_y^{(h)} = -Z_h\frac{\varepsilon_3}{\sqrt{\gamma^2 - \varepsilon_3}}, \qquad H_y^{(0)} = Z_0\frac{\varepsilon_1}{\sqrt{\gamma^2 - \varepsilon_1}}. \tag{5.3}$$

In accordance with (3.6), (3.7) inside the layer

$$-Z'(x) + \gamma X(x) = \gamma^{-1}\left(\varepsilon_{2x} + bX^2(x) + aZ^2(x)\right)X(x). \tag{5.4}$$

Then for $x = h$, using (5.2), we obtain from (5.4)

$$\gamma^{-1}\left[\varepsilon_{2x} + bX_h^2 + aZ_h^2\right]X_h = H_y^{(h)}. \tag{5.5}$$

From (5.5) we obtain the equation with respect to X_h:

$$X_h^3 + b^{-1}\left(\varepsilon_{2x} + aZ_h^2\right)X_h - b^{-1}\gamma H_y^{(h)} = 0. \tag{5.6}$$

Under taken assumptions (in regard to ε_2 and a) the value $a^{-1}(\varepsilon_2 + aZ_h^2) > 0$. Hence, this equation has at least one real root, which is considered (the root can be find explicitly by using Cardanus-Ferrari formula [35].

Using first integral (4.6) at $x = h$, we find the value $C_h^X := C|_{x=h}$ from the equation

$$C_h^X = b^2X_h^6 + 2abX_h^4Z_h^2 + a^2X_h^2Z_h^4 + 2^{-1}\left(4\varepsilon_{2x} - 3\gamma^2\right)bX_h^4 + \left(2\varepsilon_{2x} - \gamma^2\right)aX_h^2Z_h^2 + 2^{-1}\gamma^2bZ_h^4$$
$$+ \gamma^2\left(\varepsilon_{2x} - \gamma^2\right)X_h^2 + \left(\varepsilon_{2x} - \gamma^2\right)^2X_h^2 + \gamma^2\varepsilon_{2z}Z_h^2. \tag{5.7}$$

In order to find the values X_0 and Z_0 it is necessary to solve the following system (this system is obtained using formula (5.4) at $x = 0$ and the first integral at the same point):

$$\gamma\varepsilon_1Z_0 = \sqrt{\gamma^2 - \varepsilon_1}\left[\varepsilon_{2x} + bX_0^2 + aZ_0^2\right]X_0,$$
$$C_h^X = b^2X_0^6 + 2abX_0^4Z_0^2 + a^2X_0^2Z_0^4 + 2^{-1}\left(4\varepsilon_{2x} - 3\gamma^2\right)bX_0^4$$
$$+ \left(2\varepsilon_{2x} - \gamma^2\right)aX_0^2Z_0^2 + 2^{-1}\gamma^2bZ_0^4 + \gamma^2\left(\varepsilon_{2x} - \gamma^2\right)X_0^2 + \left(\varepsilon_{2x} - \gamma^2\right)^2X_0^2 + \gamma^2\varepsilon_{2z}Z_0^2. \tag{5.8}$$

It is easy to see from the second equation of this system that the values X_0 and Z_0 can have arbitrary signs. At the same time from the first equation of this system we can see that X_0 and Z_0 must be positive or negative simultaneously.

Normal components of electromagnetic field are known to be discontinues at media interfaces. And it is the discontinuity of the first kind. In this case the normal component is E_x. It is also known that the value εE_x is continuous at media interfaces. From the above and from the continuity of the tangential component E_z it follows that the transmission conditions for the functions εX and Z are

$$[\varepsilon X]_{x=0} = 0, \qquad [\varepsilon X]_{x=h} = 0, \qquad [Z]_{x=0} = 0, \qquad [Z]_{x=h} = 0, \tag{5.9}$$

where $[f]_{x=x_0} = \lim_{x\to x_0-0} f(x) - \lim_{x\to x_0+0} f(x)$ denotes a jump of the function f at the interface.

We also suppose that functions $X(x)$ and $Z(x)$ satisfy the condition

$$X(x) = O\left(\frac{1}{|x|}\right), \qquad Z(x) = O\left(\frac{1}{|x|}\right) \quad \text{as } |x| \longrightarrow \infty. \tag{5.10}$$

6. Dispersion Equation

Introduce the new variables

$$\tau(x) = \frac{\varepsilon_{2x} + bX^2(x) + aZ^2(x)}{\gamma^2}, \qquad \eta(x) = \gamma\frac{X(x)}{Z(x)}\tau(x). \tag{6.1}$$

Using new variables rewrite system (4.4),

$$\frac{d\tau}{dx} = 2\frac{\tau\eta(\gamma^2\tau - \varepsilon_{2x})}{b\eta^2 + a\gamma^2\tau^2} \times \frac{(b\eta^2 + a\gamma^2\tau^2)(b\varepsilon_{2z} - a\gamma^2\tau(\tau-1)) + (a\eta^2 + b\gamma^2\tau^2)(\gamma^2\tau - \varepsilon_{2x})}{\gamma^2\tau(b\eta^2 + a\gamma^2\tau^2) + 2b\eta^2(\gamma^2\tau - \varepsilon_{2x})},$$

$$\frac{d\eta}{dx} = \frac{\tau - 1}{\tau}\eta^2 + \varepsilon_{2z} + \left(\gamma^2\tau - \varepsilon_{2x}\right)\frac{a\eta^2 + b\gamma^2\tau^2}{b\eta^2 + a\gamma^2\tau^2}, \tag{6.2}$$

and (4.6)

$$\frac{\gamma^2\tau - \varepsilon_{2x}}{b\eta^2 + a\gamma^2\tau^2}\left[\eta^2\left(\left(\gamma^2\tau - \varepsilon_{2x}\right)^2 + \varepsilon_{2x}\left(\varepsilon_{2x} - \gamma^2\right)\right) + \gamma^4\varepsilon_{2z}\tau^2\right] + \frac{\left(\gamma^2\tau - \varepsilon_{2x}\right)^2}{2\left(b\eta^2 + a\gamma^2\tau^2\right)^2}$$
$$\times\left[\left(4\varepsilon_{2x} - 3\gamma^2\right)b\eta^4 + 2\left(2\varepsilon_{2x} - \gamma^2\right)a\gamma^2\tau^2\eta^2 + \gamma^6 b\tau^4\right] = C, \tag{6.3}$$

where constant C is equal to the constant C in (4.6).

In order to obtain the DE for the propagation constants it is necessary to find the values $\eta(0), \eta(h)$.

It is clear that $\eta(0) = \gamma(X(0)/Z(0))\tau(0)$, $\eta(h) = \gamma(X(h)/Z(h))\tau(h)$. Taking into account that $\gamma^2 X(x)\tau(x) = \varepsilon X(x)$ and using formulas (5.2), (5.3), it is easy to obtain that

$$\eta(0) = \frac{\varepsilon_1}{\sqrt{\gamma^2 - \varepsilon_1}} > 0, \qquad \eta(h) = -\frac{\varepsilon_3}{\sqrt{\gamma^2 - \varepsilon_3}} < 0. \tag{6.4}$$

It is easy to see that the right-hand side of the second equation of system (VI) is strictly positive. This means that the function $\eta(x)$ monotonically increases on interval $(0, h)$. Taking into account (6.4) we obtain that the function $\eta(x)$ cannot be differentiable on the entire interval $(0, h)$. This means that the function $\eta(x)$ has a break point. Let $x^* \in (0, h)$ be the break point. From (6.3) it is obvious that x^* is such that $\tau^* = \tau(x^*)$ is a root of the equation $C_h^\tau + 3(\tau^*)^2 - 2(\tau^*)^3 - 2\tau_0(2 - \tau^*)\tau^* = 0$. In addition $\eta(x^* - 0) \to +\infty$ and $\eta(x^* + 0) \to -\infty$.

It is natural to suppose that the function $\eta(x)$ on interval $(0, h)$ has several break points $x_0, x_1, \ldots, x_N$. The properties of function $\eta(x)$ imply

$$\eta(x_i - 0) = +\infty, \qquad \eta(x_i + 0) = -\infty, \qquad \text{where } i = \overline{0, N}. \tag{6.5}$$

Let

$$\frac{1}{w} := \frac{\tau - 1}{\tau}\eta^2 + \varepsilon_{2z} + \left(\gamma^2\tau - \varepsilon_{2x}\right)\frac{a\eta^2 + b\gamma^2\tau^2}{b\eta^2 + a\gamma^2\tau^2}, \tag{6.6}$$

where $w = w(\eta)$; $\tau = \tau(\eta)$ is expressed from (5.4).

Taking into account our hypothesis we will seek the solutions on each interval $[0, x_0), (x_0, x_1), \ldots, (x_N, h]$:

$$\begin{aligned} -\int_{\eta(x)}^{\eta(x_0)} w d\eta &= x + c_0, \\ \int_{\eta(x_i)}^{\eta(x)} w d\eta &= x + c_i, \\ \int_{\eta(x_N)}^{\eta(x)} w d\eta &= x + c_N, \end{aligned} \tag{6.7}$$

where $0 \le x \le x_0$, $x_i \le x \le x_{i+1}$, and $x_N \le x \le h$, respectively, and $i = \overline{0, N-1}$.

Substituting $x = 0$, $x = x_{i+1}$, and $x = x_N$ into equations in (6.7) (into the first, the second, and the third, resp.,) and taking into account (6.5), we find constants $c_1, c_2, \ldots, c_{N+1}$:

$$\begin{aligned} c_0 &= -\int_{\eta(0)}^{+\infty} w d\eta, \\ c_{i+1} &= \int_{-\infty}^{+\infty} w d\eta - x_{i+1}, \\ c_{N+1} &= \int_{-\infty}^{\eta(h)} w d\eta - h, \end{aligned} \tag{6.8}$$

where $i = \overline{0, N-1}$.

Using (6.8) we can rewrite (6.7) in the following form:

$$\begin{aligned} \int_{\eta(x)}^{\eta(x_0)} w d\eta &= -x + \int_{\eta(0)}^{+\infty} w d\eta, \\ \int_{\eta(x_i)}^{\eta(x)} w d\eta &= x + \int_{-\infty}^{+\infty} w d\eta - x_{i+1}, \\ \int_{\eta(x_N)}^{\eta(x)} w d\eta &= x + \int_{-\infty}^{\eta(h)} w d\eta - h, \end{aligned} \tag{6.9}$$

where $0 \le x \le x_0$, $x_i \le x \le x_{i+1}$, $x_N \le x \le h$, respectively, and $i = \overline{0, N-1}$.

Introduce the notation $T := \int_{-\infty}^{+\infty} w d\eta$. It follows from formula (6.9) that $0 < x_{i+1} - x_i = T < h$, where $i = \overline{0, N-1}$. This implies the convergence of the improper integral (it will be proved in other way below). Now consider x in (6.9) such that all the integrals on the left side vanish (i.e., $x = x_0$, $x = x_i$, and $x = x_N$), and sum all equations in (6.9). We obtain

$$0 = -x_0 + \int_{\eta(0)}^{+\infty} w d\eta + x_0 + T - x_1 + \cdots + x_{N-1} + T - x_N + x_N + \int_{-\infty}^{\eta(h)} w d\eta - h. \qquad (6.10)$$

Finally we obtain

$$-\int_{\eta(h)}^{\eta(0)} w d\eta + (N+1)T = h, \qquad (6.11)$$

where $\eta(0)$, $\eta(h)$ are defined by formulas (6.4).

Expression (6.11) is the DE, which holds for any finite h. Let γ be a solution of DE (6.11) and an eigenvalue of the problem. Then, there are eigenfunctions X and Z, which correspond to the eigenvalue γ. The eigenfunction Z has $N+1$ zeros on the interval $(0, h)$.

Notice that improper integrals in DE (6.11) converge. Indeed, function $\tau = \tau(\eta)$ is bounded as $\eta \to \infty$ since $\tau = \gamma^{-2}(\varepsilon_{2x} + bX^2 + aZ^2)$, and X, Z are bounded.

Then

$$|w| \le \frac{1}{\alpha\eta^2 + \beta}, \qquad (6.12)$$

where $\alpha > 0$, $\beta > 0$ are constants. It is obvious that improper integral $\int_{-\infty}^{+\infty} (d\eta/(\alpha\eta^2 + \beta))$ converges. Convergence of the improper integrals in (6.11) in inner points results from the requirement that the right-hand side of the second equation of system (VI) is positive.

The first equation of system (VI) jointly with the first integral can be integrated in hyperelliptic functions. The solution is expressed in implicit form by means of hyperelliptic integrals. This is the simple example of Abelian integrals. The inversion of these integrals is hyperelliptic functions and they are solutions of system (VI). Hyperelliptic functions are Abelian functions, which are meromorphic and periodic functions. Since function η is expressed algebraically through τ, therefore, η is a meromorphic periodic function. This means that the break point x^* is a pole of function η.

7. Generalized Dispersion Equation

Here we derive the generalized DE, which holds for any real values ε_2. In addition the sign of the right-hand side of the second equation in system (VI) and condition $\max(\varepsilon_1, \varepsilon_3) < \gamma^2 < \varepsilon_2$ are not taken into account. These conditions appear in the case of a linear layer and are used for derivation of DE (6.11). Though on the nonlinear case it is not necessary to limit the value γ^2 from the right side, at the same time it is clear that γ is limited from the left side, since this limit appears from the solutions in the half-spaces.

Now we assume that γ satisfies the following two-sided inequality:

$$\max(\varepsilon_1, \varepsilon_3) < \gamma^2 < +\infty. \tag{7.1}$$

Using first integral (6.3) it is possible to integrate formally any of the equations of system (VI). As earlier we integrate the second equation, we cannot obtain the solution on the entire interval $(0, h)$, since function $\eta(x)$ can have break points, which belong to $(0, h)$. It is known that function $\eta(x)$ has break points only of the second kind (η is an analytical function).

Assume that function $\eta(x)$ on interval $(0, h)$ has $N + 1$ break points $x_0, \ x_1, \ \ldots, \ x_N$. It should be noticed that

$$\eta(x_i - 0) = \pm\infty \qquad \eta(x_i + 0) = \pm\infty, \tag{7.2}$$

where $i = \overline{0, \ N}$, and signs $\pm$ are independent and unknown.

Taking into account the previous, solutions are sought on each interval $[0, x_0), (x_0, x_1), \ldots, (x_N, h]$:

$$\begin{aligned} -\int_{\eta(x)}^{\eta(x_0-0)} w d\eta &= x + c_0, \\ \int_{\eta(x_i+0)}^{\eta(x)} w d\eta &= x + c_{i+1}, \\ \int_{\eta(x_N+0)}^{\eta(x)} w d\eta &= x + c_{N+1}, \end{aligned} \tag{7.3}$$

where $0 \le x \le x_0$, $x_i \le x \le x_{i+1}$, and $x_N \le x \le h$, respectively, and $i = \overline{0, N-1}$.

From (7.3), substituting $x = 0$, $x = x_{i+1}$, and $x = x_N$ into the first, the second, and the third equations in (7.3), respectively, we find required constants $c_1, c_2, \ldots, c_{N+1}$:

$$\begin{aligned} c_0 &= -\int_{\eta(0)}^{\eta(x_0-0)} w d\eta, \\ c_{i+1} &= \int_{\eta(x_i+0)}^{\eta(x_{i+1}-0)} w d\eta - x_{i+1}, \\ c_{N+1} &= \int_{\eta(x_N+0)}^{\eta(h)} w d\eta - h, \end{aligned} \tag{7.4}$$

where $i = \overline{0, N-1}$.

Using (7.4), (7.3) take the form

$$\begin{aligned}
\int_{\eta(x)}^{\eta(x_0-0)} w d\eta &= -x + \int_{\eta(0)}^{\eta(x_0-0)} w d\eta, \\
\int_{\eta(x_i+0)}^{\eta(x)} w d\eta &= x + \int_{\eta(x_i+0)}^{\eta(x_{i+1}-0)} w d\eta - x_{i+1}, \\
\int_{\eta(x_N+0)}^{\eta(x)} w d\eta &= x + \int_{\eta(x_N+0)}^{\eta(h)} w d\eta - h,
\end{aligned} \tag{7.5}$$

where $0 \le x \le x_0$, $x_i \le x \le x_{i+1}$, and $x_N \le x \le h$, respectively, and $i = \overline{0, N-1}$.

From formulas (7.5) we obtain that

$$x_{i+1} - x_i = \int_{\eta(x_i+0)}^{\eta(x_{i+1}-0)} w d\eta, \qquad i = \overline{0, N-1}. \tag{7.6}$$

Expressions $0 < x_{i+1} - x_i < h < \infty$ imply that under the assumption about the break point existence the integral on the right side converges and $\int_{\eta(x_i+0)}^{\eta(x_{i+1}-0)} w d\eta > 0$. In the same way, from the first and the last equations of (7.5) we obtain that $x_0 = \int_{\eta(0)}^{\eta(x_0-0)} w d\eta$ and $0 < x_0 < h$ then

$$0 < \int_{\eta(0)}^{\eta(x_0-0)} w d\eta < h < \infty, \tag{7.7}$$

and $h - x_N = \int_{\eta(x_N+0)}^{\eta(h)} w d\eta$ and $0 < h - x_N < h$ then

$$0 < \int_{\eta(0)}^{\eta(x_0-0)} w d\eta < h < \infty. \tag{7.8}$$

These considerations yield that the function $w(\eta)$ has no nonintegrable singularities for $\eta \in (-\infty, \infty)$. And also this proves that the assumption about finite number break points is true.

Now, setting $x = x_0$, $x = x_i$, and $x = x_N$ into the first, the second, and the third equations in (7.5), respectively, we have that all the integrals on the left sides vanish. We add all the equations in (7.5) to obtain

$$\begin{aligned}
0 = &-x_0 + \int_{\eta(0)}^{\eta(x_0-0)} w d\eta + x_0 + \int_{\eta(x_0+0)}^{\eta(x_1-0)} w d\eta - x_1 + \cdots + x_{N-1} \\
&+ \int_{\eta(x_{N-1}+0)}^{\eta(x_N-0)} w d\eta - x_N + x_N + \int_{\eta(x_N+0)}^{\eta(h)} w d\eta - h.
\end{aligned} \tag{7.9}$$

From (7.9) we obtain

$$\int_{\eta(0)}^{\eta(x_0-0)} w d\eta + \int_{\eta(x_N+0)}^{\eta(h)} w d\eta + \sum_{i=0}^{N-1} \int_{\eta(x_i+0)}^{\eta(x_{i+1}-0)} w d\eta = h. \quad (7.10)$$

It follows from formulas (7.6) that

$$\eta(x_i + 0) = \pm\infty, \quad \eta(x_i - 0) = \mp\infty, \quad \text{where } i = \overline{0, N}, \quad (7.11)$$

and it is necessary to choose the infinities of different signs.

Thus we obtain that

$$\int_{\eta(x_0+0)}^{\eta(x_1-0)} w d\eta = \cdots = \int_{\eta(x_{N-1}+0)}^{\eta(x_N-0)} w d\eta =: T'. \quad (7.12)$$

Hence $x_1 - x_0 = \cdots = x_N - x_{N-1}$.

Now we can rewrite (7.10) in the following form:

$$\int_{\eta(0)}^{\eta(x_0-0)} w d\eta + \int_{\eta(x_N+0)}^{\eta(h)} f d\eta + NT' = h. \quad (7.13)$$

Let $T \equiv \int_{-\infty}^{+\infty} w d\eta$; then we finally obtain

$$-\int_{\eta(h)}^{\eta(0)} w d\eta \pm (N+1)T = h, \quad (7.14)$$

where $\eta(0)$, $\eta(h)$ are defined by formulas (6.4).

Expression (7.14) is the DE, which holds for any finite h. Let γ be a solution of DE (7.14) and an eigenvalue of the problem. Then, there are eigenfunctions X and Z, which correspond to the eigenvalue γ. The eigenfunction Z has $N + 1$ zeros on the interval $(0, h)$. It should be noticed that for every number $N+1$ it is necessary to solve two DEs: for $N+1$ and for $-(N+1)$.

Note 1. If there is a certain value γ_*^2, such that some of the integrals in DEs (6.11) or (7.14) diverge at certain inner points this simply means that the value γ_*^2 is not a solution of chosen DE and the value γ_*^2 is not an eigenvalue of the problem.

Note 2. It is necessary to emphasize that this boundary eigenvalue (transmission) problem essentially depends on the initial condition Z_h. The transmission problem for a linear layer does not depend on the initial condition. If the nonlinearity function is a specific one, then in some cases it will be possible to normalize the Maxwell equations in such a way that the transmission problem does not depend on initial condition Z_h explicitly (it is possible e.g., for Kerr nonlinearity in a layer and in a circle cylindrical waveguide). Once more we stress the fact that the opportunity of such normalization is an exceptional case. What is more, in spite of the fact that in certain cases this normalization is possible it does not mean that the normalized transmission problem is independent of the initial condition. In this case one of the problem's parameter depends on the initial condition.

8. Passage to the Limit in the Generalized Dispersion Equation

In this section we assume that $\varepsilon_{2x} = \varepsilon_{2z} = \varepsilon_2$ and $b = a$. Now consider the passage to the limit as $a \to 0$. The value $a = 0$ corresponds to the case of a linear medium in the layer. Two cases are possible:

(a) $\varepsilon_2 > 0$,

(b) $\varepsilon_2 < 0$ (metamaterial case).

Let us examine case (a). The DE for a linear case is well known [10] and has the form

$$tg\left(h\sqrt{\varepsilon_2-\gamma^2}\right) = \frac{\varepsilon_2\sqrt{\varepsilon_2-\gamma^2}\left(\varepsilon_1\sqrt{\gamma^2-\varepsilon_3}+\varepsilon_3\sqrt{\gamma^2-\varepsilon_1}\right)}{\varepsilon_1\varepsilon_3(\varepsilon_2-\gamma^2)-\varepsilon_2^2\sqrt{\gamma^2-\varepsilon_3}\sqrt{\gamma^2-\varepsilon_1}}. \tag{8.1}$$

Let

$$f = \frac{\tau}{\gamma^2\tau^2+\eta^2(\tau-1)}, \qquad f_1 = \frac{\varepsilon_2}{\varepsilon_2-\gamma^2}\frac{1}{\varepsilon_2^2/(\varepsilon_2-\gamma^2)+\eta^2}. \tag{8.2}$$

Using passage to the limit as $a \to 0$ we obtain the function f_1 from the function f. We seek bounded solutions $X(x)$ and $Z(x)$. This implies that the denominator of the function f_1 cannot vanish. What is more, the function f as $a \to 0$ tends to the function f_1 uniformly on $x \in [0, h]$. It is possible to pass to the limit under integral sign as $a \to 0$ in (7.14) using results of classical analysis

$$h = -\frac{\varepsilon_2}{\varepsilon_2-\gamma^2}\int_{\eta(h)}^{\eta(0)}\frac{1}{\varepsilon_2^2/(\varepsilon_2-\gamma^2)+\eta^2}d\eta + \frac{\varepsilon_2}{\varepsilon_2-\gamma^2}(N+1)\int_{-\infty}^{+\infty}\frac{1}{\varepsilon_2^2/(\varepsilon_2-\gamma^2)+\eta^2}d\eta, \tag{8.3}$$

where $\eta(0)$, $\eta(h)$ are defined by formulas (6.4).

The integrals in (8.3) are calculated analytically. Calculating these integrals we obtain

$$h\sqrt{\varepsilon_2-\gamma^2} = \operatorname{arctg}\frac{\varepsilon_2\sqrt{\varepsilon_2-\gamma^2}\left(\varepsilon_1\sqrt{\gamma^2-\varepsilon_3}+\varepsilon_3\sqrt{\gamma^2-\varepsilon_1}\right)}{\varepsilon_1\varepsilon_3(\varepsilon_2-\gamma^2)-\varepsilon_2^2\sqrt{\gamma^2-\varepsilon_3}\sqrt{\gamma^2-\varepsilon_1}} + (N+1)\pi. \tag{8.4}$$

Expression (8.4) can be easily transformed into expression (8.1).

Let us examine (b) case. We have $\varepsilon_2 < 0$ (metamaterial) and the DE for the linear case has the form [31]

$$e^{2h\sqrt{\gamma^2-\varepsilon_2}} = \frac{\varepsilon_1\sqrt{\gamma^2-\varepsilon_2}-\varepsilon_2\sqrt{\gamma^2-\varepsilon_1}}{\varepsilon_1\sqrt{\gamma^2-\varepsilon_2}+\varepsilon_2\sqrt{\gamma^2-\varepsilon_1}}\,\frac{\varepsilon_3\sqrt{\gamma^2-\varepsilon_2}-\varepsilon_2\sqrt{\gamma^2-\varepsilon_3}}{\varepsilon_3\sqrt{\gamma^2-\varepsilon_2}+\varepsilon_2\sqrt{\gamma^2-\varepsilon_3}}, \tag{8.5}$$

where $\gamma^2 - \varepsilon_1 > 0$, $\gamma^2 - \varepsilon_2 > 0$, and $\gamma^2 - \varepsilon_3 > 0$.

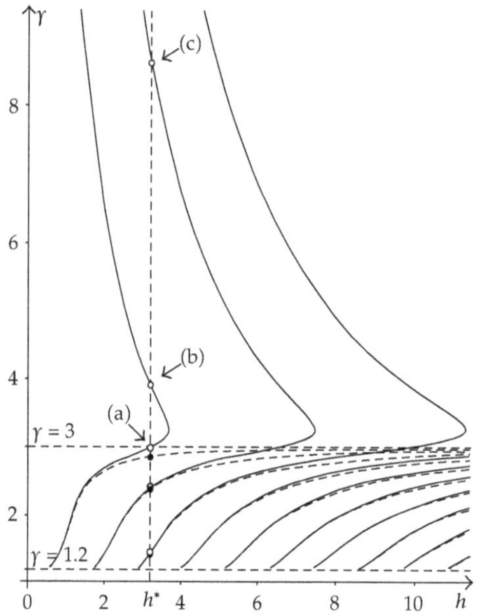

Figure 2: Plot of $\gamma(h)$. The first few dispersion curves are shown. Solid curves for the nonlinear case (solutions of (7.14)); dashed curves for the linear case (solutions of (8.1)). The following parameters are used for both cases: $\varepsilon_1 = 1.44$, $\varepsilon_2 = 9$, $\varepsilon_3 = 1$, and for the nonlinear case $a = 0.1$, and $E_z^{(h)} = 1$. Dashed lines are described by formulas: $h^* = 3.206$ (thickness of the layer), $\gamma = 1.2$ (lower bound for γ), and $\gamma = 3$ (upper bound for γ in the case of linear medium in the layer).

In the same way as above, passing to the limit in the function f as $a \to 0$ we obtain $f_2 = |\varepsilon_2|/(\gamma^2 - \varepsilon_2)(1/\eta^2 - \varepsilon_2^2/(\gamma^2 - \varepsilon_2))$. Passing to the limit in equation (7.14) as $a \to 0$ and integrating the function f_2, after simple calculations, we obtain formula(8.5)

The results here show that it is possible to pass to the limit as $a \to 0$. DE in (7.14) for the nonlinear case turns into (8.1) or (8.5) for the linear case as $a \to 0$.

9. Numerical Results

The way of solution to the DE in (7.14) is the following: we choose the segment on γ then cut this segment into p pieces with nods $\gamma_i, i = 0, 1, \ldots, p$. Then for each γ_i we can calculate all necessary values in (7.14). The integral in (7.14) can be calculated using any method of numerical integration. In order to calculate the value of the integrand at the point γ_i it is necessary to use first integral (6.3). For each value γ_i we calculate value h_i, so we obtain the grid $\{\gamma_i, h_i\}$, $i = 0, 1, \ldots, p$. Choosing reasonably dense grid on γ we can plot dependence γ on h, as it is done below.

Dispersion curves (DC) calculated from (7.14), (8.1), and (8.5) are shown in Figures 2 and 4. Eigenmodes for eigenvalues indicated in Figures 3 and 5 are shown in Figures 3 and 5.

As it is known and it is shown in Figure 2, the line $\gamma = 3$ is an asymptote for DCs in the linear case. It should be noticed that in the linear case there are no DCs in the region $\gamma^2 \geq \varepsilon_2$. It can be proved that function $h \equiv h(\gamma)$ defined from equation (7.14) is continuous at the neighborhood $\gamma^2 = \varepsilon_2$ when $a \neq 0$ (see Figure 2). This is the important distinction between linear and nonlinear cases.

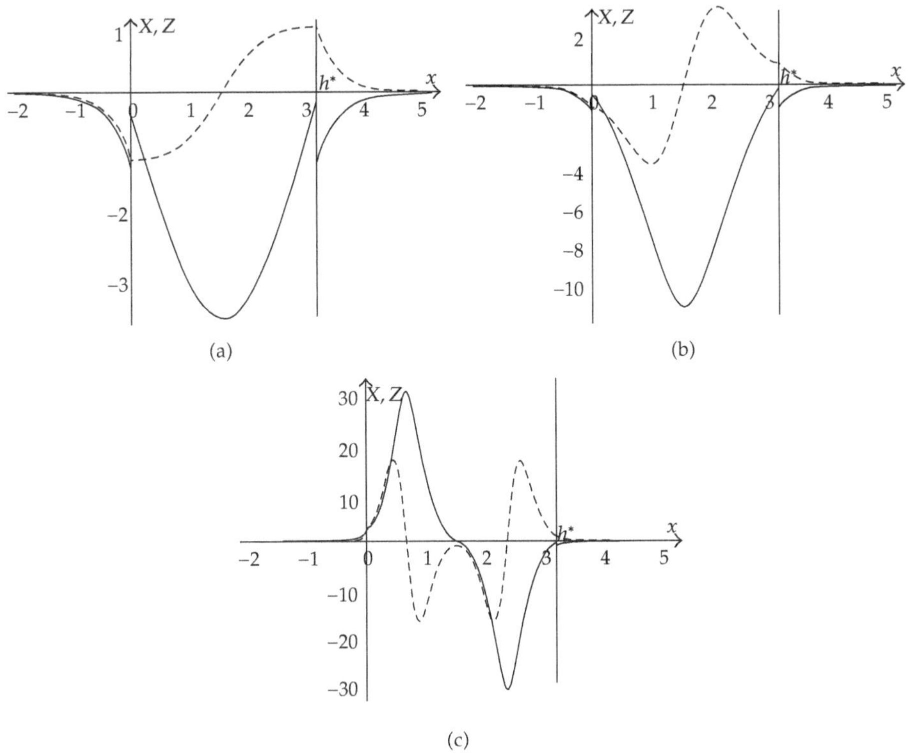

Figure 3: Eigenfunctions (fields) for the nonlinear problem are shown. Solid curves for X; dashed curves for Z. The same parameters as in Figure 2 are used. For (a), γ = 2.994; for (b), γ = 3.892: for (c), γ = 8.657, and h = 3.206 is used for all three cases. The eigenvalues are marked in Figure 2.

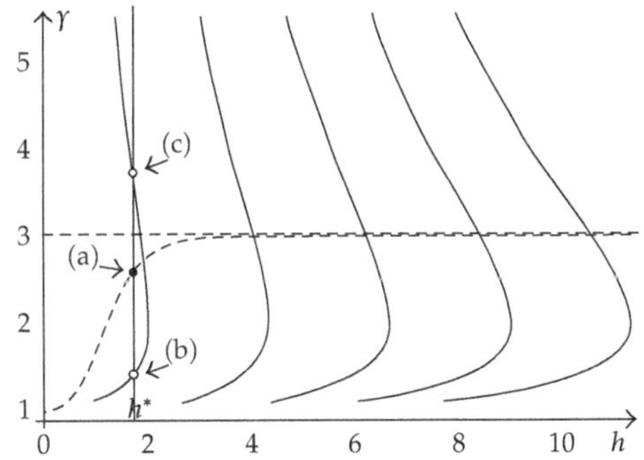

Figure 4: Plot of $\gamma(h)$. The first few dispersion curves are shown. Solid curves for the nonlinear case (solutions of (7.14)); dashed curve for the linear case (solutions of (8.5)). The following parameters are used for both cases: $\varepsilon_1 = 1$, $\varepsilon_2 = -1.5$, $\varepsilon_3 = 1$, and for the nonlinear case $a = 5.2$ and $E_z^{(h)} = 1$. Dashed lines are described by formulas: $h^* = 1.71$ (thickness of the layer), $\gamma = 1$ (lower bound for γ) and $\gamma = 3$ (upper bound for γ in the case of linear medium in the layer).

Further, it can be proved that function $h \equiv h(\gamma)$ defined from equation (7.14) when $a \neq 0$ has the following property:

$$\lim_{\gamma^2 \to +\infty} h(\gamma) = 0. \tag{9.1}$$

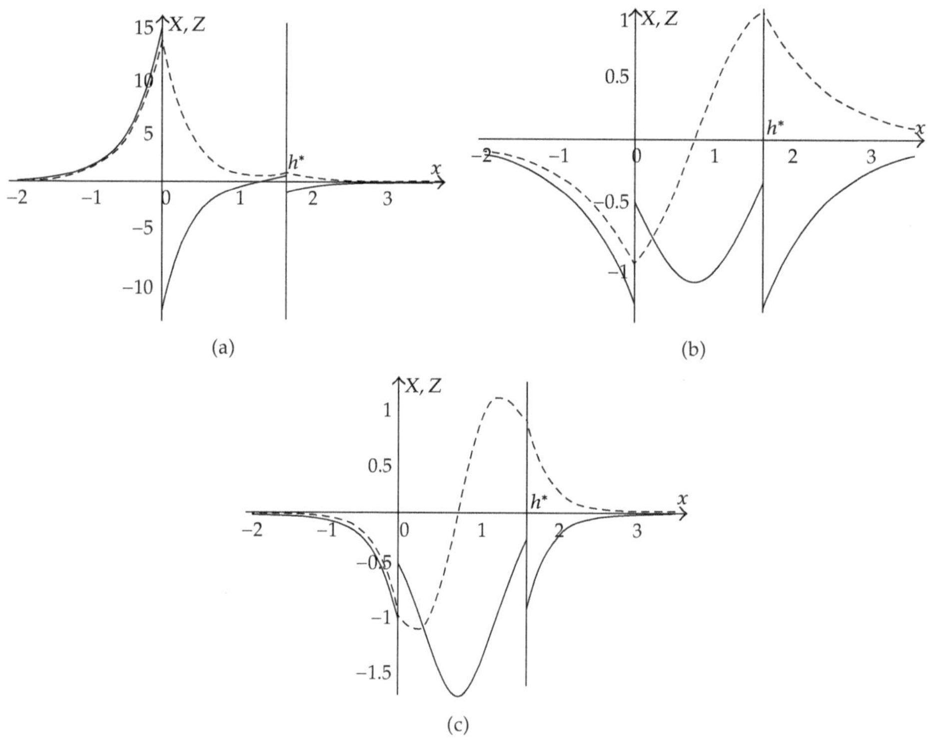

Figure 5: Eigenfunctions (fields) for the nonlinear problem are shown. Solid curves for X; dashed curves for Z. The same parameters as in Figure 4 are used. For (a), $\gamma = 2.620$, and $a = 0$ (see (8.5)); for (b), $\gamma = 1.565$; for (c), $\gamma = 3.481$, and $h = 1.71$ is used for all three cases. The eigenvalues are marked in Figure 4.

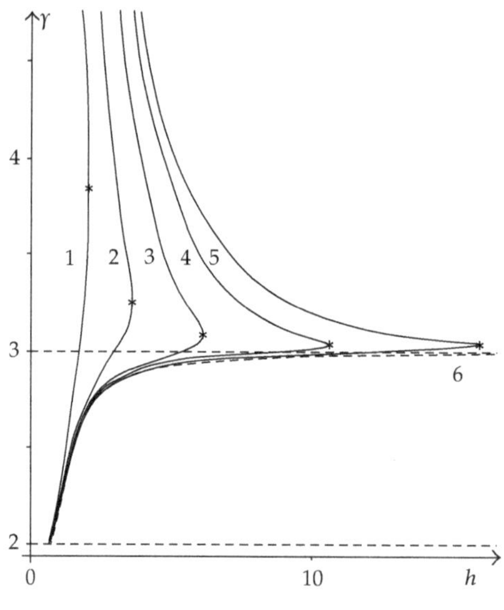

Figure 6: Plot of $\gamma(h)$ for the different values of a: 1 – $a = 1$; 2 – $a = 0.1$; 3 – $a = 0.01$; 4 – $a = 0.001$; 5 – $a = 0.0001$; 6 – $a = 0$ (linear case). The following parameters are used for both cases: $\varepsilon_1 = 4$, $\varepsilon_2 = 9$, $\varepsilon_3 = 1$, and for the nonlinear case $E_z^{(h)} = 1$. Dashed lines are described by formulas: $\gamma = 2$ (lower bound for γ), $\gamma = 3$ (upper bound for γ in the case of linear medium in the layer). Curves (solid) 1–5 are solutions of (7.14), and curve 6 (dashed) is solutions of (8.1).

In Figure 2 for $h = 3.206$ in the case of a linear layer there are 3 eigenvalues (black dots where the line $h = 3.206$ intersects with DCs). These eigenvalues correspond to 3 eigenmodes. In the case of a nonlinear layer in Figure 2 are shown 5 eigenvalues (uncolored dots). These eigenvalues correspond to 5 eigenmodes. Taking into account the last paragraph's statement it is clear that in this case there is infinite number of eigenvalues.

It is easy to see in Figure 6 that the less nonlinearity coefficient a is the more stretched DCs in the nonlinear case. The maximum points of the curves $h(\gamma)$ (in Figure 6 they are marked by asterisks) move to the right. The parts of the DCs that locate below the maximum points asymptotically tend to the DCs for the linear case as $a \to 0$.

It should be noticed that in the case of Kerr nonlinearity in a layer and TE waves there are strong constraints on the value a depending on the value ε_2 (for details see [16]). It is natural to suppose that there are constraints of the kind in the case under consideration.

As far as we know experiments to observe the new nonlinear eigenmodes were not carried out. So the question if the modes corresponding to the new eigenvalues exist (in an experiment) stays open! We should like to emphasize that it is interesting to observe purely nonlinear modes that do not arise in a linear limiting case. If to see Figure 2 then points (b) and (c) correspond to these purely nonlinear modes.

Acknowledgments

The authors gratefully acknowledge support of Russian Foundation for Basic Research (RFBR) Grant no. 11-07-00330-A and Russian Federation President Grant no. MK-2074.2011.1.

References

[1] S. A. Akhmanov and R. V. Khokhlov, *Problems of Non-linear Optics*, VINITI, Moscow, Russia, 1964, (in Russian).

[2] N. Bloembergen, *Nonlinear Optics*, W. A. Benjamin, New York, NY, USA, 1965.

[3] N. N. Akhmediev and A. Ankevich, *Solitons, Nonlinear Pulses and Beams*, Chapman and Hall, London, UK, 1997.

[4] A. D. Boardman, P. Egan, F. Lederer, U. Langbein, and D. Mihalache, "Third-order nonlinear electromagnetic TE and TM guided," in *Nonlinear Surface Electromagnetic Phenomena*, H.-E. Ponath and G. I. Stegeman, Eds., Elsevier, 1991.

[5] R. Y. Chiao, E. Garmire, and C. Townes, "Self-trapping of optical beams," *Physical Review Letters*, vol. 13, no. 15, pp. 479–482, 1964.

[6] V. M. Eleonskii, L. G. Oganes'yants, and V. P. Silin, "Cylindrical nonlinear waveguides," *Soviet Journal of Experimental and Theoretical Physics*, vol. 35, p. 44, 1972.

[7] V. M. Eleonskii and V. P. Silin, "Nonlinear theory of penetration of p-polarized waves into a conductor," *Soviet Journal of Experimental and Theoretical Physics*, vol. 33, p. 1039, 1971.

[8] D. V. Valovik and Y. G. Smirnov, "Calculation of the propagation constants and fields of polarized electromagnetic TM waves in a nonlinear anisotropic layer," *Journal of Communications Technology and Electronics*, vol. 54, no. 4, pp. 411–417, 2009.

[9] D. V. Valovik and Y. G. Smirnov, "Calculation of the propagation constants of TM electromagnetic waves in a nonlinear layer," *Journal of Communications Technology and Electronics*, vol. 53, no. 8, pp. 934–940, 2008.

[10] A. Snyder and J. Love, *Optical Waveguide Theory*, Chapman and Hall, London, UK, 1983.

[11] R. I. Joseph and D. N. Christodoulides, "Exact field decomposition for TM waves in nonlinear media," *Optics Letters*, vol. 12, no. 10, pp. 826–828, 1987.

[12] K. M. Leung and R. L. Lin, "Scattering of transverse-magnetic waves with a nonlinear film: formal field solutions in quadratures," *Physical Review B*, vol. 44, no. 10, pp. 5007–5012, 1991.

[13] K. M. Leung, "p-polarized nonlinear surface polaritons in materials with intensity-dependent dielectric functions," *Physical Review B*, vol. 32, no. 8, pp. 5093–5101, 1985.

[14] H. W. Schürmann and R. Schmoldt, "Optical response of a nonlinear absorbing dielectric film," *Optics Letters*, vol. 21, no. 6, pp. 387–389, 1996.

[15] H. W. Schürmann, V. S. Serov, Y. V. Shestopalov et al., "TE-polarized waves guided by a lossless nonlinear three-layer structure," *Physical Review E*, vol. 58, no. 1, pp. 1040–1050, 1998.

[16] D. V. Valovik, "Propagation of electromagnetic waves in a nonlinear metamaterial layer," *Journal of Communications Technology and Electronics*, vol. 56, no. 5, pp. 544–556, 2011.

[17] A. D. Boardman, A. A. Maradudin, G. I. Stegeman, T. Twardowski, and E. M. Wright, "Exact theory of nonlinear p-polarized optical waves," *Physical Review A*, vol. 35, no. 3, pp. 1159–1164, 1987.

[18] C. Qin and Z. H. Wang, "Exact dispersion relations for TM waves guided by thin dielectric films bounded by nonlinear media," *Optics Letters*, vol. 18, no. 1, pp. 260–262, 1993.

[19] A. D. Boardman and P. Egan, "S-polarized waves in a thin dielectric film asymmetrically bounded by optically nonlinear media," *IEEE Journal of Quantum Electronics*, vol. 21, no. 10, pp. 1701–1713, 1958.

[20] C. T. Seaton, J. Valera, R. Shoemaker et al., "Calculations of nonlinear TE waves guided by thin dielectric films bounded by nonlinear media," *IEEE Journal of Quantum Electronics*, vol. 21, no. 7, pp. 774–783, 1985.

[21] V. M. Agranovich, V. S. Babichenko, and V. Ya Chernyak, "Nonlinear surface polaritons," *JETP Letters*, vol. 32, pp. 532–535, 1980.

[22] C. T. Seaton, J. D. Valera, B. Svenson, and G. I. Stegeman, "Comparison of solutions for TM-polarized nonlinear guided waves," *Optics Letters*, vol. 10, no. 3, pp. 149–150, 1985.

[23] R. A. Sammut and C. Pask, "Gaussian and equivalent-step-index approximations for nonlinear waveguides," *Journal of the Optical Society of America B*, vol. 8, no. 2, pp. 395–402, 1991.

[24] A. E. Kaplan, "Theory of hysteresis reflection and refraction of light by a boundary of a nonlinear medium," *Soviet Physics, JETP*, vol. 72, pp. 896–905, 1977.

[25] U. Langbein, F. Lederer, T. Peschel, and H.-E. Ponath, "Nonlinear guided waves in saturable nonlinear media," *Optics Letters*, vol. 10, no. 11, pp. 571–573, 1985.

[26] W. J. Tomlinson, "Surface wave at a nonlinear interface," *Optics Letters*, vol. 5, no. 7, pp. 323–325, 1980.

[27] D. V. Valovik and Yu. G. Smirnov, "A nonlinear boundary eigenvalue problem for TM-polarized electromagnetic waves in a nonlinear layer," *Russian Mathematics*, vol. 52, no. 10, pp. 60–63, 2008.

[28] D. V. Valovik and Yu. G. Smirnov, "On the propagation of TM-polarized electromagnetic waves in a nonlinear layer with the nonlinearity expressed by the Kerr law," *Journal of Communications Technology and Electronics*, vol. 48, no. 12, pp. 2186–2194, 2008.

[29] Y. G. Smirnov and D. V. Valovik, "Boundary eigenvalue problem for maxwell equations in a nonlinear dielectric laye," *Applied Mathematics*, vol. 1, no. 1, pp. 29–36, 2010.

[30] Y. G. Smirnov, *Mathematical Methods for Electromagnetics Problems*, Penza State University Press, Penza, Russia, 2010.

[31] Y. G. Smirnov and D. V. Valovik, *Electromagnetic Wave Propagation in Nonlinear Waveguide Structures*, Penza State University Press, Penza, Russia, 2011.

[32] H. W. Schürmann, Y. G. Smirnov, and Y. V. Shestopalov, "Propagation of TE waves in cylindrical nonlinear dielectric waveguides," *Physical Review E*, vol. 71, no. 1, part 2, Article ID 016614, 2005.

[33] Y. G. Smirnov, H. W. Schürmann, and Y. U. Shestopalov, "Integral equation approach for the propagation of TE-waves in a nonlinear dielectric cylindrical waveguide," *Journal of Nonlinear Mathematical Physics*, vol. 11, no. 2, pp. 256–268, 2004.

[34] N. N. Bautin and E. A. Leontovich, *Metody i Priemy Kachestvennogo Issledovaniya Dinamicheskikh Sistem na Ploskosti*, Nauka, Moscow, Russia, 1990, (in Russian).

[35] G. A. Korn and T. M. Korn, *Mathematical Handbook for Scientists and Engineers*, McGraw-Hill Book, New York, NY, USA, 1968.

11

The Asymptotic Synchronization Analysis for Two Kinds of Complex Dynamical Networks

Ze Tang and Jianwen Feng

College of Mathematics and Computational Sciences, Shenzhen University, Shenzhen 518060, China

Correspondence should be addressed to Ze Tang, tangze0124@126.com

Academic Editor: Teoman Özer

We consider a class of complex networks with both delayed and nondelayed coupling. In particular, we consider the situation for both time delay-independent and time delay-dependent complex dynamical networks and obtain sufficient conditions for their asymptotic synchronization by using the Lyapunov-Krasovskii stability theorem and the linear matrix inequality (LMI). We also present some simulation results to support the validity of the theories.

1. Introduction

A complex dynamical network is a large set of interconnected nodes that represent the individual elements of the system and their mutual relationships. Owing to their immense potential for applications to various fields, complex networks have been intensively investigated in the past decade in areas as diverse as mathematics, physics, biology, engineering, and even the social sciences [1–3]. The synchronization problem for complex networks was first posed by Saber and Murray [4, 5] who also introduced a theoretical framework for their investigation by viewing them as the adjustments of the rhythms of their interaction states [6] and many different kinds of synchronization phenomena and models have also been discovered such as complete synchronization, phase synchronization, lag synchronization, antisynchronization, impulsive synchronization, and projective synchronization.

Time delays are an important consideration for complex networks although these were usually ignored in early investigations of synchronization and control problems [6–11]. To make up for this deficiency, uniformly distributed time delays have recently been incorporated into network models [12–25] and Wang et al. [18] even considered networks with both delayed and nondelayed couplings and obtained sufficient conditions for asymptotic stability. Similarly, Wu and Lu [19] investigated the exponential synchronization of general weighted delay and nondelay coupled complex dynamical networks with different topological structures. There remains, however, much room for improvement in both the scope of the systems considered by Wang and Xu as well as in their methods of proofs.

The main contributions of this paper are two-fold. Firstly, we present a more general model for networks with both delayed and nondelayed couplings and derive criteria for their asymptotical synchronization. Secondly, we apply the Lyapunov-Krasovskii theorem and the LMIs to ensure the inevitable attainment of the required synchronization.

The rest of the paper is organized as follows. In Section 2, we present the general complex dynamical network model under consideration and state some preliminary definitions and results. In Section 3, we present the main results of this paper. In particular, we consider the situation for both time delay-independent and time delay-dependent complex dynamical networks and derive sufficient conditions for their asymptotic synchronization by using the Lyapunov-Krasovskii stability theorem and the linear matrix inequality (LMI). In Section 4, we present some numerical simulation results that verify our theoretical results. The paper concludes in Section 5.

2. Preliminaries and Model Description

In general, a linearly coupled ordinary differential equation system (LCODES) can be described as follows:

$$\frac{dx_i(t)}{dt} = f(x_i(t)) + c_1 \sum_{j\neq i,j=1}^{N} b_{ij} A x_j(t) + c_2 \sum_{j\neq i,j=1}^{N} b'_{ij} A' x_j(t-\tau). \tag{2.1}$$

Since $x_i - x_i = 0$ for all $i = 1,\dots,N$, we can choose any values for a_{ii} in the above equations. Hence, letting $b_{ii} = -\sum_{j\neq i,j=1}^{N} b_{ij}$ and $b'_{ii} = -\sum_{j\neq i,j=1}^{N} b'_{ij}$, the above equations can be rewritten as follows:

$$\frac{dx_i(t)}{dt} = f(x_i(t)) + c_1 \sum_{j=1}^{N} b_{ij} A x_j(t) + c_2 \sum_{j=1}^{N} b'_{ij} A' x_j(t-\tau), \tag{2.2}$$

where N is the number of nodes, $x^i(t) = (x_{i1}, x_{i2},\dots,x_{iN})^T \in R^n$ are the state variables of the ith node, $t \in [0,+\infty)$ and $f : R^n \to R^n$ is a continuously differentiable function. The constants c_1 and c_2 (possibly distinct) are the coupling strengths, $b_{ij} \geq 0$, $b'_{ij} > 0$ (for $i, j = 1,\dots,N$), $A, A' \in R^{n\times n}$ are inner-coupled matrices, $B, B' \in R^{n\times n}$ are coupled matrices with zero-sum rows with $b_{ij}, b'_{ij} \geq 0$ for $i \neq j$ that determines the topological structure of the network. We assume that B and B' are symmetric and irreducible matrices so that there are no isolated nodes in the system.

If all the eigenvalues of a matrix $A \in R^{n\times n}$ are real, then we denote its ith eigenvalue by $\lambda_i(A)$ and sort them by $\lambda_1(A) \leq \lambda_2(A) \leq \cdots \leq \lambda_n(A)$. A symmetric real matrix A is positive definite (semidefinite) if $x^T A x > 0$ (≥ 0) for all nonzero x and denoted by $A > 0$ $(A \geq 0)$. Finally, I stands for the identity matrix and the dimensions of all vectors and matrices should be clear in the context.

Definition 2.1. A complex network with delayed and nondelayed coupling (2.2) is said to achieve asymptotic synchronization if

$$x_1(t) = x_2(t) = \cdots = x_N(t) = s(t), \quad t \longrightarrow +\infty, \tag{2.3}$$

where $s(t)$ is a solution of the local dynamics of an isolated node satisfying $\dot{s(t)} = f(s(t))$.

Definition 2.2. A matrix $L = (l_{ij})_{i,j=1}^{N}$ is said to belong to the class $A1$, denoted by $L \in A1$ if

(1) $l_{ij} \leq 0, i \neq j, l_{ii} = -\sum_{j=1,j\neq i}^{N} l_{ij}, i = 1, 2, \ldots, N,$
(2) L *is irreducible.*

If $L \in A1$ is symmetrical, then we say that L belongs to the class $A2$, denoted by $L \in A2$.

Lemma 2.3 (see [26]). *If* $L \in A1$, *then* $\operatorname{rank}(L) = N - 1$, *that is, 0 is an eigenvalue of* L *with multiplicity 1, and all the nonzero eigenvalues of* L *have positive real part.*

Lemma 2.4 (Wang and Chen [11]). *If* $G = (g_{ij})_{N\times N}$ *satisfies the above conditions, then there exists a unitary matrix* $\Phi = (\phi_1, \ldots, \phi_N)$ *such that*

$$G^T \phi_k = \lambda_k \phi_k, \quad k = 1, 2, \ldots, N, \tag{2.4}$$

where λ_i, $i = 1, 2, \ldots, N$, *are the eigenvalues of* G.

Lemma 2.5 (Schur complement [22]). *The linear matrix inequality (LMI)*

$$\begin{pmatrix} Q(x) & S(x) \\ S(x)^T & R(x) \end{pmatrix} > 0, \tag{2.5}$$

where $Q(x)$ *and* $R(x)$ *are symmetric matrices and* $S(x)$ *is a matrix with suitable dimensions is equivalent to one of the following conditions:*

(i) $Q(x) > 0,\ R(x) - S(x)^T Q(x)^{-1} S(x) > 0;$
(ii) $R(x) > 0,\ Q(x) - S(x) R(x)^{-1} S(x) > 0.$

Lemma 2.6 ((the Lyapunov-Krasovskii stability theorem). (Kolmanovskii and Myshkis, Hale and Verduyn Lunel [16])). *Consider the delayed differential equation*

$$x(t) = \dot{f}(t, x(t)), \tag{2.6}$$

where $f : R\times C \to R^n$ *is continuous and takes* $R\times$ *(bounded subsets of* C*) into bounded subsets of* R^n, *and let* $u, v, w : R^+ \to R^+$ *be continuous and strictly monotonically nondecreasing functions with* $u(s), v(s), w(s)$ *being positive for* $s > 0$ *and* $u(0) = v(0) = 0$. *If there exists a continuous functional* $V : R \times C \to R$ *such that*

$$\begin{gathered} u(\|x\|) \leq V(t, x) \leq v(\|x\|), \\ \dot{V}(t, x(t, x(t))) \leq -w(\|x(t)\|), \end{gathered} \tag{2.7}$$

where $\dot{V}$ is the derivative of V along the solutions of the above delayed differential equation, then the solution $x = 0$ of this equation is uniformly asymptotically stable.

Remark 2.7. The functional V is called a Lyapunov-Krasovskii functional.

Lemma 2.8 (Moon et al. [22]). *Let $a(\cdot) \in R^{n_a}$, $b(\cdot) \in R^{n_b}$ and $M(\cdot) \in R^{n_a \times n_b}$ be defined on an interval Ω. Then, for any matrices $X \in R^{n_a \times n_a}$, $Y \in R^{n_a \times n_b}$, and $Z \in R^{n_b \times n_b}$, one has*

$$-2\int_{\Omega} a(x)^T Mb(x)dx \leq \int_{\Omega} \begin{bmatrix} a(x) \\ b(x) \end{bmatrix}^T \begin{bmatrix} X & Y - M \\ Y^T - M^T & Z \end{bmatrix} \begin{bmatrix} a(x) \\ b(x) \end{bmatrix} dx, \tag{2.8}$$

where

$$\begin{bmatrix} X & Y \\ Y^T & Z \end{bmatrix} \geq 0. \tag{2.9}$$

Lemma 2.9. *For all positive-definite matrices P and vectors x and y, one has*

$$-2x^T y \leq \inf_{P>0}\left\{x^T Px + y^T P^{-1} y\right\}. \tag{2.10}$$

Lemma 2.10 (see [16]). *Consider the delayed dynamical network (2.2). Let*

$$\begin{gathered} 0 = \lambda_1 > \lambda_2 \geq \lambda_3 \geq \cdots \geq \lambda_N, \\ 0 = \mu_1 > \mu_2 \geq \mu_3 \geq \cdots \geq \mu_N \end{gathered} \tag{2.11}$$

be the eigenvalues of the outer-coupling matrices B and B', respectively. If the n-dimensional linear time-delayed and nontime delayed system

$$\dot{w}_i(t) = J(t)w_i(t) + c_1\lambda_i A w_i(t) + c_2\mu_i A' w_i(t - \tau), \quad k = 2, 3, \ldots, N, \tag{2.12}$$

of $N - 1$ differential equations is asymptotically stable about their zero solutions for some Jacobian matrix $J(t) \in R^{n \times n}$ of $f(x(t))$ at $s(t)$, then the synchronized states (2.3) are asymptotically stable.

3. The Criteria for Asymptotic Synchronization

In this section, we derive the conditions for the asymptotic synchronization of time-delayed coupled dynamical networks when they are either time-dependent or time-independent.

3.1. *Case 1: The Time Delay-Independent Stability Criterion*

Theorem 3.1. *Consider the general time delayed and non-time delayed complex dynamical network (2.2). If there exist two positive definite matrices P and $Q > 0$ such that*

$$\begin{bmatrix} J(t)^T P + PJ(t) + 2c_1\lambda_i A + Q & c_2\mu_N PA' \\ c_2\mu_N A'P & -Q \end{bmatrix} > 0, \tag{3.1}$$

then the synchronization manifold (2.3) of network (2.2) can be asymptotically synchronized for all fixed time delay $\tau > 0$.

Proof. For each fixed $i = 1, 2, \dots, N$, choose the Lyapunov-Krasovskii functional

$$V_i(t) = w_i(t)^T P w_i(t) + \int_{t-\tau}^{t} w_i(s)^T Q w_i(s) ds \tag{3.2}$$

for some matrices $P > 0$ and $Q > 0$ to be determined. Then the derivative of $V_i(t)$ along the trajectories of (3.2) is

$$\frac{dV_i(t)}{dt} = \dot{w}_i(t)^T P w_i(t) + w_i(t)^T P \dot{w}_i(t) + w_i(t)^T Q w_i(t) - w_i(t-\tau)^T Q w_i(t-\tau) \tag{3.3}$$

which, upon substitution of (2.12), gives

$$\begin{aligned} \dot{V}_i(t) = {} & \left[J(t)w_i(t) + c_1\lambda_i A w_i(t) + c_2\mu_i A' w_i(t-\tau)\right]^T P w_i(t) + w_i(t)^T P \\ & \times \left[J(t)w_i(t) + c_1\lambda_i A w_i(t) + c_2\mu_i A' w_i(t-\tau)\right] + w_i(t)^T Q w_i(t) - w_i(t-\tau)^T Q w_i(t-\tau) \\ = {} & w_i(t)^T J(t)^T w_i(t) + w_i(t)^T \lambda_i A^T P w_i(t) + w_i(t-\tau)^T \mu_i A' P w_i(t) \\ & + w_i(t)^T P J(t) w_i(t) + w_i(t)^T c_1\lambda_i A w_i(t) + w_i(t)^T P c_2\mu_i A' w_i(t-\tau) \\ & + w_i(t)^T Q w_i(t) - w_i(t-\tau)^T Q w_i(t-\tau) \\ = {} & w_i(t)^T \left[J(t)^T P + PJ(t) + 2c_1\lambda_i A + Q\right] w_i(t) + 2w_i(t)^T c_2\mu_i PA' w_i(t-\tau) \\ & - w_i(t-\tau)^T Q w_i(t-\tau). \end{aligned} \tag{3.4}$$

Now, by using the inequality in Lemma 2.9, we have

$$2w_i(t)^T c_2\mu_i PA'^{w_i}(t-\tau) \le w_i(t-\tau)^T Q w_i(t-\tau) + w_i(t)^T c_2{}^2 \mu_i{}^2 PA'QA'Pw_i(t), \tag{3.5}$$

which, upon substituting (3.5) into (3.2), gives

$$\dot{V}_i(t) \le w_i(t)^T \left[J(t)^T P + PJ(t) + 2c_1\lambda_i A + c_2\mu_i{}^2 w_i(t)^T PA'QA'Aw(t) + Q\right] w_i(t). \tag{3.6}$$

It therefore follows from the Schur complement (Lemma 2.5) and the linear matrix inequality (3.1) that $\dot{V}_i(t) < 0$ for all the $N-1$ equations in the general time delayed and non-time delayed system (2.12) and hence the system (2.12) is asymptotically synchronized by the Lyapunov-Krasovskii stability theorem. So, by Theorem 3.1, the synchronization manifold (2.3) of the network (2.2) is asymptotically synchronized. This completes the proof of the theorem. □

The following corollaries follow immediately from the above theorem.

Corollary 3.2. *Consider the general non-time delayed complex dynamical network*

$$\dot{x}_i(t) = f(x_i(t)) + c_1 \sum_{j=1}^{N} b_{ij} A x_j(t). \tag{3.7}$$

If there exists a positive definite matrix $P > 0$ *such that*

$$J(t)^T P + c_1 \lambda_i A P < 0, \tag{3.8}$$

then the synchronization manifold (2.3) *of network* (3.7) *can be asymptotically synchronized.*

Proof. From Lemma 2.10, we have

$$\dot{w}_i(t) = J(t) w_i(t) + c_1 \lambda_i A w_i(t) \tag{3.9}$$

and the result follows by choosing the Lyapunov functional $V_i(t) = (1/2) w_i(t)^T P w_i(t)$. □

Corollary 3.3. *Consider the general time delayed complex dynamical network*

$$\dot{x}_i(t) = f(x_i(t)) + c_2 \sum_{j=1}^{N} b'_{ij} A' x_j(t - \tau). \tag{3.10}$$

If there exist two positive definite matrices $P > 0$ *and* $Q > 0$ *such that*

$$\begin{bmatrix} J(t)^T P + P J(t) + Q & c_2 \mu_N P A' \\ c_2 \mu_N A' P & -Q \end{bmatrix} < 0, \tag{3.11}$$

then the synchronization manifold (2.3) *of network* (3.10) *can be asymptotically synchronized.*

Remark 3.4. The results of [16] are obtainable as particular cases of Theorem 3.1.

Remark 3.5. The above analysis is applicable to a general system with arbitrary time delays. A simpler synchronization scheme, however, could be applied to systems with time delays that are already known and are small in value.

3.2. Case 2: The Criterion for Time Delay-Dependent Stability

Theorem 3.6. *Consider the general time delayed and non-time delayed complex dynamical network* (2.2) *with a fixed time delay* $\tau \in (0, h]$ *for some small* h. *If there exist three positive definite matrices* $P, Q, Z > 0$, *such that*

$$\begin{bmatrix} (1,1) & Pc_2\mu_i A' - Y + h(J(t) + c_1\lambda_i A)^T Z c_2\mu_i A' \\ c_2\mu_i A'^T P - Y^T + hc_2\mu_i A'^T Z(J(t) + c_1\lambda_i A) & hc_2\mu_i{}^2 A'^T Z A' - Q \end{bmatrix} < 0 \tag{3.12}$$

with

$$\begin{aligned} (1,1) = {} & \left(J(t) + c_1\lambda_i A + c_2\mu_i A'\right)^T P + P(J(t) + c_1\lambda_i A + c_2\mu_i A') + hX \\ & + Y^T + Y + Q + h(J(t) + c_1\lambda_i A)^T (J(t) + c_1\lambda_i A) \end{aligned} \tag{3.13}$$

$$\begin{bmatrix} X & Y \\ Y^T & Z \end{bmatrix} \geq 0,$$

then the synchronization manifold (2.3) *of network* (2.2) *can be asymptotically synchronized.*

Proof. For each fixed $i = 1, 2, \ldots, N$, choose the Lyapunov-Krasovskii functional

$$V_i(t) = w_i(t)^T P w_i(t) + \int_{t-\tau}^{t} w_i(s)^T Q w_i(s) + \int_{-\tau}^{0}\int_{t+\beta}^{t} \dot{w}_i(s)^T Z \dot{w}_i(s) ds d\beta \tag{3.14}$$

for some matrices $P, Q, Z > 0$ to be determined and let

$$\begin{gathered} V_1 = \dot{w}_i(t)^T P w_i(t), \qquad V_2 = \int_{t-\tau}^{t} w_i(s)^T Q w_i(s), \\ V_3 = \int_{-\tau}^{0}\int_{t+\beta}^{t} \dot{w}_i(s)^T Z \dot{w}_i(s) ds d\beta. \end{gathered} \tag{3.15}$$

Then, $V_i(t) = V_1 + V_2 + V_3$ and it follows from the Newton-Leibniz equation that

$$\int_{t-\tau}^{t} \dot{w}_i(\xi) d\xi = w_i(t) - w_i(t - \tau) \tag{3.16}$$

so that (2.12) can be transformed into

$$\dot{w}_i(t) = \left(J(t) + c_1\lambda_i A + c_2\mu_i A'\right) w_i(t) - c_2\mu_i A'^{\int_{t-\tau}^{t} \dot{w}_i}(s) ds. \tag{3.17}$$

Hence

$$\begin{aligned} V_1 &= \dot{w}_i(t)^T P w_i(t) + w_i(t)^T P \dot{w}_i(t) \\ &= w_i(t)^T \left[\left(J(t) + c_1\lambda_i A + c_2\mu_i A' \right)^T P + P\left(J(t) + c_1\lambda_i A + c_2\mu_i A' \right) \right] w_i(t) \\ &\quad - 2w_i(t)^T P c_2\mu_i A' \int_{t-\tau}^{t} \dot{w}_i(s) ds \end{aligned} \tag{3.18}$$

and so, by Lemma 2.9, we have

$$\begin{aligned} & -2w_i(t)^T P c_2\mu_i A' \int_{t-\tau}^{t} \dot{w}_i(s) ds \\ &= -2\int_{t-\tau}^{t} w_i(t)^T (P c_2\mu_i A') \dot{w}_i(s) ds \\ &\leq \int_{t-\tau}^{t} \begin{bmatrix} w_i(t) \\ \dot{w}_i(s) \end{bmatrix}^T \begin{bmatrix} X & Y - P c_2\mu_i A' \\ Y^T - A'^T c_2\mu_i P & Z \end{bmatrix} \begin{bmatrix} w_i(t) \\ \dot{w}_i(s) \end{bmatrix} dx \\ &= \int_{t-\tau}^{t} w_i(t)^T X w_i(t) ds + \int_{t-\tau}^{t} \dot{w}_i(s)^T Z \dot{w}_i(s) ds + 2\int_{t-\tau}^{t} w_i(t)^T (Y - P c_2\mu_i A') \dot{w}_i(s) ds \\ &= \tau w_i(t)^T X w_i(t) + 2w_i(t)^T (Y - P c_2\mu_i A') \int_{t-\tau}^{t} \dot{w}_i(s) ds + \int_{t-\tau}^{t} \dot{w}_i(s)^T Z \dot{w}_i(s) \\ &= \tau w_i(t)^T X w_i(t) + 2w_i(t)^T (Y - P c_2\mu_i A') w_i(t) - 2w_i(t)^T (Y - P c_2\mu_i A') w_i(t-\tau) \\ &\quad + \int_{t-\tau}^{t} \dot{w}_i(s) Z \dot{w}_i(s) \end{aligned} \tag{3.19}$$

and so

$$\begin{aligned} V_1 \leq\, & w_i(t)^T \left[\left(J(t) + c_1\lambda_i A + c_2\mu_i A' \right)^T P + P\left(J(t) + c_1\lambda_i A + c_2\mu_i A' \right) \right] w_i(t) \\ & + 2w_i(t)^T (P c_2\mu_i A' - Y) w_i(t-\tau) + \int_{t-\tau}^{t} \dot{w}_i(s)^T Z w_i(s) ds. \end{aligned} \tag{3.20}$$

Similarly, we have

$$
\begin{aligned}
V_2 &= w_i(t)^T Q w_i(t) - w_i(t-\tau)^T Q w_i(t-\tau),\\
V_3 &= \tau \dot{w}_i(t)^T Z \dot{w}_i(t) - \int_{t-\tau}^{t} \dot{w}_i(s)^T Z \dot{w}_i(s)ds \le h\left[(J(t)+c_1\lambda_i A)w_i(t) + c_2\mu_i A' w_i(t-\tau)\right]^T\\
&\quad \times Z\left[(J(t)+c_1\lambda_i A)w_i(t) + c_2\mu_i A' w_i(t-\tau)\right] - \int_{t-\tau}^{t} \dot{w}_i(s)^T Z \dot{w}_i(s)ds\\
&= h w_i(t)^T (J(t)+c_1\lambda_i A)^T Z (J(t)+c_1\lambda_i A) w_i(t) + h w_i(t)^T\\
&\quad \times (J(t)t + c_1\lambda_i A)^T Z c_2\mu_i A' w_i(t-\tau) + h w_i(t-\tau)^T c_2 \mu_i A'^T Z^T\\
&\quad \times (J(t)+c_1\lambda_i A) w_i(t) + h w_i(t-\tau)^T \mu_i^2 A'^T Z A' w_i(t-\tau) - \int_{t-\tau}^{t} \dot{w}_i(s)^T Z \dot{w}_i(s)ds\\
&= h w_i(t)^T (J(t)+c_1\lambda_i A)^T Z (J(t)+c_1\lambda_i A) w_i(t) + 2h w_i(t)^T\\
&\quad \times (J(t)+c_1\lambda_i A)^T Z c_2\mu_i A' w_i(t-\tau) + h w_i(t-\tau)^T \mu_i^2 A'^T Z A' w_i(t-\tau)\\
&\quad - \int_{t-\tau}^{t} \dot{w}_i(s)^T Z \dot{w}_i(s)ds
\end{aligned}
\tag{3.21}
$$

and so

$$
\begin{aligned}
\dot{V}_i(t) &= V_1 + V_2 + V_3\\
&\le w_i(t)^T\Big[(J(t)+c_1\lambda_i A + c_2\mu_i A')^T P + P(J(t)+c_1\lambda_i A + c_2\mu_i A')\\
&\qquad + hX + Y^T + Y + Q + h(J(t)+c_1\lambda_i A)^T (J(t)+c_1\lambda_i A)\Big] w_i(t)\\
&\quad + w_i(t-\tau)^T\left[h\mu_i^2 A'^T Z A' - Q\right] w_i(t-\tau)\\
&\quad + 2w_i(t)^T\left[(P c_2\mu_i A' - Y) + h(J(t)+c_1\lambda_i)^T Z c_2 \mu_i A'\right] w_i(t-\tau).
\end{aligned}
\tag{3.22}
$$

Finally, we have

$$
\dot{V}_i(t) \le \begin{bmatrix} w_i(t) \\ w_i(t-\tau) \end{bmatrix}^T \begin{bmatrix} (1,1) & (1,2) \\ (2,1) & hc_2\mu^2 A'^T Z A'_Q \end{bmatrix} \begin{bmatrix} w_i(t) \\ w_i(t-\tau) \end{bmatrix}, \tag{3.23}
$$

where

$$(1,1) = (J(t) + c_1\lambda_i A + c_2\mu_i A')^T P + P(J(t) + c_1\lambda_i A + c_2\mu_i A') + hX$$
$$+ Y^T + Y + Q + h(J(t) + c_1\lambda_i A)^T (J(t) + c_1\lambda_i A)$$
$$(1,2) = Pc_2\mu_i A' - Y + h(J(t) + c_1\lambda_i A)^T Zc_2\mu_i A' \tag{3.24}$$
$$(2,1) = c_2\mu_i A'^T P - Y^T + hc_2\mu_i A'^T Z(J(t) + c_1\lambda_i A)$$
$$\begin{bmatrix} X & Y \\ Y^T & Z \end{bmatrix} \geq 0.$$

It now follows from Lemma 2.5 that the conditions of the theorem are equivalent to $\dot{V}_i(t) < 0$ and that by the Lyapunov-Krasovskii Stability Theorem all the nodes of the system (2.12) are asymptotically stable when (3.12) and (3.13) hold for $i = 1, 2, \ldots, N$. This completes the proof of Theorem 3.6. □

Corollary 3.7. *Consider the general time delayed complex dynamical network* (3.10) *with a fixed time delay* $\tau \in (0, h]$

$$\dot{x}_i(t) = f(x_i(t)) + c_2 \sum_{j=1}^{N} b'_{ij} A' x_j(t - \tau) \tag{3.25}$$

for some $h < +\infty$. *If there exist two positive definite matrices,* $P, Q > 0$, X, Y, *and* Z *such that*

$$\begin{bmatrix} (1,1) & c_2\mu_i PA' - Y + hJ(t)^T Zc_2\mu_i A' \\ c_2\mu_A'^T P - Y^T + hc_2\mu_i A'^T ZJ(t) & hc_2^2\mu_i^2 A'^T ZA' - Q \end{bmatrix} < 0, \tag{3.26}$$

where $(1,1) = PJ(t) + J(t)^T + hX + Y^T + Y + Q + hJ(t^T ZJ(t))$, *then the synchronization manifold* (2.3) *of network* (3.10) *is asymptotic synchronization.*

Remark 3.8. The proof can be found in [16]. Those are the two results of general complex dynamical network with fixed time-invariant delay $\tau \in (0, h]$ for some $h < +\infty$; the conclusions are less conservative than the time-independent delay. The delay-dependent stability is another method applying to the delayed system. And it could provide a useful and meaningful upper bound of the delay h, which could ensure the delayed system achieves asymptotic synchronization only if the time delay is less than h.

4. Numerical Simulations

The above criteracould be applied to networks with different topologies and different size. We put two examples to illustrate the validity of the theories.

Example 4.1. We use a three-dimensional stable nonlinear system as an example to illustrate the main results, Theorem 3.1, of our paper; this is the time delay-independent situation. The model could be described as follows:

$$\begin{bmatrix} \dot{x}_{i1} \\ \dot{x}_{i2} \\ \dot{x}_{i3} \end{bmatrix} = \begin{bmatrix} -x_{i1} + x_{i2}^2 \\ -2x_{i2} \\ -3x_{i3} + x_{i2}x_{i3} \end{bmatrix}, \quad i = 1,2,3. \tag{4.1}$$

The solution of the 3-dimensional stable nonlinear system equations can be written as

$$x_{i1} = c_1 e^{-t} - c e^{-4t}, \qquad x_{i2} = c_2 e^{-2t}, \qquad x_{i3} = \frac{c_3 e^{-3t} - c_2 e^{2t}}{2}, \tag{4.2}$$

which is asymptotically stable at the equilibrium point of the system $s(t) = 0$, where $c = -c_2^2/3$ and c_1, c_2, c_3 are all constants. It is easy to see that the Jacobian matrix is $J = \mathrm{diag}\{-1, -2, -3\}$. We assume the inner-coupling matrices A, A' are all identity matrices, namely, $A = A' = \mathrm{diag}\{1, 1, 1\}$, and the outer coupling configuration matrices

$$B = B' = \begin{bmatrix} -2 & 1 & 0 & 1 \\ 1 & -2 & 1 & 0 \\ 0 & 1 & -2 & 1 \\ 1 & 0 & 1 & 2 \end{bmatrix}. \tag{4.3}$$

The eigenvalues of the coupling matrices are $\lambda(B) = \lambda(B') = \{0, -1.5, -1.5\}$. We choose the coupling strength $c_1 = 0.5$, $c_2 = 1$. By using Theorem 3.1 and the LMI Toolbox in MATLAB, we obtained the following common two positive-definite matrices:

$$P = \mathrm{diag}\{1.1204, 12.3091, 10.165\}, \qquad Q = \mathrm{diag}\{2.3710, 22.5713, 26.0849\}. \tag{4.4}$$

According to the conditions in Theorem 3.1, we know the synchronized state $s(t)$ is global asymptotically stable for any fixed delay. The quantity

$$E(t) = \sqrt{\left(\frac{\sum_{i=1}^{N} |x_i(t) - s(t)|^2}{N} \right)} \tag{4.5}$$

is used to measure the quality of the synchronization process. We plot the evolution of $E(t)$ in the upper part in Figure 1. For the time delay here we choose $\tau = 0.1$. The lower subplot indicates the synchronization results of the network.

Example 4.2. We use a 4-nodes networks model as another example to illustrate the Theorem 3.6; this is the time delay-dependent situation. The model could be described as follows:

$$\begin{bmatrix} \dot{x}_{i1} \\ \dot{x}_{i2} \\ \dot{x}_{i3} \end{bmatrix} = \begin{bmatrix} -x_{i1} \\ -2x_{i2} + x_{i3}^2 \\ -3x_{i3} + x_{i2}x_{i3} \end{bmatrix}, \quad i = 1,2,3,4. \tag{4.6}$$

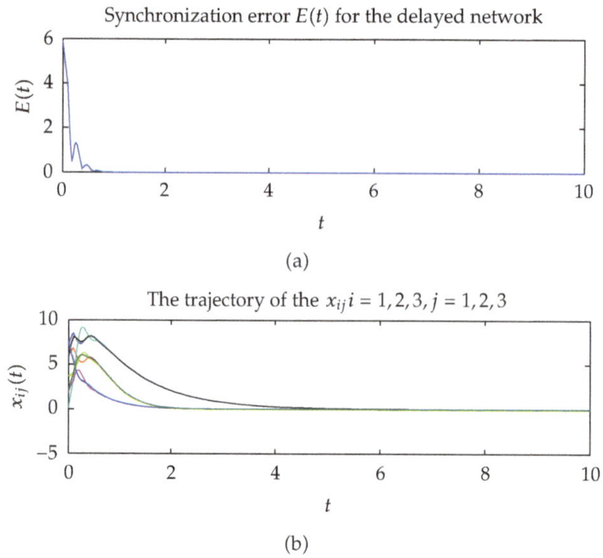

Figure 1: Synchronization evolution $E(t)$ for the delay-independent network with $c_1 = 0.5$, $c_2 = 1$, and $\tau = 0.1$.

We choose the same coupling strength $c_1 = 0.5$, $c_2 = 1$; the eigenvalues of the coupling matrices are $\lambda(B) = \lambda(B') = \{0, -2, -2, -4\}$. By using Theorem 3.6 and the LMI Toolbox in MATLAB, we obtained the following matrices:

$$\begin{aligned} P &= \operatorname{diag}\{1.4078, 1.4057, 1.4054\}, & Q &= \{1.4157, 1.4157, 1.4157\}, \\ Z &= \{-1.4218, -1.4303, -1.4367\}, & X &= \{18.3024, 19.6808, 21.0815\}, \\ Y &= \{0.0221, 0.0208, 0.0196\}. \end{aligned} \tag{4.7}$$

By using Theorem 3.6 in this paper, it is found that the maximum delay bound for the complex dynamical network to form asymptotic synchronization is $h = 1$. $E(t)$ are defined the same as in the example. We plot the evolution of $E(t)$ in upper part in Figure 2. The lower subplot indicates the synchronization results of the network. It can be seen from the figures that the network in this example can achieve asymptotic synchronization.

5. Conclusion

This paper considered a class of complex networks with both time delayed and non-time delayed coupling. We derived, respectively, a sufficient criterion for time delay-dependent and time delay-independent asymptotic synchronization which are more general than those obtained in previous works. These asymptotic synchronization results were obtained by using the Lyapunov-Krasovskii stability theorem and the linear matrix inequality. Two simple examples were also used to validate the theoretical analysis.

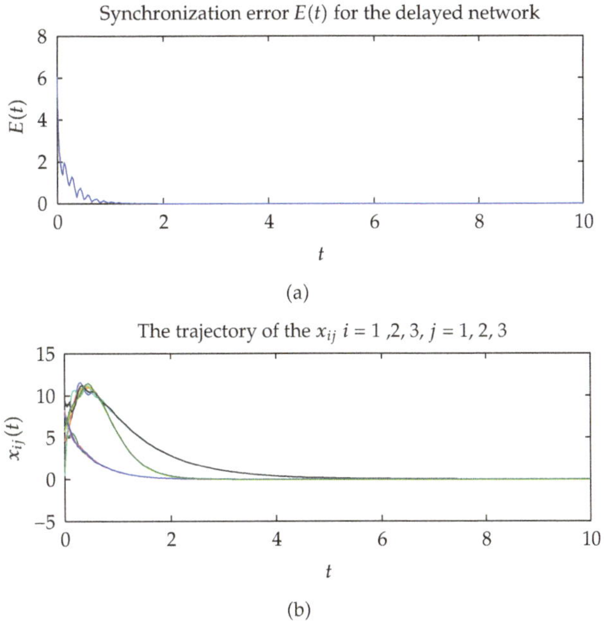

Figure 2: Synchronization evolution $E(t)$ for the delay-dependent network with $c_1 = 0.5$, $c_2 = 1$, and $\tau = 0.1$.

Acknowledgments

The authors thank the referees and the editor for their valuable comments on this paper. This work is supported by the National Natural Science Foundation of China (Grant no. 61273220), Guandong Education University Industry Cooperation Projects (Grant no. 2009B090300355), and the Shenzhen Basic Research Project (JC201006010743A, JC200903120040A).

References

[1] S. H. Strogatz, "Exploring complex networks," *Nature*, vol. 410, no. 6825, pp. 268–276, 2001.

[2] A.-L. Barabási and R. Albert, "Emergence of scaling in random networks," *Science*, vol. 286, no. 5439, pp. 509–512, 1999.

[3] X. F. Wang, "Complex networks: topology, dynamics and synchronization," *International Journal of Bifurcation and Chaos in Applied Sciences and Engineering*, vol. 12, no. 5, pp. 885–916, 2002, Chaos control and synchronization (Shanghai, 2001).

[4] R. O. Saber and R. M. Murray, "Flocking with Obstacle Avoidance: cooperation with Limited Communication in Mobile Networks," in *Proceedings of the 42nd IEEE Conference on Decision and Control*, pp. 2022–2028, December 2003.

[5] R. O. Saber and R. M. Murray, "Graph rigidity and distributed formation stabilization of multi-vehicle systems," in *Proceedings of the 41st IEEE Conference on Decision and Control*, pp. 2965–2971, usa, December 2002.

[6] A. Pikovsky, M. Rosenblum, and J. Kurths, *Synchronization: A Universal Concept in Nonlinear Sciences*, vol. 12 of *Cambridge Nonlinear Science Series*, Cambridge University Press, Cambridge, 2001.

[7] X. F. Wang and G. Chen, "Synchronization in scale-free dynamical networks: robustness and fragility," *IEEE Transactions on Circuits and Systems. I. Fundamental Theory and Applications*, vol. 49, no. 1, pp. 54–62, 2002.

[8] X. F. Wu, C. Xu, and J. Feng, "Mean synchronization of pinning complex networks with linearly and nonlinearly time-delay coupling," *International Journal of Digital Content Technology and its Applications*, vol. 5, no. 3, pp. 33–46, 2011.

[9] X. S. Yang, J. D. Cao, and J. Q. Lu, "Synchronization of coupled neural networks with random coupling strengths and mixed probabilistic time-varying delays," *International Journal of Robust and Nonlinear Control*. In press.

[10] J. Cao, Z. Wang, and Y. Sun, "Synchronization in an array of linearly stochastically coupled networks with time delays," *Physica A*, vol. 385, no. 2, pp. 718–728, 2007.

[11] X. F. Wang and G. Chen, "Synchronization in scale-free dynamical networks: robustness and fragility," *IEEE Transactions on Circuits and Systems. I*, vol. 49, no. 1, pp. 54–62, 2002.

[12] W. Sun, F. Austin, J. Lü, and S. Chen, "Synchronization of impulsively coupled complex systems with delay," *Chaos*, vol. 21, no. 3, Article ID 033123, 7 pages, 2011.

[13] W. He, F. Qian, J. Cao, and Q.-L. Han, "Impulsive synchronization of two nonidentical chaotic systems with time-varying delay," *Physics Letters A*, vol. 375, no. 3, pp. 498–504, 2011.

[14] J. Tang, J. Ma, M. Yi, H. Xia, and X. Yang, "Delay and diversity-induced synchronization transitions in a small-world neuronal network," *Physical Review E*, vol. 83, no. 4, Article ID 046207, 2011.

[15] C. W. Wu and L. O. Chua, "Synchronization in an array of linearly coupled dynamical systems," *IEEE Transactions on Circuits and Systems. I*, vol. 42, no. 8, pp. 430–447, 1995.

[16] C. Li and G. Chen, "Synchronization in general complex dynamical networks with coupling delays," *Physica A. Statistical Mechanics and its Applications*, vol. 343, no. 1–4, pp. 263–278, 2004.

[17] D. Xu and Z. Su, "Synchronization criterions and pinning control of general complex networks with time delay," *Applied Mathematics and Computation*, vol. 215, no. 4, pp. 1593–1608, 2009.

[18] L. Wang, H.-P. Dai, and Y.-X. Sun, "Random pseudofractal networks with competition," *Physica A*, vol. 383, no. 2, pp. 763–772, 2007.

[19] X. Wu and H. Lu, "Exponential synchronization of weighted general delay coupled and non-delay coupled dynamical networks," *Computers & Mathematics with Applications*, vol. 60, no. 8, pp. 2476–2487, 2010.

[20] S. Wen, S. Chen, and W. Guo, "Adaptive global synchronization of a general complex dynamical network with non-delayed and delayed coupling," *Physics Letters A*, vol. 372, no. 42, pp. 6340–6346, 2008.

[21] W. Lu and T. Chen, "New approach to synchronization analysis of linearly coupled ordinary differential systems," *Physica D*, vol. 213, no. 2, pp. 214–230, 2006.

[22] S. Boyd, L. El Ghaoui, E. Feron, and V. Balakrishnan, *Linear Matrix Inequalities in System and Control Theory*, SIAM, Philadelphia, 1994.

[23] Y. Kuramoto, *Chemical Oscillations, Waves, and Turbulence*, vol. 19 of *Springer Series in Synergetics*, Springer, Berlin, Germany, 1984.

[24] A. Turing, "The chemical basis of morphogenesis," *Philosophical Transactions of the Royal Society B*, vol. 237, p. 37, 1952.

[25] F. C. Hoppensteadt and E. M. Izhikevich, "Pattern recognition via synchronization in phase-locked loop neural networks," *IEEE Transactions on Neural Networks*, vol. 11, no. 3, pp. 734–738, 2000.

[26] W. Lu and T. Chen, "Synchronization of coupled connected neural networks with delays," *IEEE Transactions on Circuits and Systems. I*, vol. 51, no. 12, pp. 2491–2503, 2004.

12

Creation of Two-Particle Entanglement in Open Macroscopic Quantum Systems

M. Merkli,[1] G. P. Berman,[2] F. Borgonovi,[3,4] and V. I. Tsifrinovich[5]

[1] *Department of Mathematics and Statistics, Memorial University of Newfoundland, St. John's, NL, Canada A1C 5S7*
[2] *Theoretical Division, Los Alamos National Laboratory, MS B213, Los Alamos, NM 87545, USA*
[3] *Dipartimento di Matematica e Fisica, and Interdisciplinary Laboratories for Advanced Materials Physics, Università Cattolica, Via Musei 41, 25121 Brescia, Italy*
[4] *Istituto Nazionale di Fisica Nucleare, Sezione di Pavia, Via Bassi 6, 27100 Pavia, Italy*
[5] *Department of Applied Physics, Polytechnic Institute of NYU, 6 MetroTech Center, Brooklyn, NY 11201, USA*

Correspondence should be addressed to M. Merkli, merkli@mun.ca

Academic Editor: P. G. Kevrekidis

We consider an open quantum system of N not directly interacting spins (qubits) in contact with both local and collective thermal environments. The qubit-environment interactions are energy conserving. We trace out the variables of the thermal environments and $N - 2$ qubits to obtain the time-dependent reduced density matrix for two arbitrary qubits. We numerically simulate the reduced dynamics and the creation of entanglement (concurrence) as a function of the parameters of the thermal environments and the number of qubits, N. Our results demonstrate that the two-qubit entanglement generally decreases as N increases. We show analytically that, in the limit $N \to \infty$, no entanglement can be created. This indicates that collective thermal environments cannot create two-qubit entanglement when many qubits are located within a region of the size of the environment coherence length. We discuss possible relevance of our consideration to recent quantum information devices and biosystems.

1. Introduction

In open many-body systems, such as solid-state and biological ones, quantum behavior reveals itself in many ways. Often the quantitative parameter used to measure "quantumness" (possibly even of macroscopic order) is entanglement. The presence of entanglement implies that the wave function (or the reduced density matrix) cannot be represented as a product of the corresponding objects for the individual qubits. It is important to note that to produce and to measure entanglement in such systems, one does not necessarily need to know much detail about the system, possibly not even its Hamiltonian [1]. The questions then are how useful is entanglement as a measure of quatumness, what can it add to our knowledge of system properties and behavior, and how can it be utilized? Indeed, just knowing that the system is entangled (knowledge of complicated quantum behavior) is not sufficient to imply that its quantum properties are useful for specific applications. Fortunately, however, in some situations entanglement provides very useful properties, including additional exponential resources for quantum computation [2] and a possible enhancement of photosynthesis in bio-systems [3] (See also references therein.)

Entanglement could be produced by direct interaction between the qubits. This interaction should be of a "conditional" nature, mixing an initial product state (disentangled) in such a way that the final state becomes correlated in a quantum way. Many aspects of entanglement creation are widely discussed in the literature (See, e.g., [4–11] and references therein.)

More recently, interest appeared in the possibility to create entanglement in the absence of direct interactions between qubits (or when the latter are very small). Entanglement can then still be created merely by the indirect interaction of noninteracting qubits through a collective thermal bath. In [4] this situation was considered for a model of two noninteracting spins 1/2 (qubits) interacting only with a collective thermal bosonic environment. It was demonstrated numerically in [4] that for some initially unentangled two-qubit states and under some conditions on the thermal bath, measurable entanglement between the two qubits is created for intermediate times. The model of [4] is energy conserving, ignoring relaxation processes for the qubits and including only the effects of decoherence. In [11] these results were extended to a more general model having (i) both local and collective thermal environments (at the same temperature) and (ii) energy-conserving and energy exchange interactions between qubits and their environments. The conditions for entanglement creation were discussed and analyzed numerically in [11]. It was concluded that, in spite of the competition between the local thermal environments (which destroy entanglement) and the collective thermal environment (tending to create entanglement), the creation of measurable entanglement can be realized for some finite times. In both papers, [4, 11], only two qubits are analyzed hence no direct connection to many-body systems was made. As was recently shown in [12], the presence of a large number of indirectly interacting qubits interacting only with their common collective thermal environment could significantly modify the effective single-qubit characteristics including their relaxation and decoherence rates.

In the present paper, we consider a model of N not directly interacting spins 1/2 (qubits) placed in a constant effective magnetic field (oriented in the z-direction). The qubits interact with both local and collective thermal environments (all at the same temperature). The collective interaction introduces an indirect qubit interaction. In the total density matrix of all qubits and environments, we trace over the variables of the environments and $N-2$ qubits. This gives us the time-dependent reduced density matrix for two arbitrary qubits. In the z-representation it is represented by a time-dependent 4×4 matrix. It is important to notice that the matrix elements, $[\rho_t]_{n,m}$, $n, m = 1, \ldots, 4$, depend not only on the parameters of the thermal environments but also on the total number of qubits, N. We study numerically the concurrence $C(t)$ of the reduced two qubit density matrix and its dependence on the parameters of the system and on N. To realize and study this situation in an experiment, one must have access to the two selected qubits (such as their particular frequencies), in order to manipulate them and prepare the initial state. Our main result is that the amplitude of concurrence, $C_{\max}(t)$, generally decreases as N increases. This means that one should not expect that the collective thermal environment can create by itself measurable entanglement even of two qubits, in the presence of many other qubits within the range of the collective environment coherence length.

1.1. *Outline of Main Results*

The initial state of the entire system is disentangled, a product state in which each of the N spins is in a state ρ_j, $j = 1, \dots, N$; all local reservoir states are thermal and so is that of the collective reservoir, at a fixed common temperature (A generalization to a nonequilibrium situation where each local and the collective reservoir have different individual temperatures is immediate.).

1.1.1. *Analytic Results*

(i) Explicit Dynamics

As the spins interact with the reservoirs via energy-conserving couplings only, the reduced two-spin dynamics can be calculated explicitly; see Proposition 2.1. Consequences of the energy conservation are that populations, that is, the diagonal density matrix elements, are time independent and that the off-diagonal elements evolve independently. As an example, we discuss here the dynamics of the (1, 2) matrix element:

$$[\rho_t]_{12} = [\rho_0]_{12} e^{i\omega_2 t}\, e^{i\varkappa_c^2 S(t)}\, e^{-\varkappa_\ell^2 \Gamma_\ell(t) - \varkappa_c^2 \Gamma_c(t)}\, P_N(t). \tag{1.1}$$

The other matrix elements have similar behavior. Each factor on the r.h.s. has an interpretation.

(i) $[\rho_0]_{12}$ is the initial condition of the matrix element in question. None of the other initial matrix elements are involved (energy-conserving coupling).

(ii) $e^{i\omega_2 t}$ is the uncoupled dynamics (no interaction with environments);

(iii) $e^{i\varkappa_c^2 S(t)}$ is a dephasing factor with a time-dependent phase $S(t) \le 0$ becoming linear for large t (for the considered infrared behavior $|k|^{1/2}$ of the coupling constants in three dimensions, see (2.10)); it represents a "Lamb shift" contribution to the real part of the effective energy; this term is generated by the collective reservoir, but it is independent of the presence of the $N-2$ traced-out spins (the term would be the same if only two spins were coupled to the reservoirs).

(iv) $e^{-\varkappa_\ell^2 \Gamma_\ell(t) - \varkappa_c^2 \Gamma_c(t)}$ is a decaying factor with time-dependent decay rates, $\Gamma(t) \ge 0$, becoming linear for large t (see (2.11)). Both the local and collective reservoirs contribute; however, the term is independent of the $N-2$ traced-out spins (again, it would be the same if only two spins were coupled to the reservoirs);

(v) $P_N(t)$ is a product of $N-2$ oscillating terms encoding the effect of all the traced-out spins (see (2.9)). It is important to notice that $P_N(t)$ *only depends on the diagonal density matrix elements of the initial states of the* $N-2$ *traced-out spins*. (This is so since the dynamics is energy-conserving, and tracing out any of the spins involves only the diagonal of the initial (time zero) density matrix. (See also Remark 2.2(1) after Proposition 2.1.)). Consequently, the two-qubit state does not depend on the initial off-diagonal density matrix elements of the $N-2$ traced-out "background" spins. Typically, we expect those spins to be initially in (close to) equilibrium, corresponding to vanishing off-diagonals.

Some general properties of $P_N(t)$ can be explained easily for the case in which all $N-2$ spins are initially in the high-temperature equilibrium state $(1/\sqrt{2})(|+\rangle + |-\rangle)$. Then $P_N(t) = [\cos(\varkappa_c^2 S(t))]^{N-2}$ and its magnitude oscillate between zero and one. Due to the large power $(N-2)$, the peaks of the function $|P_N(t)|$, centered at the discrete times t_p satisfying $S(t_p) \in (\pi/\varkappa_c^2)\mathbb{Z}$, are of very narrow width $O(1/(\varkappa_c^2\sqrt{N}))$ for large N. Consequently, in the limit $N \to \infty$, with $\varkappa_c$ held fixed, $|P_N(t)|$ is zero for all t except for $t = t_p$, where $|P_N(t_p)| = 1$. But the density matrix becomes very simple if $P_N(t) = 0$, because many entries vanish (c.f. Proposition 2.1) and the corresponding concurrence is zero. It follows that, in the large N limit, concurrence is zero for all times (except possibly for some isolated instances, t_p).

(ii) N-Dependent Scaling of the Interaction $\varkappa_c$

The above analysis suggests that one cannot generate two-spin entanglement for N large at fixed interaction strength $\varkappa_c$. However, the width of the peaked function $P_N(t)$ which is of order $O(1/(\varkappa_c^2\sqrt{N}))$ becomes appreciable if $\varkappa_c \gtrsim N^{1/4}$. Hence we consider an N-dependent scaling of the coupling, replacing $\varkappa_c$ by $\varkappa_c/N^{\eta}$, for some $\eta > 0$. According to the above discussion, the borderline case is $\eta = 1/4$.

Starting from the explicit expressions (Proposition 2.1) and using the scaling $\varkappa_c/N^{\eta}$, we calculate the limit $N \to \infty$ of ρ_t, for $t \in \mathbb{R}$ fixed. The analytic expressions we obtain for $0 < \eta < 1/4$ and $\eta > 1/4$ show that the limiting dynamics does not create entanglement, for any time t. While we are able to obtain explicit expressions for concurrence in the regime of $N \to \infty$, we are not so for N finite. (The reduced density matrix is given explicitly for all N and all t, but calculating from it explicitly the concurrence is more difficult.) However, since no entanglement is generated in the limiting case, $N \to \infty$, but we know entanglement is created for $N = 2$ (see, e.g., [4, 11]), we expect that entanglement creation decays with increasing N. We study this decay numerically.

1.1.2. Numerical Results

We introduce ν_c, the highest frequency at which spin-reservoir interactions occur and call it the cut-off frequency. In the simulations, we take ν_c of the order of the thermal frequency $\nu_T = k_B T/h$. In the infra-red regime, our coupling is proportional to $\sqrt{|k|}$ (see after (2.11)).

(i) For $N = 2$, concurrence creation is maximal if both spins start out in their high-temperature state $(1/\sqrt{2})(|+\rangle + |-\rangle)$; see Figure 1. Consequently, in the subsequent

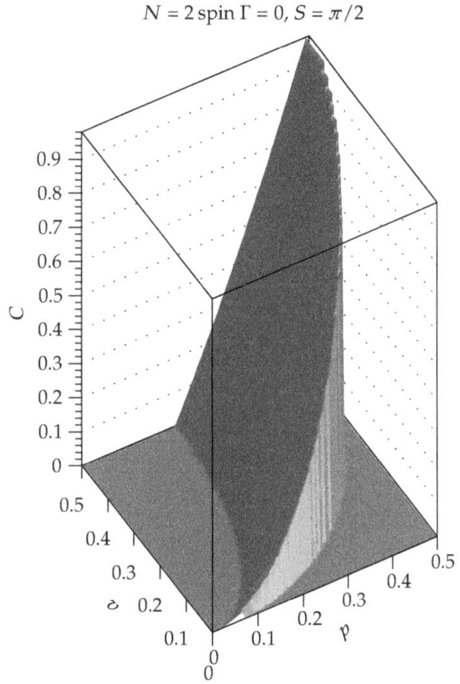

Figure 1: Maximal concurrence as a function of p and v, for fixed $\Gamma(t) = 0$ ans $S(t) = \pi/2$. Here, $N = 2$ spins is considered.

simulations, we take initial states of the two not traced-out qubits very close to this state, and we take the diagonals of the initial states of the $N - 2$ traced-out quibts to be constant $1/2$ (remember that the off-diagonals of these qubits do not influence the dynamics at all). In Figure 4 we modify the initial state of the two not traced-out qubits and check that maximal concurrence is indeed obtained when both qubits are in the above state, even for large N.

(ii) For general N, entanglement evolves according to a rescaled time $t \mapsto \varkappa_c^2 \nu_c t$, see Figure 2. This figure shows that a reduction of $\varkappa_c$ diminishes the created concurrence in a moderate way. For instance, decreasing $\varkappa_c$ by a factor 10 only decreases concurrence by less than $1/3$.

(iii) In Figure 3 we show that the maximum of created concurrence decays with increasing N. For intermediate values of N (with the current parameters N~10–150) the decrease is exponential; for smaller and larger values of N, it is superexponential.

(iv) In the same Figure 3(c), we study the dependence of the maximal time, τ_c, (before recurrence) at which the concurrence is not zero. We have found that this time decays exponentially in the number of spins, N, for sufficiently large N.

(v) Results on the rescaled model $\varkappa_c \mapsto \varkappa_c/N^\eta$ are shown in Figure 5. We find a decrease of maximal concurrence with increasing N for all $\eta \geq 0$. The critical value, $\eta = 1/4$ (see analytic results above), divides the concurrence decay into two regimes. In the range, $\eta > 1/4$, the maximal concurrence decreases exponentially in N, for intermediate values of N (between 10 and 180), with a *universal decay rate* (i.e., not depending on η). For $\eta < 1/4$ the decay is superexponential and varies with η. We conclude that no scaling $\varkappa_c \mapsto \varkappa_c/N^\eta$ can compensate the decay of created concurrence for large N.

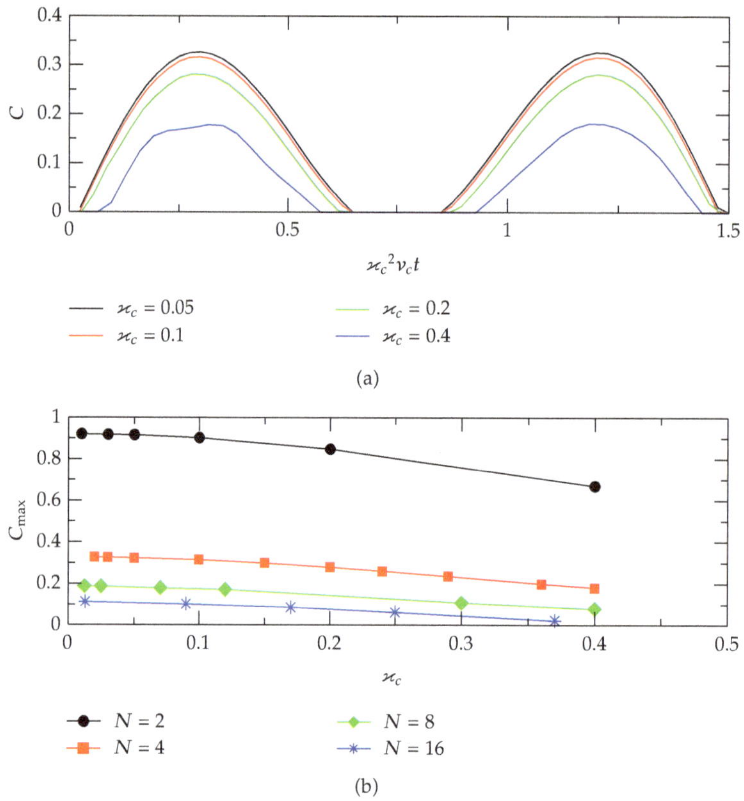

Figure 2: (a) Concurrence as a function of the rescaled time, for different $\varkappa_c$ and fixed $\epsilon = \nu_c/\nu_T = 1$, being respectively the cut-off and thermal frequency ($\nu_T = k_B T/h$ with $T = 300\,\text{K}$) and the form factor $f_c(k) = \sqrt{|k|}$. Other data are $p = 0.5, v = 0.48$, $N = 4$. (b) Plot of the concurrence at the peak (obtained from (a)) as a function of the coupling strength, $\varkappa_c$, for different N values as indicated in the legend. Other values are the same as in (a).

2. Model and Reduced Density Matrix

The full Hamiltonian of the N noninteracting spins 1/2 coupled by energy conserving interactions to local and collective bosonic heat reservoirs is given by

$$H = -\hbar \sum_{n=1}^{N} \omega_n S_n^z + \sum_{n=1}^{N} H_{\mathrm{R}_n} + H_{\mathrm{R}} \tag{2.1}$$

$$+ \sum_{n=1}^{N} \varkappa_n S_n^z \otimes \phi_c(f_c) + \sum_{n=1}^{N} \nu_n S_n^z \otimes \phi_n(f_n). \tag{2.2}$$

Below we use dimensionless variables and parameters. To do so, we introduce a characteristic frequency, ω_0, typically of the order of spin transition frequency. The total Hamiltonian, energies of spin states, and temperature are measured in units $\hbar\omega_0$. The frequencies of spins, $\omega_n > 0$, bosonic excitations, $\omega(k) = c|\vec{k}|$ (where c is the speed of

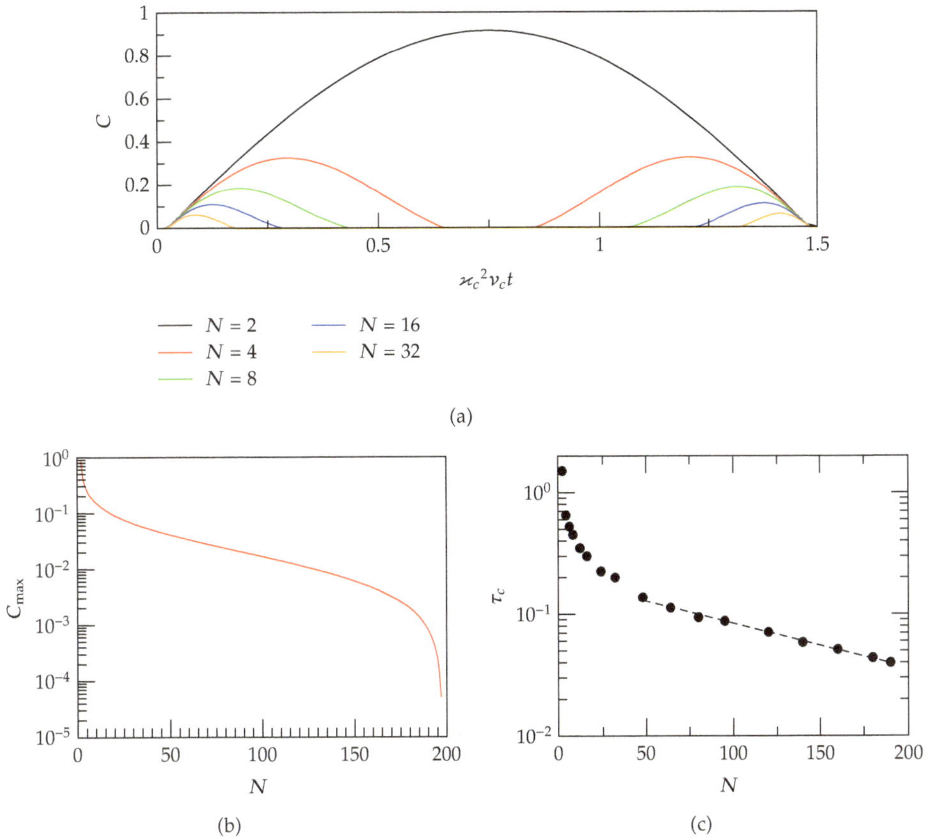

Figure 3: (a) Concurrence as a function of the rescaled time, for fixed $\epsilon = \nu_c/\nu_T = 1$, being respectively the cut-off and thermal frequency ($\nu_T = k_B T/h$ with $T = 300\,K$) and the form factor, $f_c(k) = \sqrt{|k|}$. Other data are $p = 0.5$, $v = 0.48$, $\varkappa_c = 0.05$. (b) Plot of the concurrence at the peak (obtained from (a)) as a function of the number of spins N. (c) Plot of the collapse time, τ_c, as a function of the number of spins, N. The dashed line is the best fit $\exp(-\alpha N)$ with $\alpha = 0.0838 \pm 0.0002$.

light), the wave vectors of bosonic excitations are normalized by ω_0/c, and all constants of interactions are measured in units of ω_0. The dimensionless time is defined as $\omega_0 t$.

In (2.1) and (2.2), $\omega_n > 0$ is the frequency of spin n,

$$S^z = \frac{1}{2}\begin{bmatrix} 1 & 0 \\ 0 & -1 \end{bmatrix}, \tag{2.3}$$

and S_n^z denotes the S^z of spin n. H_{R} is the Hamiltonian of the bosonic collective reservoir,

$$H_{\mathrm{R}} = \int_{\mathbb{R}^3} |k| a^*(k) a(k) \mathrm{d}^3 k, \tag{2.4}$$

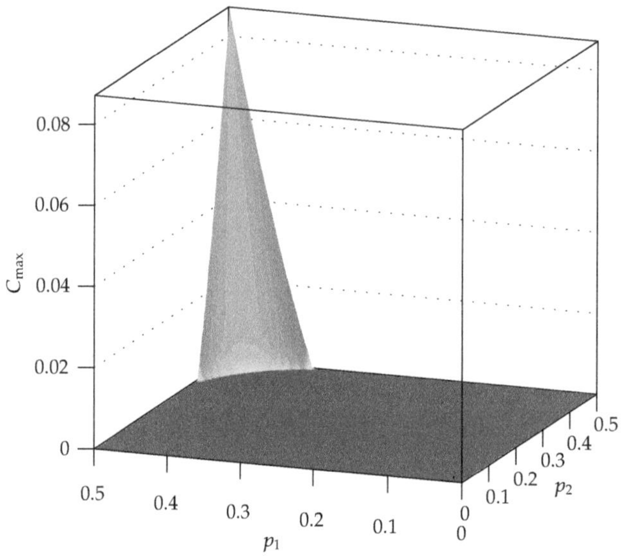

Figure 4: Maximal concurrence as a function of the independent parameters, $p_1 = v_1$ and $p_2 = v_2$. As one can see the maximal concurrence is realized at the external corner, that is $p_1 = p_2 = 1/2$. Here is $\epsilon = 1$, $N = 40$ and $p = 1/2$ for all other spins.

and H_{R_n} is that same Hamiltonian for the nth individual reservoir. For a square-integrable *form factor* $h(k)$, $k \in \mathbb{R}^3$, $\phi(h)$ is given by

$$\phi(h) = \frac{1}{\sqrt{2}} \int_{\mathbb{R}^3} \{h(k)a^*(k) + h(k)^* a(k)\} \mathrm{d}^3k. \tag{2.5}$$

The real numbers, $\varkappa_n$ and ν_n, are coupling constants, measuring the strengths of the energy-conserving collective coupling and the energy-conserving local coupling, respectively.

Since the spins interact with the reservoirs only through energy-conserving channels, this model is *exactly solvable*. For simplicity of exposition, we take

$$\begin{aligned} \varkappa_n &= \varkappa_c \quad \forall n \text{ (collective)}, \\ \nu_n &= \nu_\ell \quad \forall n \text{ (local)}. \end{aligned} \tag{2.6}$$

We also take, for simplicity, all local form factors equal (f_ℓ) and all collective ones also (f_c).

Fix any pair of spins, and (re-)label their frequencies by ω_1 and ω_2; see (2.1). We write the reduced density matrix, ρ_t, of the two fixed spins as a 4×4 matrix $[\rho_t]_{ij}$ in the ordered energy basis

$$\Phi_1 = \varphi_1 \otimes \varphi_1, \qquad \Phi_2 = \varphi_1 \otimes \varphi_{-1}, \qquad \Phi_3 = \varphi_{-1} \otimes \varphi_1, \qquad \Phi_4 = \varphi_{-1} \otimes \varphi_{-1}, \tag{2.7}$$

(Another equivalent notation is: $\Phi_1 = |++\rangle, \Phi_2 = |+-\rangle, \Phi_3 = |-+\rangle, \Phi_4 = |--\rangle$.) where $S^z\varphi_{\pm 1} = \pm(1/2)\varphi_{\pm 1}$. For instance, $[\rho_t]_{2,4} = \langle \Phi_2, \rho_t \Phi_4 \rangle$.

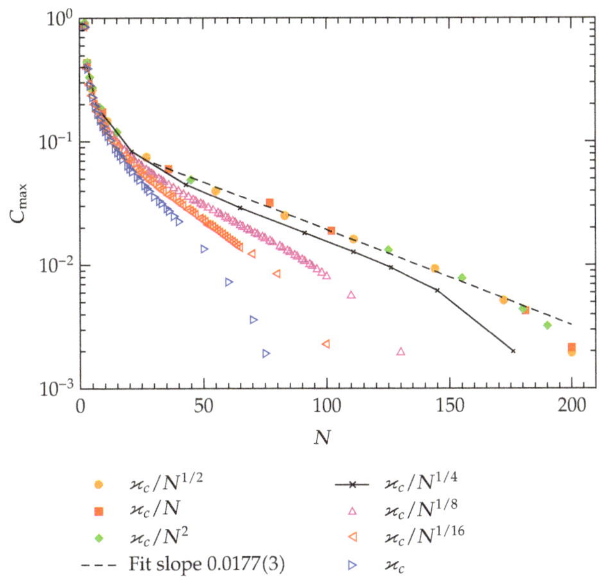

Figure 5: Maximal concurrence as a function of the number of spins, for different power law scaling, as indicated in the legend. Here is $f_c(k) = \sqrt{|k|}$, $p = 0.5$, $v = 0.48$, $\epsilon = \nu_c/\nu_T = 1$, $T = 300\,\text{K}$, $\varkappa_c = 0.2$. The dashed line indicates a fitting exponential for the cases $\eta > 1/4$. The solid curve indicates the case, $\eta = 1/4$.

The *initial state* of the spins is a product state of the form $\rho_{S_1,0} \otimes \cdots \otimes \rho_{S_N,0}$, where

$$\rho_{S_j,0} = \begin{bmatrix} p_j & v_j \\ v_j^* & 1-p_j \end{bmatrix}, \tag{2.8}$$

with $0 \le p_j \le 1$ and $|v_j|^2 \le p_j(1-p_j)$. The upper bound on the off-diagonal guarantees that the eigenvalues of $\rho_{S_j,0}$ are nonnegative.

We introduce the quantities:

$$P_N(t) = \prod_{j=3}^{N} \left[p_j e^{i\varkappa_c^2 S(t)} + (1-p_j) e^{-i\varkappa_c^2 S(t)} \right], \tag{2.9}$$

$$S(t) = -\frac{1}{2} \int_{\mathbb{R}^3} |f_c(k)|^2 \frac{|k|t - \sin(|k|t)}{|k|^2} d^3k, \tag{2.10}$$

$$\Gamma_{e,c}(t) = \int_{\mathbb{R}^3} |f_{e,c}(k)|^2 \coth\left(\frac{\beta|k|}{2}\right) \frac{\sin^2(|k|t/2)}{|k|^2} d^3k. \tag{2.11}$$

The integrals in (2.10) and (2.11) are made to converge introducing a suitable cut-off wavenumber, $|k_c|$, or cut-off frequency, $\nu_c = |k_c|/2\pi$. (Here we use dimensionless units.) For instance, for numerical simulations, we choose as form factor the function $f_c(k) = \sqrt{|k|}\,\chi_{|k|\le|k_c|}$, where $\chi_{|k|\le|k_c|} = 1$ if $|k| \le |k_c|$ and $\chi_{|k|\le|k_c|} = 0$ otherwise.

We also define $\widetilde{P}_N(t)$ to be the same as $P_N(t)$, but with $\varkappa_c^2$ replaced by $2\varkappa_c^2$. With this notation, we have the following result.

Proposition 2.1 (explicit dynamics of the reduced density matrix). *The evolution of the density matrix is given by*

$$\begin{aligned}
[\rho_t]_{12} &= [\rho_0]_{12} e^{i\omega_2 t} e^{i\varkappa_c^2 S(t)} e^{-\varkappa_e^2 \Gamma_e(t) - \varkappa_c^2 \Gamma_c(t)}\ P_N(t),\\
[\rho_t]_{13} &= [\rho_0]_{13}\ e^{i\omega_1 t}\ e^{i\varkappa_c^2 S(t)}\ e^{-\varkappa_c^2 \Gamma_c(t)}\ P_N(t),\\
[\rho_t]_{14} &= [\rho_0]_{14} e^{i(\omega_1+\omega_2)t} e^{-2\varkappa_e^2 \Gamma_e(t) - 4\varkappa_c^2 \Gamma_c(t)}\ \widetilde{P}_N(t),\\
[\rho_t]_{23} &= [\rho_0]_{23} e^{i(\omega_1-\omega_2)t} e^{-2\varkappa_e^2 \Gamma_e(t)},\\
[\rho_t]_{24} &= [\rho_0]_{24} e^{i\omega_1 t} e^{-i\varkappa_c^2 S(t)} e^{-\varkappa_e^2 \Gamma_e(t) - \varkappa_c^2 \Gamma_c(t)}\ P_N(t),\\
[\rho_t]_{34} &= [\rho_0]_{34} e^{i\omega_2 t} e^{-i\varkappa_c^2 S(t)} e^{-\varkappa_e^2 \Gamma_e(t) - \varkappa_c^2 \Gamma_c(t)}\ P_N(t),
\end{aligned} \tag{2.12}$$

and the populations are constant, $[\rho_t]_{jj} = [\rho_0]_{jj}$, *for* $j = 1, \ldots, 4$ *and* $t \in \mathbb{R}$.

The proof of this proposition is a rather simple calculation. One can proceed as in [13–15] (proof of Proposition 7.4 in [13]).

Remark 2.2. (1) The effect of spins $3, \ldots, N$ is contained entirely in the factors $P_N(t)$ and $\widetilde{P}_N(t)$. They only depend on the initial populations p_j, $j = 3, \ldots, N$ (see (2.9)), but not on the off-diagonals, v_j. This is explained by the fact that when tracing over a single spin, $j \geq 3$, we perform the operation $\mathrm{Tr}_{\mathrm{spin}\, j}\ U\rho_{S_j,0}V$, where U, V are operators commuting with S_2^z (energy-conserving interactions only!). Clearly the latter trace only involves the diagonal of $\rho_{S_j,0}$.

(2) The oscillatory phases, $e^{i\omega t}$, in (2.12) represent the free, uncoupled dynamics of the spins. Consider the modified two-spin density matrix:

$$\rho_t' = e^{it(-\omega_1 S_1^z - \omega_2 S_2^z)} \rho_t e^{-it(-\omega_1 S_1^z - \omega_2 S_2^z)}, \tag{2.13}$$

("interaction picture" dynamics of ρ_t). Because ρ_t' and ρ_t are related by conjugation of a unitary operator of the product form, $e^{it(-\omega_1 S_1^z)} \otimes e^{it(-\omega_2 S_2^z)}$, *the concurrences of* ρ_t *and* ρ_t' *are the same*. In other words, when examining concurrence of ρ_t, we may use formulas (2.12) with $\omega_1 = \omega_2 = 0$.

2.1. *Concurrence*

Recall that the concurrence of the reduced density matrix, ρ_t, is unchanged when we pass to the interaction picture ρ_t' (see the remark explaining (2.13)). In the basis (2.7), the evolution of ρ_t' (2.13) is given by (2.12) with $\omega_1 = \omega_2 = 0$ and where the initial condition is ($\rho_0' = \rho_0$):

$$[\rho_0] = \begin{bmatrix} p_1p_2 & p_1v_2 & v_1p_2 & v_1v_2 \\ p_1v_2^* & p_1(1-p_2) & v_1v_2^* & v_1(1-p_2) \\ v_1^*p_2 & v_1^*v_2 & (1-p_1)p_2 & (1-p_1)v_2 \\ v_1^*v_2^* & v_1^*(1-p_2) & (1-p_1)v_2^* & (1-p_1)(1-p_2) \end{bmatrix}. \tag{2.14}$$

2.1.1. Variation of N Dependence

For homogeneous initial conditions, $p_j = p$ for $j = 3, \ldots, N$, we have

$$P_N(t) = \left[pe^{i\varkappa_c^2 S(t)} + (1-p)e^{-i\varkappa_c^2 S(t)}\right]^{N-2}. \tag{2.15}$$

Unless $p = 0, 1$, $|P_N(t)|$ oscillates in t between its minimum value $|2p-1|^{N-2}$ (when $\cos(2\varkappa_c^2 S(t)) = -1$) and its maximum value 1 (when $\cos(2\varkappa_c^2 S(t)) = 1$) (We have $|pe^{i\varkappa_c^2 S(t)} + (1-p)e^{-i\varkappa_c^2 S(t)}|^2 = p^2 + 2p(1-p)\cos(2\varkappa_c^2 S(t)) + (1-p)^2$.) The width of the oscillations becomes very narrow with increasing N. In the limit of large N, $P_N(t)$ is zero for all times, except for the discrete set of $t \in \mathbb{R}$ satisfying $\cos(2\varkappa_c^2 S(t)) = 1$, in which case $|P_N(t)| = 1$.

This implies that, for large N, all off-diagonal density matrix elements of ρ_t vanish with the exception of $[\rho_t]_{23} = e^{i(\omega_1-\omega_2)t}e^{-2\varkappa_e^2 \Gamma_e(t)}$ (and $[\rho_t]_{32}$, of course) for almost all values of t. This suppression of off-diagonals comes from the large number of particles and is mediated through the collective energy-conserving interaction. (For $\varkappa_c = 0$ we have $P_N(t) = 1$.)

In order to try to have a nontrivial dynamics for large N, one may scale the collective conserving coupling constant as

$$\varkappa_c \longrightarrow \frac{\varkappa_c}{N^\eta}, \quad \text{some } \eta > 0. \tag{2.16}$$

Then (2.15) becomes

$$P_N(t) = \left[pe^{i\varkappa_c^2 S(t)/N^{2\eta}} + (1-p)e^{-i\varkappa_c^2 S(t)/N^{2\eta}}\right]^{N-2}. \tag{2.17}$$

An expansion in large N yields

$$P_N(t) = e^{-i\varkappa_c^2 S(t)[1-2p]N^{1-2\eta}}e^{-2\varkappa_c^4 S^2(t)N^{1-4\eta}[p(1-p)+O(N^{-2\eta})]}. \tag{2.18}$$

Thus as $N \to \infty$,

$$P_N(t) \longrightarrow \begin{cases} e^{-i\varkappa_c^2 S(t)[1-2p]N^{1/2}}e^{-2\varkappa_c^4 S^2(t)p(1-p)N^{1-4\eta}}, & 0 < \eta \le \frac{1}{4} \ (p \neq 0,1), \\ e^{-i\varkappa_c^2 S(t)[1-2p]N^{1-2\eta}}, & \frac{1}{4} < \eta. \end{cases} \tag{2.19}$$

Remark 2.3. (1) By replacing, in these limits, $\varkappa_c^2$ by $2\varkappa_c^2$, we obtain the corresponding limits for $\widetilde{P}_N(t)$.

(2) For $0 < \eta < 1/4$, $P_N(t)$ vanishes as $N \to \infty$.

(3) The rapid oscillating phases disappear if $p = 1/2$ (any $\eta > 0$) or $\eta = 1/2$ (any p).

2.1.2. *Asymptotic Concurrence* ($N \rightarrow \infty$)

(*i*) $0 < \eta < 1/4$

The reduced two-spin density matrix (in the interaction picture) at time t is

$$[\rho_t'] = \begin{bmatrix} p_1 p_2 & 0 & 0 & 0 \\ 0 & p_1(1-p_2) & v_1 v_2^* e^{-2\varkappa_e^2 \Gamma_e(t)} & 0 \\ 0 & v_1^* v_2 e^{-2\varkappa_e^2 \Gamma_e(t)} & (1-p_1)p_2 & 0 \\ 0 & 0 & 0 & (1-p_1)(1-p_2) \end{bmatrix}, \tag{2.20}$$

from which we obtain the concurrence

$$C(\rho_t) = \max\left\{0, -2\left[\sqrt{p_1(1-p_1)p_2(1-p_2)} - |v_1||v_2|e^{-2\varkappa_e^2 \Gamma_e(t)}\right]\right\} = 0. \tag{2.21}$$

Remember that $|v_j|^2 \leq p_j(1-p_j)$. This shows that the $N \rightarrow \infty$ asymptotic dynamics cannot create entanglement at any time.

(*ii*) $\eta > 1/4$

Call the r.h.s. of (2.19) $P_\infty(t)$ (a quantity still depending on N unless $\eta = 1/2$). By replacing $\varkappa_c^2$ by $2\varkappa_c^2$ in (2.19), we obtain the limit of $\widetilde{P}_N(t)$, which we call $\widetilde{P}_\infty(t)$. For $\eta > 1/4$ we have the relation $\widetilde{P}_\infty(t) = [P_\infty(t)]^2$. The reduced two-spin density matrix (in the interaction picture) at time t is, for $N \rightarrow \infty$,

$$[\rho_t'] = \begin{bmatrix} p_1 p_2 & p_1 v_2 D_e P_\infty(t) & v_1 p_2 D_e P_\infty(t) & v_1 v_2 D_e^2 P_\infty(t)^2 \\ p_1 v_2^* D_e P_\infty(t)^* & p_1(1-p_2) & v_1 v_2^* D_e^2 & v_1(1-p_2) D_e P_\infty(t) \\ v_1^* p_2 D_e P_\infty(t)^* & v_1^* v_2 D_e^2 & (1-p_1)p_2 & (1-p_1) v_2 D_e P_\infty(t) \\ v_1^* v_2^* D_e^2 \left[P_\infty(t)^2\right]^* & v_1^*(1-p_2) D_e P_\infty(t)^* & (1-p_1) v_2^* D_e P_\infty(t)^* & (1-p_1)(1-p_2) \end{bmatrix}, \tag{2.22}$$

where

$$D_e = D_e(t) = e^{-\varkappa_e^2 \Gamma_e(t)}, \qquad P_\infty(t) = e^{-i\varkappa_c^2 S(t)[1-2p]N^{1-2\eta}}. \tag{2.23}$$

The density matrix (2.22) is of the product form:

$$\begin{bmatrix} p_1 & v_1 D_e(t) P_\infty(t) \\ v_1^* D_e(t) P_\infty(t)^* & 1-p_1 \end{bmatrix} \otimes \begin{bmatrix} p_2 & v_2 D_e(t) P_\infty(t) \\ v_2^* D_e(t) P_\infty(t)^* & 1-p_2 \end{bmatrix}. \tag{2.24}$$

This shows that the $N \rightarrow \infty$ asymptotic dynamics is factorizable and cannot create entanglement at any time.

3. Numerical Results

(i) Let us first consider the case of two spins only, $N = 2$. In (2.12) we put for simplicity $\Gamma_\ell = \Gamma_c = \Gamma$ and regard Γ and S as two independent parameters. Taking both spins initially in the same state given by p, v, see (2.8), we examine the maximal concurrence, as a function of S and Γ, for arbitrary fixed values of p and v.

We find that, for fixed p, v, the maximal concurrence is given at $S = \pi/2$, $\Gamma = 0$. Having such values fixed and plotting the concurrence as a function of p, v, the maximal concurrence is realized when $p = v = 1/2$; see Figure 1, where a plot of the concurrence as a function of p, v is shown. Maximal generation of concurrence is thus obtained starting from pure-state initial conditions $(1/\sqrt{2})(|+\rangle + |-\rangle)$ for each spin.

(ii) Let us now consider the case of N spins. For concreteness we choose $p = 1/2$ for all spins, (the traced-out ones and the two not traced-out ones). For the two not traced-out spins we take off-diagonals $v = 0.48$. (Then p is close to v which favors larger entanglement creation.) Recall that the dynamics is independent of the off-diagonals of the $N-2$ traced-out spins (i.e., we do not have to specify the v of the $N-2$ traced-out spins).

As mentioned after (2.11), we choose the form factor, $f_c(k) = \sqrt{|k|}\ \chi_{|k|\le 2\pi\nu_c}$, with the cut-off frequency equal to the thermal frequency, $\nu_c = \nu_T = k_B T/h$, at room temperature, $T = 300$ K.

In Figure 2 we investigate the effects of an increase in the coupling parameter, $\varkappa_c$. The first effect is a time shift for the concurrence evolution, described by a scaling, $t \to \varkappa_c^2 \nu_c t$; see Figure 2(a), where ν_c is the cut-off frequency. The second effect is a reduction of the maximal concurrence in a smooth way; see Figure 2(b). As one can see, the effective decrease in amplitude for not too strong coupling strength, $\varkappa_c$, is relatively small. For instance, changing $\varkappa_c$ for $N = 2$ by one order of magnitude from 0.04 to 0.4 changes the amplitude by only 27%. The percentage change is almost the same for larger N values; see Figure 2(b).

In Figure 3 we show that the creation of concurrence decreases with the number of spins. In (a) we plot the concurrence as a function of (rescaled) time for various values of $N = 2, \ldots, 32$. As one can see, the same time rescaling is also valid for $N > 2$. Moreover, the maximum concurrence created, $C_{\max}$, reported in (b), decreases exponentially in N in the range $10 < N < 150$ and faster than exponentially outside this range. For larger N, the concurrence decays superexponentially in N. For N exceeding 200, the concurrence becomes too small to be significant (of the order 10^{-4}).

It is also interesting to note that the graph of concurrence shows collapses and revivals and that the revival times for $N > 2$ are always less than the revival time for $N = 2$. It is also interesting to consider how the collapse time, τ_c, defined as the first time at which concurrence drops abruptly to zero, depends on the number of spins, N. This study has been reported in Figure 3(c) and shows that the rescaled collapse time decays exponentially with the number of spins: $\tau_c = \varkappa_c^2 \nu_c t_c \simeq \exp(-\alpha N)$, where $\alpha = 0.0838 \pm 0.0002$. (See dashed line in Figure 3(c).)

(iii) One can also vary the initial conditions for the spins by choosing independent $p_{1,2}$ and $v_{1,2}$, while all other spins have the same value $p_j = 1/2$, $j = 3, \ldots, N$ (their off-diagonals v_j do not influence the dynamics at all).

In order to simplify the problem, we also set $p_1 = v_1$ and $p_2 = v_2$ and consider the maximal concurrence as a function of two independent parameters, p_1 and p_2, only.

An example of the 3D plot obtained is reported in Figure 4: the maximal concurrence is realized for $p_1 = p_2 = v_1 = v_2 = 1/2$, independently of N (in the picture $N = 40$, but similar plots are obtained for other values of N).

(iv) The numerical analysis of the rescaled model with $\varkappa_c$ replaced by $\varkappa_c/N^\eta$ shows that the concurrence is always a decreasing function of N and that the maximum of the created concurrence is a universal function of the number of spins N, independent of η for $\eta > 1/4$.

Results are shown in Figure 5, where the dashed line is the best exponential fit $\exp(-aN)$, with $a = 0.0177 \pm 0.0003$, the best fitting value, for the cases $\eta > 1/4$. The same figure shows that when $\eta \leq 1/4$, the decay is superexponential and no universality occurs.

This suggests that no power law scaling with N of the coupling strength can compensate the rapid decay of concurrence with the number of spins.

Qualitatively similar results, not reported here, can be obtained by changing the ratio between the thermal and cut-off frequency in the range $(0.5, 4)$.

4. Conclusion

We have analyzed the two-qubit (two effective spin) entanglement in an N-qubit open system, where individual qubits interact through a collective thermal environment. We have demonstrated that concurrence (a measure of two-qubit entanglement) quickly decays with increasing number of surrounding qubits. It follows from our consideration that, for creation of entanglement by implementing our approach, one has to use a small number of qubits collectively interacting with the thermal environment, preferably only two qubits. In this paper, we consider a quantum noise generated by a thermal bosonic environment, interacting in a purely dephasing way with qubits, but no Markovian approximation has been made. In contrast, it has been shown [16–18] that an N-qubit register can be driven to entangled target states (within a wide class of states), under an appropriately *engineered* Markovian dissipative dynamics in Lindblad form and that the convergence rate to the target state is independent of N. This entirely different noise has thus a completely different effect on entanglement.

The results of this paper are important for a better understanding and a characterization of the collective thermal environment and its ability to create and destroy entanglement in many-particle open quantum systems. Any quantum information processor could be used for the study of entanglement induced by a thermal environment, as, for example, two (or more) ions in an ion trap quantum computer [2] or two (or more) superconducting qubits [19] or an NMR quantum information processor [2] operating at room temperature. In the latter case, the experiment is performed with an ensemble of many identical molecules, each having a similar number of weakly interacting nuclear spins. One can study in this system, for example, how the collective vibrational modes of a molecule (thermal environment) influence the two-spin "effective entanglement" in the ensemble of molecules, depending on the number of spins in a molecule. Important and relevant systems are represented by photosynthetic complexes in which quantum effects for exciton dynamics and primary charge separation have been recently discovered [20–23]. Here, the effective spins are exciton or electron sites (pigments), and the thermal environment represents the vibrational protein modes.

Acknowledgment

This work Supported by NSERC under Discovery Grant 205247 (M. Merkli). This work was carried out under the auspices of the NNSA of the U. S. DOE at LANL under Contract No. DEAC52-06NA25396. Supported by the LDRD program at LANL (G. P. Berman).

References

[1] M. Cramer, M. B. Plenio, and H. Wunderlich, "Measuring entanglement in condensed matter systems," *Physical Review Letters*, vol. 106, no. 2, Article ID 020401, 2011.

[2] M. A. Nielsen and I. L. Chuang, *Quantum Computation and Quantum Information*, Cambridge University Press, Cambridge, UK, 2000.

[3] M. Sarovar, A. Ishizaki, G. R. Fleming, and K. B. Whaley, "Quantum entanglement in photosynthetic light-harvesting complexes," *Nature Physics*, vol. 6, no. 6, pp. 462–467, 2010.

[4] D. Braun, "Creation of entanglement by interaction with a common heat bath," *Physical Review Letters*, vol. 89, no. 27, pp. 2779011–2779014, 2002.

[5] T. Yu and J. H. Eberly, "Finite-time disentanglement via spontaneous emission," *Physical Review Letters*, vol. 93, no. 14, Article ID 140404, pp. 1–4, 2004.

[6] T. Yu and J. H. Eberly, "Sudden death of entanglement," *Science*, vol. 323, no. 5914, pp. 598–601, 2009.

[7] T. Yu and J. H. Eberly, "Sudden death of entanglement: classical noise effects," *Optics Communications*, vol. 264, no. 2, pp. 393–397, 2006.

[8] B. Bellomo, R. Lo Franco, and G. Compagno, "Non-Markovian effects on the dynamics of entanglement," *Physical Review Letters*, vol. 99, no. 16, Article ID 160502, 2007.

[9] J. H. Huang and S. Y. Zhu, "Sudden death time of two-qubit entanglement in a noisy environment," *Optics Communications*, vol. 281, no. 8, pp. 2156–2159, 2008.

[10] J. P. Paz and A. J. Roncaglia, "Dynamics of the entanglement between two oscillators in the same environment," *Physical Review Letters*, vol. 100, no. 22, Article ID 220401, 2008.

[11] M. Merkli, G. P. Berman, F. Borgonovi, and K. Gebresellasie, "Evolution of entanglement of two qubits interacting through local and collective environments," *Quantum Information & Computation*, vol. 11, no. 5-6, pp. 390–419, 2011.

[12] M. Merkli, G. P. Berman, and A. Redondo, "Application of resonance perturbation theory to dynamics of magnetization in spin systems interacting with local and collective bosonic reservoirs," *Journal of Physics A*, vol. 44, no. 30, Article ID 305306, 2011.

[13] M. Merkli, I. M. Sigal, and G. P. Berman, "Resonance theory of decoherence and thermalization," *Annals of Physics*, vol. 323, no. 2, pp. 373–412, 2008.

[14] M. Merkli, G. P. Berman, and I. M. Sigal, "Dynamics of collective decoherence and thermalization," *Annals of Physics*, vol. 323, no. 12, pp. 3091–3112, 2008.

[15] M. Merkli, I. M. Sigal, and G. P. Berman, "Decoherence and thermalization," *Physical Review Letters*, vol. 98, no. 13, Article ID 130401, 2007.

[16] F. Verstraete, M. M. Wolf, and J. Ignacio Cirac, "Quantum computation and quantum-state engineering driven by dissipation," *Nature Physics*, vol. 5, no. 9, pp. 633–636, 2009.

[17] B. Kraus, H. P. Büchler, S. Diehl, A. Kantian, A. Micheli, and P. Zoller, "Preparation of entangled states by quantum Markov processes," *Physical Review A*, vol. 78, no. 4, Article ID 042307, 2008.

[18] S. Diehl, A. Micheli, A. Kantian, B. Kraus, H. P. Büchler, and P. Zoller, "Quantum states and phases in driven open quantum systems with cold atoms," *Nature Physics*, vol. 4, no. 11, pp. 878–883, 2008.

[19] J. M. Martinis, "Superconducting phase qubits," *Quantum Information Processing*, vol. 8, no. 2-3, pp. 81–103, 2009.

[20] H. Lee, Y. C. Cheng, and G. R. Fleming, "Coherence dynamics in photosynthesis: protein protection of excitonic coherence," *Science*, vol. 316, no. 5830, pp. 1462–1465, 2007.

[21] G. Panitchayangkoon, D. Hayes, K. A. Fransted et al., "Long-lived quantum coherence in photosynthetic complexes at physiological temperature," *Proceedings of the National Academy of Sciences of the United States of America*, vol. 107, no. 29, pp. 12766–12770, 2010.

[22] E. Collini, C. Y. Wong, K. E. Wilk, P. M. G. Curmi, P. Brumer, and G. D. Scholes, "Coherently wired light-harvesting in photosynthetic marine algae at ambient temperature," *Nature*, vol. 463, no. 7281, pp. 644–647, 2010.

[23] G. L. Celardo, F. Borgonovi, M. Merkli, V. I. Tsifrinovich, and G. P. Berman, "Superradiance transition in photosynthetic light-harvesting complexes," Submitted, http://arxiv.org/abs/1111.5443.

13

Spacetime Junctions and the Collapse to Black Holes in Higher Dimensions

Filipe C. Mena

Centro de Matemática, Universidade do Minho, Campus de Gualtar, 4710-057 Braga, Portugal

Correspondence should be addressed to Filipe C. Mena, fmena@math.uminho.pt

Academic Editor: Burak Polat

We review recent results about the modelling of gravitational collapse to black holes in higher dimensions. The models are constructed through the junction of two exact solutions of the Einstein field equations: an interior collapsing fluid solution and a vacuum exterior solution. The vacuum exterior solutions are either static or containing gravitational waves. We then review the global geometrical properties of the matched solutions which, besides black holes, may include the existence of naked singularities and wormholes. In the case of radiating exteriors, we show that the data at the boundary can be chosen to be, in some sense, arbitrarily close to the data for the Schwarzschild-Tangherlini solution.

1. Introduction

Higher dimensional black holes have recently been the subject of an increasing number of research works (see [1] for a review) and play an important role in theoretical physics, particularly in String Theory. However, the study of the actual dynamical process of collapse resulting in black hole formation has deserved little attention from the mathematical point of view.

Mathematically, black hole formation can be analysed by constructing appropriate matched spacetimes which settle down through gravitational collapse to black hole solutions. The junction (or matching) of two spacetimes requires the equality of the respective first and second fundamental forms at some matching boundary hypersurface. These conditions amount to a set of differential equations that the spacetime metric functions need to satisfy at that matching boundary. This is often not an easy problem as the metric functions are also required to satisfy the respective Einstein field equations (EFEs) on both sides of the matching surface. In what follows, we review some results about models of black hole formation with emphasis on models resulting from the spacetime matching with an exterior which is either vacuum or a cosmological background.

For a zero cosmological constant Λ, there are well known spherically symmetric examples, one of the simplest being the Oppenheimer-Snyder model which results from the matching of a Friedman-Lemaître-Robertson-Walker (FLRW) dust metric to a Schwarzschild exterior. For $\Lambda \neq 0$ spherical cases, the results are less well-known: the matching of Λ-FLRW with the Λ-Schwarzschild solution (also called Kottler solution) has been investigated by Balbinot et al. [2], Nakao [3], and Markovic and Shapiro [4] and its Λ-Lemaître-Tolman-Bondi counterpart by Lake [5]. The matching of the dust Λ-Szekeres solution with Kottler was recently studied by Debnath et al. [6] (following a work of Bonnor [7] for $\Lambda = 0$), while the matching of a collapsing fluid with tangential pressure and $\Lambda \neq 0$ to Kottler has been investigated by Madhav et. al. [8]. The above examples give rise to black hole solutions with $\Lambda \neq 0$ in spherical symmetry.

The collapse to nonspherical black holes has been less studied. Smith and Mann [9] have shown that one can match a collapsing $k = -1$ FLRW spacetime to an asymptotically anti-de Sitter (AdS) exterior, as a model of gravitational collapse to higher genus asymptotically AdS black holes. In a related work, Lemos [10] matched a flat FLRW metric to a radiating Vaidya exterior whose collapse results in a toroidal black hole. Subsequently, there has been considerable interest in models of toroidal and higher genus black holes (also called topological black holes), shells, and horizons, see, for example, [11–14], partly due to the existence of a "landscape" of vacua states in String theory with Λ positive, negative, and zero (see, e.g., [15]). More recently, [16] generalised the earlier work and constructed models for the collapse of inhomogeneous and anisotropic fluids to topological asymptotically AdS black holes in planar and hyperbolic symmetry.

There is also a vast literature about the formation of black holes in cosmological models. There are, essentially, three types of models which are geometrically different, namely, models with (i) junctions with no shells where two metrics are involved, for example, swiss-cheese-type models (see, e.g., [17] and references therein); (ii) junctions with shells (these can be thin or thick), for example, models with domain walls and bubbles mentioned in the previous paragraph; (iii) no junctions, where a single metric is involved in the problem, for example, models of collapsing over-densities resulting from perturbations on cosmological backgrounds.

In the context of primordial black hole (PBH) formation, the gravitational collapse due to first-order density perturbations in FLRW models of the early universe radiative phase was first studied by Zeldovich and Novikov [18] as well as Hawking [19]. These works were then generalised [20, 21], in particular, to include black hole formation from the collapse of cosmic strings produced in phase transitions with symmetry breaking [22]. The collision of bubble walls formed at phase transitions was also considered as sources of PBHs in [23–25]. Different scenarios were subsequently studied and a review is in [26]. More recently, cosmological black hole formation due to QCD and electroweak phase transitions were studied by, for example, [27] following earlier works of [28] and bubble collision has been investigated in detail (for various potentials and using planar, hyperbolic, and spherical symmetry in both dS and AdS phases) by, for example, [11, 29]. Furthermore, PBH formation in cosmological models with $\Lambda \neq 0$ has very recently been analysed in [30], which contains an interesting review of numerical results up to the year 2011 (for a review on the $\Lambda = 0$ case see [31]). In turn, the collapse to black holes in FLRW models have also been investigated using nonlinear scalar metric perturbations on radiative [32] and scalar field backgrounds [33], with those recent works establishing interesting observational bounds on the cosmological spectra nongaussianity.

In this paper, we will be mostly interested in cases which involve spacetime junctions with no shells and, in particular, we will focus on nonspherical spacetimes containing a non-zero Λ. We start by reviewing the theory of spacetime junctions and we then summarise recent results about the gravitational collapse to black holes in higher dimensions [34] which generalise the results of [16]. We start by considering a family of solutions to the Λ-vacuum Einstein equations in $n+2$ dimensions which contains black hole solutions. We find some possible interior collapsing solutions with dust as source and study the corresponding matching problem. We then analyse the global geometrical properties of the matched spacetime and find conditions for which the spacetimes have no naked singularities.

We also review the interesting case of a model of radiating gravitational collapse. In particular, we take the anisotropic Bizoń-Chmaj-Schmidt (BCS) solution in 4 + 1 dimensions and prove that it can be matched to some collapsing dust interiors. The BCS solution is known [35] to settle down via radiation to the Schwarzschild-Tangherlini (SchT) solution. In turn, the latter is known to be stable [35, 36] and we show that, for some interiors, the data at the matching surface can be chosen to be arbitrarily close to the data for a SchT exterior [34]. So, the resulting spacetime models the gravitational collapse of a fluid in five dimensions with an exterior emitting gravitational waves which settles down to the SchT solution.

2. Geometric Theory of Spacetime Junctions in Brief

Let (M^+, g^+) and (M^-, g^-) be two n-dimensional C^3 space times with oriented boundaries Ω^+ and Ω^-, respectively, such that Ω^+ and Ω^- are diffeomorphic. The matched space time (M, g) is the disjoint union of $M^\pm$ with the points in $\Omega^\pm$ identified such that the junction conditions are satisfied (Israel [37], Clarke and Dray [38] and Mars and Senovilla [39]). Since $\Omega^\pm$ are diffeomorphic, one can then view those boundaries as diffeomorphic to a 3-dimensional oriented manifold Ω which can be embedded in M^+ and M^-. Let $\{\xi^\alpha\}$ and $\{x^{\pm i}\}$ be coordinate systems on Ω and $M^\pm$, respectively, where $\alpha, \beta = 1, 2, \ldots, n-1$ and $i, j = 1, 2, \ldots, n$. The two embeddings are given by the following C^3 maps:

$$\Phi^\pm : \Omega \longrightarrow M^\pm, \tag{2.1}$$

$$\xi^\alpha \longmapsto x^{i\pm} = \Phi^{i\pm}(\xi^\alpha), \tag{2.2}$$

such that $\Omega^\pm \equiv \Phi^\pm(\Omega) \subset M^\pm$. The diffeomorphism from Ω^+ to Ω^- is $\Phi^- \circ \Phi^{+^{-1}}$.

Given the basis $\{\partial/\partial\xi^\alpha|_p\}$ of the tangent plane $T_p\Omega$ at some $p \in \Omega$, the push forwards $d\Phi^\pm|_p$ map $\{\partial/\partial\xi^\alpha|_p\}$ into three linearly independent vectors at $\Phi^\pm(p)$ represented by $\vec{e}^{\,\pm}_\alpha|_{\Phi^\pm(p)}$:

$$d\Phi^\pm\left(\left.\frac{\partial}{\partial\xi^\alpha}\right|_\Omega\right) = \left.\frac{\partial\Phi^{\pm i}}{\partial\xi^\alpha}\frac{\partial}{\partial x^{\pm i}}\right|_{\Omega^\pm} \equiv \left.\vec{e}^{\,\pm}_\alpha\right|_{\Omega^\pm} = \left.e^{\pm i}_\alpha \vec{e}^{\,\pm}_\alpha \frac{\partial}{\partial x^{\pm i}}\right|_{\Omega^\pm}. \tag{2.3}$$

On the other hand, using the pull backs $\Phi^{\pm *}$ of the maps $\Phi^\pm$, the metrics $g^\pm$ can be mapped to Ω given two symmetric 2-covariant tensors $\overline{g}^+$ and $\overline{g}^-$ whose components in the basis $\{d\xi^\alpha\}$ are

$$\overline{g}^\pm_{\alpha\beta} \equiv e^{\pm i}_\alpha e^{\pm j}_\beta \left.g_{ij}\right|_{\Omega^\pm} = \left.\left(\vec{e}^{\,\pm}_\alpha \cdot \vec{e}^{\,\pm}_\beta\right)\right|_{\Omega^\pm}. \tag{2.4}$$

The *first matching conditions* are given by the equality of the first fundamental forms (Israel [37])

$$\overline{g}^{+}_{\alpha\beta} = \overline{g}^{-}_{\alpha\beta}. \tag{2.5}$$

The bases $\{\vec{e}^{+}_{\alpha}|_p\}$ and $\{\vec{e}^{-}_{\alpha}|_p\}$ can then be identified,

$$d\Phi^{+}\left(\frac{\partial}{\partial\xi^{\alpha}}\bigg|_{\Omega}\right) = d\Phi^{-}\left(\frac{\partial}{\partial\xi^{\alpha}}\bigg|_{\Omega}\right), \tag{2.6}$$

as can the hypersurfaces $\Omega^{+} \equiv \Omega^{-}$, so henceforth we represent both $\Omega^{\pm}$ by Ω.

We now define a 1-form n, normal to the hypersurface Ω, as

$$n^{\pm}(\vec{e}^{\pm}_{\alpha}) = 0. \tag{2.7}$$

The vectors $\{\vec{n}^{\pm}, \vec{e}^{\pm}_{\alpha}\}$ constitute a basis on the tangent spaces to $M^{\pm}$ at $\Omega^{\pm}$. Since the first junction condition allows the identification of $\{\vec{e}^{+}_{\alpha}\}$ with $\{\vec{e}^{-}_{\alpha}\}$, we only have to ensure that both bases have the same orientation and that $n^{+}_{i}n^{+i} \overset{\Omega}{=} n^{-}_{i}n^{-i}$ is satisfied in order to identify the whole 4-dimensional tangent spaces of $M^{\pm}$ at Ω, $\{\vec{n}^{+}, \vec{e}^{+}_{\alpha}\} \equiv \{\vec{n}^{-}, \vec{e}^{-}_{\alpha}\}$.

The second fundamental forms are given by

$$K^{\pm}_{\alpha\beta} = -n^{\pm}_{i}e^{\pm j}_{\alpha}\nabla^{\pm}_{j}e^{\pm i}_{\beta} \tag{2.8}$$

and the *second matching conditions*, for nonnull surfaces, are the equality of the second fundamental forms

$$K^{+}_{\alpha\beta} = K^{-}_{\alpha\beta}. \tag{2.9}$$

We note that these matching conditions do not depend on the choice of the normal vectors. The first matching conditions ensure the continuity of the metric across Ω while the second conditions prevent infinite jump discontinuities in the Riemann tensor so that the Einstein field equations are well defined in the distributional sense.

The theory is also fully developed for the case where the matching boundary Ω is null or changes character. In that case, the normal vectors $n_{\pm}{}^{\alpha}$ are substituted by the so-called *rigging vectors*, which are vectors nonwhere tangent to $\Omega^{\pm}$ and are used to define generalised second fundamental forms, see [39] for the details.

We are interested in the particular cases for which the matching surface inherits a certain symmetry of the two space times $(M^{\pm}, g^{\pm})$. Such matching is said to *preserve the symmetry*. In practice one demands that the matching hypersurface is tangent to the orbits of the symmetry group to be preserved, see [40].

We note that if the exterior is a vacuum spacetime and the interior contains a fluid then the normal pressure of the fluid has to vanish at the boundary Ω. This is a consequence of the fact that the matching conditions imply

$$n^{\alpha}_{-}T^{-}_{\alpha\beta} \overset{\Omega}{=} n^{\alpha}_{+}T^{+}_{\alpha\beta}. \tag{2.10}$$

We also note that if the interior contains dust then the boundary Ω^- is ruled by geodesics, and we will explore this fact to solve the matching problem in what follows.

3. Black Holes in Higher Dimensions

An $(n+2)$-dimensional Lorentzian manifold (M, g_{ab}) is a solution of the *Einstein equations with cosmological constant* Λ *and energy-momentum tensor* T_{ab} if its Ricci tensor R_{ab} satisfies

$$R_{ab} = \Lambda g_{ab} + \kappa\left(T_{ab} - \frac{1}{n}T g_{ab}\right), \tag{3.1}$$

where κ is a constant, $T = T_a^a$ and the indices $a, b = 1, 2, \ldots, n+2$. We will consider the family of higher-dimensional black holes given in the following proposition (see also [41, 42]).

Proposition 3.1. *Let* $(N, d\sigma^2)$ *be an n-dimensional Riemannian Einstein manifold with Ricci scalar* $n\lambda$*, and let*

$$V(r) = \frac{\lambda}{n-1} - \frac{2m}{r^{n-1}} - \frac{\Lambda r^2}{n+1}, \tag{3.2}$$

where m, Λ *are constants. If* $J \subset \mathbb{R}$ *is an open interval where* V *is well defined and does not vanish then the* $(n+2)$*-dimensional Lorentzian manifold* (M, ds^2) *given by* $M = \mathbb{R} \times J \times N$ *and*

$$ds^2 = -V(r)dt^2 + (V(r))^{-1}dr^2 + r^2 d\sigma^2 \tag{3.3}$$

is a solution of the vacuum Einstein equations with cosmological constant Λ.

It is easy to see that the above metric generalises the Kottler metric to arbitrary dimensions. Nevertheless, black holes with the above metric do not immediately integrate into the usual intuition of a black hole in four dimensions. For instance, since the metric on the sections of null infinity $\mathcal{I}$ does not need to be a metric of constant curvature, the spacetimes are not asymptotically flat (Nor asymptotically de Sitter (dS), resp., anti de Sitter (AdS), in the case of $\Lambda > 0$, resp., $\Lambda < 0$.); however, by analysing the equations for null geodesics in M and N, one can show that the spacetimes are weakly asymptotically simple [34].

Proposition 3.2. *The metrics* (3.3) *are conformally compactifiable at infinity and are weakly asymptotically-simple.*

Therefore, by comparison with the four-dimensional case [16], the following properties hold.

(i) With $\Lambda = 0$, $\lambda > 0$ and $m > 0$, V has a single zero, corresponding to an event horizon, and asymptotes to a positive constant at large r. This is a black-hole solution, which can be thought of as generalising the Schwarzschild metric. The (degenerate) metric on the horizon is $d\sigma^2$, which is also the conformal metric on future null infinity $\mathcal{I}^+$.

(ii) With $\Lambda > 0$, $m > 0$ and large enough positive λ, $V(r)$ is positive between two zeroes, corresponding to a black-hole event horizon and a cosmological event horizon. The solution generalises the asymptotically dS Kottler solution.

(iii) With $\Lambda < 0$ and $m > 0$, V again has a single zero, corresponding to an event horizon, and the solution generalises the asymptotically AdS Kottler solution.

(iv) The solutions in the previous class with $\lambda \leq 0$ may have no global symmetries except the staticity Killing vector. This is because compact, negative scalar curvature Einstein manifolds have no global symmetries, nor does, for example, the Ricci-flat metric on $K3$ (an example with $\lambda = 0$ and $n = 4$).

3.1. *Generalised Friedman-Lemaître-Robertson-Walker Interiors*

We will fill in the solutions of the previous section with (interior) collapsing dust solutions, so that the resulting global spacetime is a model of gravitational collapse with a black hole as the end state. In order to do that, we take the following classes of generalised FLRW spacetimes.

Proposition 3.3. *The $(n+2)$-dimensional Lorentzian metric,*

$$ds^2 = -d\tau^2 + R^2(\tau)\left(d\rho^2 + f^2(\rho)d\sigma^2\right), \tag{3.4}$$

is a solution of the Einstein equations with cosmological constant Λ and energy-momentum tensor $T_{ab} = \mu u_a u_b$, corresponding to a dust fluid with density μ and velocity $u_a dx^a = d\tau$, if and only if $R(\tau)$ and $\mu(\tau)$ satisfy the conservation equation:

$$\mu R^{n+1} = \mu_0, \tag{3.5}$$

for constant μ_0, and the Friedman-like equation:

$$\frac{\dot{R}^2}{R^2} + \frac{k}{nR^2} = \frac{2\kappa\mu}{n(n+1)} + \frac{\Lambda}{n+1}. \tag{3.6}$$

We consider the matching of an interior metric (3.4) to a static exterior (3.3) at a hypersurface Ω ruled by radial time-like geodesics in (3.3), that is, a surface of constant ρ (say $\rho = \rho_0$) in (3.4). We then solve the matching conditions (2.5) and (2.9) together with the EFEs (3.6) to get the following result [34].

Theorem 3.4. *The metric (3.3) can be matched to the FLRW-like metric (3.4) at $\rho = \rho_0$ provided that $f'(\rho_0) > 0$ and $m = \kappa\mu_0 f(\rho_0)^{n+1}/n(n+1)$.*

The matching boundary is comoving with the collapsing fluid whose dynamics is given by (3.6). We call the matched spacetime of the previous theorem FLRW-Kottler spacetime. We now summarize some properties of the FLRW-Kottler spacetime in the three cases $\Lambda = 0$, $\Lambda > 0$, and $\Lambda < 0$: when the Einstein manifold N is not cobordant to a point (e.g., $\mathbb{C}P^2$) the solutions we find cannot have a regular origin, though they can be regular with spacetime wormholes or a "cusp" at the origin. When there is a singularity at the origin, it may or may not be visible from infinity. If the singularity is visible it is called *naked singularity* as it is not hidden by an horizon. The next three propositions exploit these aspects, and Figures 1, 2, and 3 show Penrose diagrams which represent the global structure of the matched spacetimes in the cases $\Lambda = 0$, $\Lambda < 0$, and $\Lambda > 0$ [34]. In particular, by comparing the conformal lifetime

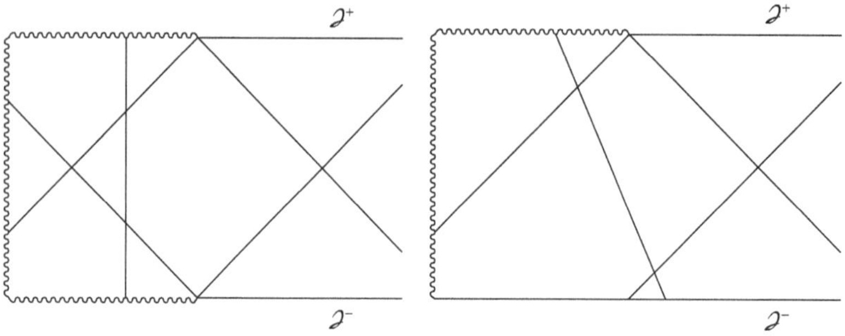

(a) (b)

Figure 1: Penrose diagram for $\Lambda > 0$ with the FLRW universe (a) recollapsing; (b) nonrecollapsing, showing the matching surfaces and the horizons.

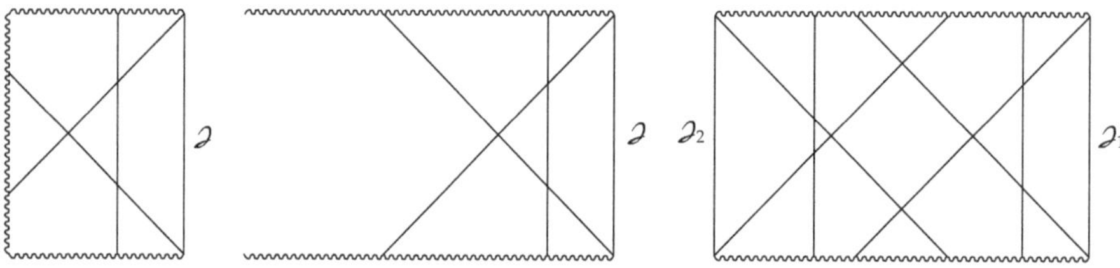

(a) (b) (c)

Figure 2: Penrose diagram for $\Lambda < 0$ and (a) $\lambda > 0$; (b) $\lambda = 0$; (c) $\lambda < 0$, showing the matching surfaces and the horizons.

of the FLRW universe with the supremum of the possible values of ρ_0 we show [34] the following.

Proposition 3.5. *If $\Lambda = 0$ (hence $\lambda > 0$) and $(N, d\sigma^2)$ is not an n-sphere then the locally naked singularity of the FLRW-Kottler spacetime at $\rho = 0$ is always visible from future null infinity $\mathcal{I}^+$ for $k \leq 0$, but can be hidden if $k > 0$ and $n \geq 4$ (space-time dimension $n + 2 \geq 6$).*

Interestingly, [43] investigated higher dimensional dust collapsing spacetimes with $\Lambda = 0$ and also found that for dimension $d = n + 2 \geq 6$ the final central singularity is always hidden by an horizon, while for $d < 6$ naked singularities may appear and thus the cosmic censorship conjecture is violated. By construction, their collapsing interiors are spherical, while here we allow interior spacetimes with other topologies.

Proposition 3.6. *If $\Lambda > 0$ (hence $\lambda > 0$) and $(N, d\sigma^2)$ is not an n-sphere then the locally naked singularity of the FLRW-Kottler spacetime at $\rho = 0$ can be always hidden except if the FLRW universe is recollapsing (hence $k > 0$) and $n < 4$.*

The case $\Lambda < 0$ has wormhole solutions that may have causal curves which cross from one $\mathcal{I}$ to the other and may violate causality. The next proposition, in particular, states conditions under which this can be avoided (see also a result of Galloway [44]). Those conditions are obtained by showing that the future (resp., past) horizons hit the matching

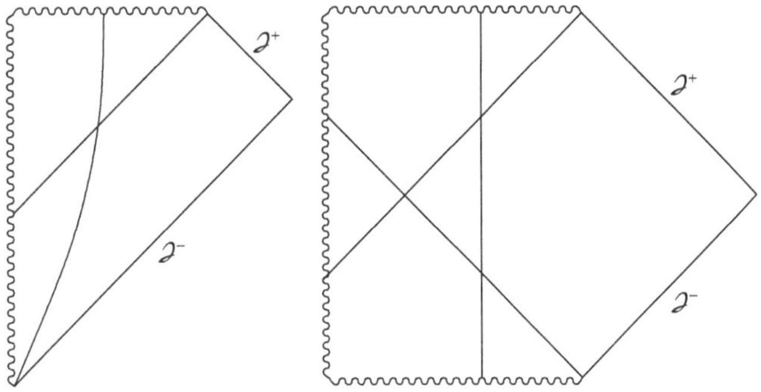

Figure 3: Penrose diagram for $\Lambda = 0$ and (a) $k \leq 0$; (b) $k > 0$, showing the matching surfaces and the horizons.

surfaces at marginally outer trapped (resp., antitrapped) surfaces which are spacelike and, in turn, that these two curves touch at $\dot{R} = \rho = 0$.

Proposition 3.7. *For $\Lambda < 0$ the FLRW-Kottler spacetime satisfies the following.*

(1) *If $\lambda > 0$ and $(N, d\sigma^2)$ is not an n-sphere then the locally naked singularity of the FLRW-Kottler spacetime at $\rho = 0$ can always be hidden.*

(2) *If $\lambda = 0$ then the cusp singularity is not locally naked.*

(3) *If $\lambda < 0$ then no causal curve can cross the wormhole from one $\mathscr{I}$ to the other.*

3.2. Generalised Lemaître-Tolman-Bondi Interiors

We consider now higher dimensional versions of the inhomogeneous Lemaître-Tolman-Bondi (LTB) solutions, generalising those of [45].

Proposition 3.8. *The $(n+2)$-dimensional Lorentzian metric,*

$$ds^2 = -d\tau^2 + A(\tau,\rho)^2 d\rho^2 + B(\tau,\rho)^2 d\sigma^2, \tag{3.7}$$

is a solution of the Einstein equations with cosmological constant Λ and energy-momentum tensor $T_{ab} = \mu u_a u_b$, corresponding to a dust fluid with density μ and velocity $u_a dx^a = d\tau$, if and only if $A(\tau,\rho)$, $B(\tau,\rho)$ and $\mu(\tau,\rho)$ satisfy

$$\begin{aligned} A &= B'(1 + w(\rho)), \\ \mu A B^n &= M'(\rho)(1 + w(\rho)), \end{aligned} \tag{3.8}$$

for some functions $w(\rho)$ and $M(\rho)$, and

$$\dot{B}^2 B^{n-1} + \left(\frac{\lambda}{n-1} - \frac{1}{(1+w(\rho))^2} - \frac{\Lambda}{n+1}B^2\right)B^{n-1} = \frac{2\kappa M(\rho)}{n}, \tag{3.9}$$

(where dot and prime denote differentiation with respect to τ and ρ).

This metric has three free functions of ρ, namely, $w(\rho)$, $M(\rho)$, and $B(0,\rho)$, one of which can be removed by coordinate freedom. We seek to match an interior represented by the metric (3.7) to a static exterior represented by the metric (3.3) at a surface Ω ruled by radial time-like geodesics in (3.3), that is, a surface of constant ρ (say $\rho = \rho_0$) in (3.7). The solutions to the matching conditions (2.5) and (2.9) allow us to prove [34].

Theorem 3.9. *The metric* (3.3) *can be matched to the LTB-like metric* (3.7) *at* $\rho = \rho_0$ *provided that* $1 + w(\rho_0) > 0$ *and* $m = (\kappa/n)M(\rho_0)$.

Note that the matching boundary is again comoving with the collapsing inhomogeneous fluid and its dynamics is now given by (3.9). We will use the term LTB-Kottler space-times for these matched solutions. The global properties of the LTB-Kottler spacetime obtained in Theorem 3.9 are much more diverse than what we find in the FLRW-Kottler spacetimes. For instance, one can easily find examples of black hole formation with wormholes inside the matter with positive λ and $\Lambda = 0$, see [34] (similar results in 4 dimensions are in [46]).

4. Collapse with Radiating Exteriors

In this section, we consider models of gravitational collapse with a gravitational wave exterior, so that the exterior metrics will be time-dependent generalisations of (3.3). This is a specially interesting problem for observational and experimental purposes [36]. We take the Bizoń-Chmaj-Schmidt (BCS) metric in (4 + 1) dimensions [36], although similar ansetz can be made in other dimensions (see [47]). We then consider three different interiors with this exterior, which are anisotropic generalisations of FLRW models [34]. Consider the metric:

$$ds^{2+} = -Ae^{-2\delta}dt^2 + A^{-1}dr^2 + \frac{r^2}{4}e^{2B}\left(\sigma_1^2 + \sigma_2^2\right) + \frac{r^2}{4}e^{-4B}\sigma_3^2, \tag{4.1}$$

where A, δ, and B are functions of t and r. The one forms σ_i are left invariant for the standard Lie group structure on S^3 and can be taken as

$$\begin{aligned}\sigma_1 &= \cos\psi d\theta + \sin\theta\sin\psi d\phi,\\ \sigma_2 &= \sin\psi d\theta - \sin\theta\cos\psi d\phi,\\ \sigma_3 &= d\psi + \cos\theta d\phi,\end{aligned} \tag{4.2}$$

where θ, ψ, ϕ are Euler angles on S^3 with $0 < \theta < \pi$, $0 < \phi < 2\pi$, and $0 < \psi < 4\pi$. The space-time with $B \neq 0$ is interpreted as containing pure gravitational waves with radial symmetry [36] and the Schwarzschild-Tangherlini (SchT) limit is obtained for $B = 0$. There is a residual coordinate freedom $t \to \hat{t} = f(t)$; $\delta \to \hat{\delta} = \delta + \log\dot{f}$ in (4.1), which one can use to set δ arbitrarily along a timelike curve. The (4 + 1) dimensional vacuum EFEs give [36]

$$\partial_r A = -\frac{2A}{r} + \frac{1}{3r}\left(8e^{-2B} - 2e^{-8B}\right) - 2r\left(e^{2\delta}A^{-1}(\partial_t B)^2 + A(\partial_r B)^2\right), \tag{4.3}$$

$$\partial_t A = -4rA(\partial_t B)(\partial_r B), \tag{4.4}$$

$$\partial_r \delta = -2r\left(e^{2\delta} A^{-2}(\partial_t B)^2 + (\partial_r B)^2\right), \tag{4.5}$$

together with the quasi-linear wave equation for B

$$\partial_t\left(e^{\delta} A^{-1} r^3 (\partial_t B)\right) - \partial_r\left(e^{-\delta} A r^3 (\partial_r B)\right) + \frac{4}{3} e^{-\delta} r\left(e^{-2B} - e^{-8B}\right) = 0. \tag{4.6}$$

In [36], the authors solve this system numerically by giving B and $\partial_t B$ at $t = 0$ with $A(0,0) = 0 = \delta(t, 0)$. Within the BCS class, they provide numerical evidence that linear perturbations of SchT decay in time so that SchT is linearly stable, in this sense. The nonlinear stability of SchT within the BCS class was later established in [35].

We will give data A, B and the normal derivative $\nabla_n B$ at the timelike boundary Ω of the collapsing interior with the gauge choice $\delta \overset{\Omega}{=} 0$. Uniqueness and local existence follow as standard. From [35], one knows that if data close to that for SchT is given on an asymptotically flat hypersurface then the solution will exist forever and stay close to the SchT solution. As we will see, in some of our cases, data on Ω can be chosen to be arbitrarily close to data for SchT.

4.1. Generalised FLRW Interiors

As interior metrics, we will consider three classes of FLRW-like solutions based on Riemannian Bianchi-IX spatial metrics which are, respectively, the Eguchi-Hanson metric (with $R_{ij} = 0$), the k-Eguchi-Hanson metric (with $R_{ij} = k g_{ij}$ excluding the case $k = 0$), and the k-Taub-NUT metric (with $R_{ij} = k g_{ij}$, including, $k = 0$ as a particular case).

We summarize now our main result and leave the details of the proof to the next three sections (see also [34]).

Theorem 4.1. *In each case, the interior metric gives consistent data for the metric* (4.1) *at a comoving time-like hypersurface. Local existence of the radiating exterior in the neighbourhood of the matching surface is then guaranteed. In the case of Eguchi-Hanson and k-Taub-NUT with* $k < 0$, *the data can be chosen to be close to the data for the Schwarzschild-Tangherlini solution.*

4.1.1. The Eguchi-Hanson Metric

Eguchi and Hanson found a class of self-dual solutions to the Euclidean Einstein equations with metric given by [48]

$$h_{\rm EH} = \left(1 - \frac{a^4}{\rho^4}\right)^{-1} d\rho^2 + \frac{\rho^2}{4}\left(\sigma_1^2 + \sigma_2^2\right) + \frac{\rho^2}{4}\left(1 - \frac{a^4}{\rho^4}\right)\sigma_3^2, \tag{4.7}$$

with σ_i given by (4.2) and a is a real constant. The generalised FLRW metric built on this Riemannian metric is

$$ds^{2-} = -d\tau^2 + R^2(\tau) h_{\rm EH}, \tag{4.8}$$

with the EFEs for a dust source reducing to

$$\mu R^4 = \mu_0, \qquad \dot{R}^2 = \frac{\kappa\mu_0}{6R^2}. \tag{4.9}$$

We match at $\rho = \rho_0$ so that Ω^- is parameterised by $\Omega^- = \{\tau, \rho = \rho_0\}$. The equality of the first and second fundamental forms at Ω then gives

$$r \overset{\Omega}{=} R\rho e^{-B}, \qquad e^{-6B} \overset{\Omega}{=} 1 - \frac{a^4}{\rho^4}, \qquad \nabla_n B \overset{\Omega}{=} -\frac{2a^4}{3R\rho^5}\left(1 - \frac{a^4}{\rho^4}\right)^{-1/2}, \qquad Ae^{-\delta}\dot{t} \overset{\Omega}{=} e^{2B}\left(1 - \frac{a^4}{3\rho^4}\right), \tag{4.10}$$

from which we calculate A, $\partial_t B$, and $\partial_r B$ on Ω in terms of quantities from the interior. At this point, we have B, $\nabla_n B$, and A on Ω, and we use the gauge freedom to set $\delta = 0$ on Ω. By (4.10), we have

$$B \overset{\Omega}{=} O\left(\frac{a^4}{\rho^4}\right), \qquad \nabla_n B \overset{\Omega}{=} (\rho R)^{-1} O\left(\frac{a^4}{\rho^4}\right), \tag{4.11}$$

and, to say that the data is close to SchT data, we want these to be small. The first term is small if $\rho \gg a$. The second term will increase without bound as R decreases to zero. If we restrict R by its value when a marginally outer trapped surface forms on Ω then, from the Friedman equation and with $\rho \gg a$, this happens when $R^2\rho^2 \sim \kappa\mu_0\rho^4$ so that we control $\nabla_n B$ on Ω by controlling μ_0. So, by choice of the location of Ω, at $\rho = \rho_0$, and choice of μ_0, we can choose data close to SchT.

4.1.2. The k-Eguchi-Hanson Metric

The k-Eguchi-Hanson metric is given by [49]

$$h_{k\mathrm{EH}} = \Delta^{-1} d\rho^2 + \frac{\rho^2}{4}\left(\sigma_1^2 + \sigma_2^2\right) + \frac{\rho^2}{4}\Delta\sigma_3^2, \tag{4.12}$$

where $\Delta = 1 - a^4/\rho^4 - k\rho^2/6$. This metric is complete for $k < 0$ and $a^4 = 4(p-2)^2(p+1)/3k^2$, $\rho > (-2(p-2)/k)^{1/2}$, where $p \geq 3$ in an integer. Since k is related to a for a complete solution, we cannot obtain the previous case from this case by taking $k \to 0$. However, the matching formulae do formally allow this limit. The generalised FLRW metric built on this metric is $ds^{2-} = -d\tau^2 + R^2(\tau)h_{k\mathrm{EH}}$ and the respective EFEs for a dust source reduce to

$$\mu R^4 = \mu_0, \qquad \dot{R}^2 + \frac{k}{3} = \frac{\kappa\mu_0}{6R^2}. \tag{4.13}$$

The matching conditions on Ω give

$$r \overset{\Omega}{=} R\rho e^{-B}, \qquad e^{-6B} \overset{\Omega}{=} \Delta, \qquad \nabla_n B \overset{\Omega}{=} -\frac{\Delta^{-1/2}}{3\rho R}\left(\frac{2a^4}{\rho^4} - \frac{k\rho^2}{6}\right), \qquad Ae^{-\delta}\dot{t} \overset{\Omega}{=} e^{2B}\left(1 - \frac{a^4}{3\rho^4} - \frac{2k\rho^2}{9}\right). \tag{4.14}$$

We can calculate A in terms of interior data, as before, and check that this is consistent with the EFEs (4.3) and (4.4). However, in this case, it is not clear that we may choose data close to SchT data as the normal derivative of B is

$$\nabla_n B \overset{\Omega}{=} \frac{k\rho}{6R}, \tag{4.15}$$

and for this to be small, we would require R to be large on Ω outside the marginally trapped surface. It is hard to see how to arrange this and so although the solution in the exterior exists locally, we do not have a good reason to think that it will settle down to SchT.

4.1.3. k-Taub-NUT

We consider the Riemannian Taub-NUT metric with a cosmological constant (k rather than Λ with our conventions) [50, 51]

$$h_{\mathrm{TN}} = \frac{1}{4}\Sigma^{-1}d\rho^2 + \frac{1}{4}\left(\rho^2 - L^2\right)\left(\sigma_1^2 + \sigma_2^2\right) + L^2\Sigma\sigma_3^2, \tag{4.16}$$

where $\Sigma = (\rho - L)(1 - (k/12)(\rho - L)(\rho + 3L))/(\rho + L)$, and use it to construct the $4+1$ interior:

$$ds^{2-} = -d\tau^2 + R^2(\tau)h_{\mathrm{TN}}. \tag{4.17}$$

The EFEs for a dust source are again (4.13). From matching the first fundamental forms we get

$$r \overset{\Omega}{=} R\left(\rho^2 - L^2\right)^{1/2} e^{-B}, \qquad e^{-6B} \overset{\Omega}{=} \frac{4L^2\Sigma}{\rho^2 - L^2}, \tag{4.18}$$

and the second matching conditions read

$$Ae^{-b}\dot{t} \overset{\Omega}{=} \frac{4R}{3r}\Sigma^{1/2}e^{4B}\left(\frac{2L^2(2\rho + L)\Sigma}{\rho^2 - L^2} - \frac{k}{6}L^2(\rho - L)\right), \qquad \nabla_n B \overset{\Omega}{=} \frac{1}{3R}\left(\frac{2\Sigma^{1/2}}{\rho + L} + \frac{k(\rho - L)}{6\Sigma^{1/2}}\right). \tag{4.19}$$

As before, the expression for A on Ω is consistent with $\dot{A}$ calculated from (4.3) and (4.4). Now note that if $kL^2 = -3$ then the metric (4.16) is precisely the 4-dimensional hyperbolic metric. In this case, B and $\nabla_n B$ vanish on Ω whatever is the value of ρ_0, so that the exterior metric is precisely SchT: this is a case from the previous section as the interior is now a standard FLRW cosmology. Consequently, if we take kL^2 close to -3 we get data close to SchT data. To see this, set $kL^2 = -3(1 + \varepsilon)$. Then $\Sigma = ((\rho^2 - L^2)/4L^2)(1 + O(\varepsilon))$, so that

$$e^{-6B} \overset{\Omega}{=} 1 + O(\varepsilon), \qquad \nabla_n B \overset{\Omega}{=} \frac{1}{LR}O(\varepsilon), \tag{4.20}$$

and the data $(B, \nabla_n B)$ can be chosen as small as desired by choosing large ρ_0 and small ε.

Acknowledgments

The work summarised in this paper was done in collaboration with José Natário and Paul Tod. The author is supported by FCT Projects PTDC/MAT/108921/2008, CERN/FP/116377/2010, and Est-C/MAT/UI0013/2011 and by CMAT, Universidade do Minho, through FCT plurianual FEDER Funds "Programa Operacional Factores de Competitividade COMPETE."

References

[1] R. Emparan and H. S. Reall, "Black holes in higher dimensions," *Living Reviews in Relativity*, vol. 11, p. 6, 2008.
[2] R. Balbinot, R. Bergamini, and A. Comastri, "Solution of the Einstein-Straus problem with a Λ term," *Physical Review D*, vol. 38, no. 8, pp. 2415–2418, 1988.
[3] K. Nakao, "The Oppenheimer-Snyder space-time with a cosmological constant," *General Relativity and Gravitation*, vol. 24, no. 10, pp. 1069–1081, 1992.
[4] D. Markovic and S. L. Shapiro, "Gravitational collapse with a cosmological constant," *Physical Review D*, vol. 61, no. 8, 084029, pp. 1–7, 2000.
[5] K. Lake, "Gravitational collapse of dust with a cosmological constant," *Physical Review D*, vol. 62, no. 2, 027301, 2000.
[6] U. Debnath, S. Nath, and S. Chakraborty, "Quasi-spherical collapse with cosmological constant," *Monthly Notices of the Royal Astronomical Society*, vol. 369, no. 4, pp. 1961–1964, 2006.
[7] W. B. Bonnor, "Non-radiative solutions of Einstein's equations for dust," *Communications in Mathematical Physics*, vol. 51, no. 3, pp. 191–199, 1976.
[8] T. A. Madhav, R. Goswami, and P. S. Joshi, "Gravitational collapse in asymptotically anti-de Sitter or de Sitter backgrounds," *Physical Review D*, vol. 72, no. 8, 084029, 2005.
[9] W. L. Smith and R. B. Mann, "Formation of topological black holes from gravitational collapse," *Physical Review D*, vol. 56, no. 8, pp. 4942–4947, 1997.
[10] J. P. S. Lemos, "Gravitational collapse to toroidal, cylindrical, and planar black holes," *Physical Review D*, vol. 57, no. 8, pp. 4600–4605, 1998.
[11] B. Freivogel, G. T. Horowitz, and S. Shenker, "Colliding with a crunching bubble," *Journal of High Energy Physics*, no. 5, p. 090, 2007.
[12] Yumei Wu, M. F A Da Silva, N. O. Santos, and Anzhong Wang, "Topological charged black holes in high dimensional spacetimes and their formation from gravitational collapse of a type II fluid," *Physical Review D*, vol. 68, no. 8, 084012, 2003.
[13] D. Ida, T. Shiromizu, and H. Ochiai, "Semiclassical instability of the brane-world: Randall-Sundrum bubbles," *Physical Review D*, vol. 65, no. 2, 023504, 2002.
[14] D. Birmingham and M. Rinaldi, "Bubbles in anti-de Sitter space," *Physics Letters B*, vol. 544, no. 3-4, pp. 316–320, 2002.
[15] M. R. Douglas and S. Kachru, "Flux compactification," *Reviews of Modern Physics*, vol. 79, no. 2, pp. 733–796, 2007.
[16] F. C. Mena, J. Natário, and P. Tod, "Gravitational collapse to toroidal and higher genus asymptotically AdS black holes," *Advances in Theoretical and Mathematical Physics*, vol. 12, no. 5, pp. 1163–1181, 2008.
[17] A. Krasinski, *Inhomogeneous Cosmological Models*, Cambridge University Press, 1997.
[18] Ya. B. Zeldovich and I. D. Novikov, "The hypothesis of cores retarded during expansion and the hot cosmological model," *Soviet Astronomy*, vol. 10, p. 602, 1967.
[19] S. Hawking, "Gravitationally collapsed objects of very low mass," *Monthly Notices of the Royal Astronomical Society*, vol. 152, p. 75, 1971.
[20] J. B. Carr and S. W. Hawking, "Black holes in the early universe," *Monthly Notices of the Royal Astronomical Society*, vol. 168, pp. 399–416, 1974.
[21] A. G. Polnarev and M. Yu. Khlopov, "Cosmology, primordial black holes, and supermassive particles," *Soviet Physics Uspekhi*, vol. 145, pp. 213–232, 1985.
[22] S. W. Hawking, "Black holes from cosmic strings," *Physics Letters B*, vol. 231, no. 3, pp. 237–239, 1989.
[23] S. W. Hawking, I. G. Moss, and J. M. Stewart, "Bubble collisions in the very early universe," *Physical Review D*, vol. 26, no. 10, pp. 2681–2693, 1982.

[24] I. G. Moss, "Singularity formation from colliding bubbles," *Physical Review D*, vol. 50, no. 2, pp. 676–681, 1994.
[25] M. Yu. Khlopov, R. V. Konoplich, S. G. Rubin, and A. S. Sakharov, "First order phase transitions as a source of black holes in the early Universe," *Gravitation & Cosmology*, vol. 23, pp. 153–156, 2000.
[26] B. J. Carr, "Primordial black holes as a probe of cosmology and high energy physics," in *Quantum Gravity*, vol. 631 of *Lecture Notes in Physics*, pp. 301–321, 2003.
[27] J. I. Kapusta and T. Springer, "Potentials for soft-wall AdS/QCD," *Physical Review D*, vol. 81, no. 8, 086009, 2010.
[28] K. Jedamzik, "Primordial black hole formation during the QCD epoch," *Physical Review D*, vol. 55, no. 10, pp. R5871–R5875, 1997.
[29] D. Hwang, B. Lee, W. Lee, and D. Yeom, "Bubble collision with gravitation," 2012, http://arxiv.org/abs/1201.6109.
[30] I. Musco, "Critical collapse and black hole formation within an expanding perfect fluid," 2011, http://arxiv.org/abs/1106.5287.
[31] B. J. Carr, T. Harada, and H. Maeda, "Can a primordial black hole or wormhole grow as fast as the universe?" *Classical and Quantum Gravity*, vol. 27, no. 18, p. 183101, 2010.
[32] M. Shibata and M. Sasaki, "Black hole formation in the Friedmann universe: formulation and computation in numerical relativity," *Physical Review D*, vol. 60, no. 8, 084002, 1999.
[33] J. C. Hidalgo, *Primordial Black Holes in non-linear perturbation theory [Ph.D. thesis]*, Queen Mary, University of London, 2009, http://arxiv.org/abs/0910.1876.
[34] F. C. Mena, J. Natário, and P. Tod, "Formation of higher-dimensional topological black holes," *Annales Henri Poincaré*, vol. 10, no. 7, pp. 1359–1376, 2010.
[35] M. Dafermos and G. Holzegel, "On the nonlinear stability of higher dimensional triaxial Bianchi-IX black holes," *Advances in Theoretical and Mathematical Physics*, vol. 10, no. 4, pp. 503–523, 2006.
[36] P. Bizoń, T. Chmaj, and B. G. Schmidt, "Critical behavior in vacuum gravitational collapse in 4 + 1 dimensions," *Physical Review Letters*, vol. 95, no. 7, p. 071102, 2005.
[37] W. Israel, "Singular hypersurfaces and thin shells in general relativity," *Il Nuovo Cimento B*, vol. 44, no. 1, pp. 1–14, 1966.
[38] C. J. S. Clarke and T. Dray, "Junction conditions for null hypersurfaces," *Classical and Quantum Gravity*, vol. 4, no. 2, pp. 265–275, 1987.
[39] M. Mars and J. M. M. Senovilla, "Geometry of general hypersurfaces in spacetime: junction conditions," *Classical and Quantum Gravity*, vol. 10, no. 9, pp. 1865–1897, 1993.
[40] R. Vera, "Symmetry-preserving matchings," *Classical and Quantum Gravity*, vol. 19, no. 20, pp. 5249–5264, 2002.
[41] G. Gibbons and S. A. Hartnoll, "Gravitational instability in higher dimensions," *Physical Review D*, vol. 66, no. 6, 064024, 2002.
[42] G. W. Gibbons, D. Ida, and T. Shiromizu, "Uniqueness and non-uniqueness of static vacuum black holes in higher dimensions," *Progress of Theoretical Physics. Supplement*, no. 148, pp. 284–290, 2002.
[43] R. Goswami and P. S. Joshi, "Cosmic censorship in higher dimensions," *Physical Review D*, vol. 69, no. 10, 104002, 2004.
[44] G. J. Galloway, "A "finite infinity" version of topological censorship," *Classical and Quantum Gravity*, vol. 13, no. 6, pp. 1471–1478, 1996.
[45] S. G. Ghosh and A. Beesham, "Higher dimensional inhomogeneous dust collapse and cosmic censorship," *Physical Review D*, vol. 64, no. 12, 124005, 2001.
[46] C. Hellaby, "A Kruskal-like model with finite density," *Classical and Quantum Gravity*, vol. 4, no. 3, pp. 635–650, 1987.
[47] P. Bizoń, T. Chmaj, A. Rostworowski, B. G. Schmidt, and Z. Tabor, "Vacuum gravitational collapse in nine dimensions," *Physical Review D*, vol. 72, no. 12, 121502, 2005.
[48] T. Eguchi and A. J. Hanson, "Asymptotically flat self-dual solutions to euclidean gravity," *Physics Letters B*, vol. 74, no. 3, pp. 249–251, 1978.
[49] H. Pedersen, "Eguchi-Hanson metrics with cosmological constant," *Classical and Quantum Gravity*, vol. 2, pp. 579–587, 1985.
[50] H. Boutaleb-Joutei, "The general Taub-NUT-de Sitter metric as a self-dual Yang-Mills solution of gravity," *Physics Letters B*, vol. 90, no. 1-2, pp. 181–184, 1980.
[51] M. M. Akbar, "Classical boundary-value problem in Riemannian quantum gravity and Taub-bolt-anti-de Sitter geometries," *Nuclear Physics B*, vol. 663, no. 1-2, pp. 215–230, 2003.

Permissions

The contributors of this book come from diverse backgrounds, making this book a truly international effort. This book will bring forth new frontiers with its revolutionizing research information and detailed analysis of the nascent developments around the world.

We would like to thank all the contributing authors for lending their expertise to make the book truly unique.They have played a crucial role in the development of this book. Without their invaluable contributions this book wouldn't have been possible. They have made vital efforts to compile up to date information on the varied aspects of this subject to make this book a valuable addition to the collection of many professionals and students.

This book was conceptualized with the vision of imparting up-to-date information and advanced data in this field. To ensure the same, a matchless editorial board was set up. Every individual on the board went through rigorous rounds of assessment to prove their worth. After which they invested a large part of their time researching and compiling the most relevant data for our readers. Conferences and sessions were held from time to time between the editorial board and the contributing authors to present the data in the most comprehensible form. The editorial team has worked tirelessly to provide valuable and valid information to help people across the globe.

Every chapter published in this book has been scrutinized by our experts. Their significance has been extensively debated. The topics covered herein carry significant findings which will fuel the growth of the discipline. They may even be implemented as practical applications or may be referred to as a beginning point for another development. Chapters in this book were first published by Hindawi Publishing Corporation; hereby published with permission under the Creative Commons Attribution License or equivalent.

The editorial board has been involved in producing this book since its inception. They have spent rigorous hours researching and exploring the diverse topics which have resulted in the successful publishing of this book. They have passed on their knowledge of decades through this book. To expedite this challenging task, the publisher supported the team at every step. A small team of assistant editors was also appointed to further simplify the editing procedure and attain best results for the readers.

Our editorial team has been hand-picked from every corner of the world. Their multi-ethnicity adds dynamic inputs to the discussions which result in innovative outcomes. These outcomes are then further discussed with the researchers and contributors who give their valuable feedback and opinion regarding the same. The feedback is then collaborated with the researches and they are edited in a comprehensive manner to aid the understanding of the subject.

Apart from the editorial board, the designing team has also invested a significant amount of their time in understanding the subject and creating the most relevant covers. They scrutinized every image to scout for the most suitable representation of the subject and create an appropriate cover for the book.

The publishing team has been involved in this book since its early stages. They were actively engaged in every process, be it collecting the data, connecting with the contributors or procuring relevant information. The team has been an ardent support to the editorial, designing and production team. Their endless efforts to recruit the best for this project, has resulted in the accomplishment of this book. They are a veteran in the field of academics and their pool of knowledge is as vast as their experience in printing. Their expertise and guidance has proved useful at every step. Their uncompromising quality standards have made this book an exceptional effort. Their encouragement from time to time has been an inspiration for everyone.

The publisher and the editorial board hope that this book will prove to be a valuable piece of knowledge for researchers, students, practitioners and scholars across the globe.

List of Contributors

Gerd Niestegge
Zillertalstrasse 39, 81373 Muenchen, Germany

Ray-Yeng Yang
Tainan Hydraulics Laboratory, Research Center of Ocean Environment and Technology, Tainan Hydraulics Laboratory, National Cheng Kung University, No.1, Ta-Hsueh Road, Tainan 701, Taiwan

Hwung Hweng Hwung
Department of Hydraulic and Ocean Engineering, Research Center of Ocean Environment and Technology, National Cheng Kung University, No.1, Ta-Hsueh Road, Tainan 701, Taiwan

A. Afsar Khan and R. Ellahi
Department of Mathematics & Statistics, FBAS, IIU, Islamabad, Pakistan

R. Ellahi and K. Vafai
Department of Mechanical Engineering, University of California Riverside, USA

B. Abdesselam
Laboratoire de Physique Theorique, Universite d'Oran Es-Senia, 31100 Oran, Algeria
Institut des Sciences et de la Technologie, Centre Universitaire d'Ain T´emouchent, 46000 Ain T´emouchent, Algeria

A. Chakrabarti
Centre de Physique Th´eorique, Ecole Polytechnique, 91128 Palaiseau Cedex, France

Sangeet Kumar
Department of Applied Sciences, Rayat Institute of Engineering & Information Technology, Railmajra, SBS Nagar, Punjab 144533, India

Rakesh Kumar
Department of Basic and Applied Sciences, University College of Engineering, Punjabi University, Patiala 147002, India

R. K. Nagaich
Department of Mathematics, Punjabi University, Patiala 147002, India

S. A. El-Wakil, Essam M. Abulwafa, E. K. El-Shewy, H. G. Abdelwahed, and Hamdi M. Abd-El-Hamid
Theoretical Physics Group, Physics Department, Faculty of Science, Mansoura University, Mansoura 35516, Egypt

Glenn Barnich and Pierre-Henry Lamber
Physique Theorique et Mathematique, Universite Libre de Bruxelles, Campus Plaine, CP 231, 1050 Bruxelles, Belgium
International Solvay Institutes, Campus Plaine, CP 231, 1050 Bruxelles, Belgium

H. T. Cho and T. R. Huang
Department of Physics, Tamkang University, Tamsui, New Taipei City 25137, Taiwan

A. S. Cornell
National Institute for Theoretical Physics, School of Physics, University of the Witwatersrand, Wits 2050, South Africa

Jason Doukas
Yukawa Institute for Theoretical Physics, Kyoto University, Kyoto 606-8502, Japan

Wade Naylor
International College and Department of Physics, Osaka University, Toyonaka, Osaka 560-0043, Japan

Hongwei Yang and Huanhe Dong
Information School, Shandong University of Science and Technology, Qingdao 266590, China

Baoshu Yin
Institute of Oceanology, China Academy of Sciences, Qingdao 266071, China
Key Laboratory of Ocean Circulation and Wave, Chinese Academy of Sciences, Qingdao 266071, China

Zhenyu Liu
Department of Mathematics, Zaozhuang College, Zaozhuang 277160, China

Yu G. Smirnov and D. V. Valovik
Department of Mathematics and Supercomputer Modeling, Penza State University, Krasnaya Street 40, Penza 440026, Russia

Ze Tang and Jianwen Feng
College of Mathematics and Computational Sciences, Shenzhen University, Shenzhen 518060, China

M. Merkli
Department of Mathematics and Statistics, Memorial University of Newfoundland, St. John's, NL, Canada A1C 5S7

G. P. Berman
Theoretical Division, Los Alamos National Laboratory, MS B213, Los Alamos, NM 87545, USA

F. Borgonovi
Dipartimento di Matematica e Fisica, and Interdisciplinary Laboratories for Advanced Materials Physics, Universit`a Cattolica, Via Musei 41, 25121 Brescia, Italy
Istituto Nazionale di Fisica Nucleare, Sezione di Pavia, Via Bassi 6, 27100 Pavia, Italy

V. I. Tsifrinovich
Department of Applied Physics, Polytechnic Institute of NYU, 6 MetroTech Center, Brooklyn, NY 11201, USA

Ze Tang and Jianwen Feng
College of Mathematics and Computational Sciences, Shenzhen University, Shenzhen 518060, China